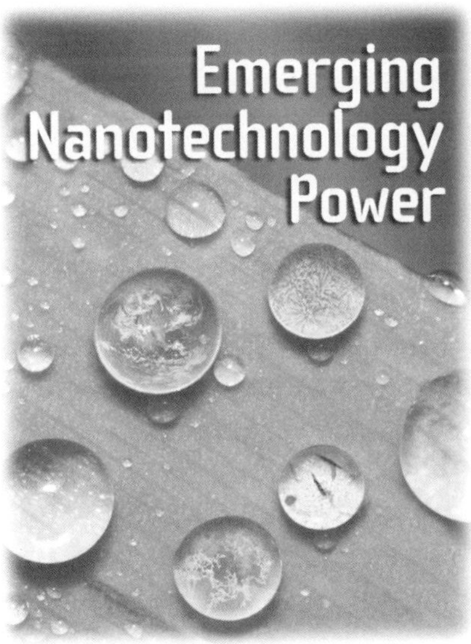

Nanotechnology R&D and Business Trends in the Asia Pacific Rim

Emerging Nanotechnology Power

Nanotechnology R&D and Business Trends in the Asia Pacific Rim

Editor

Lerwen Liu

NanoGlobe Pte Ltd, Singapore

World Scientific

NEW JERSEY • LONDON • SINGAPORE • BEIJING • SHANGHAI • HONG KONG • TAIPEI • CHENNAI

Published by

World Scientific Publishing Co. Pte. Ltd.
5 Toh Tuck Link, Singapore 596224
USA office: 27 Warren Street, Suite 401-402, Hackensack, NJ 07601
UK office: 57 Shelton Street, Covent Garden, London WC2H 9HE

British Library Cataloguing-in-Publication Data
A catalogue record for this book is available from the British Library.

EMERGING NANOTECHNOLOGY POWER: NANOTECHNOLOGY R&D AND BUSINESS TRENDS IN THE ASIA PACIFIC RIM

ISBN-13 978-981-4261-54-8 (pbk)
ISBN-10 981-4261-54-8 (pbk)

Printed in Singapore by Mainland Press Pte Ltd.

PREFACE

I have been involved in the world of nanotechnology in Asia for nearly 15 years engaging both in scientific research and in monitoring and analyzing government policy, research and development and industry trends in the field. I am now keen to share some insights into what is currently happening in this exciting part of the world.

I received my introduction to nanoscience in 1992, completing a PhD at the University of New South Wales and subsequently moving to Japan in 1995 to pursue my postdoctoral studies. In 1999 I started concentrating on analysing nanotechnology policy, R&D and industry trends. At this time, nanotechnology was still a relatively small field with few coordinated government funding programs and minimal investment by industry except in the advanced economies such as Japan, Germany, UK and USA. However, after President Clinton announced the US National Nanotechnology Initiative in 2000, the field has undergone profound changes especially in the Asia Pacific region. Public and private funding in nanotechnology has increased more than tenfold and the number of people working in industry and academia has also increased significantly.

Whereas Japan, a developed economy, has long been a leader in nanotechnology R&D and commercialization in this region, since 2003 developing economies such as Thailand and Vietnam have also started to pursue nanotechnology. Even Indonesia has started to become more active in this field and will launch a major nanotechnology initiative in 2009.

Over the last 10 years, I have been conducting research into government policy, R&D and industry trends through site visits, participating in major nanotechnology events and interviewing government policy makers, distinguished scientists, industry leaders and

My special thanks should go to the Singapore Institute of Materials Research and Engineering (IMRE) staff Dr LOW Hong Yee, Eugene LOW, TAN Wei Wei, and Jofelyn LYE as well as ItN Nanovation AG for providing the cover page pictures; and the World Scientific team Rajesh Babu, Jimmy Low, and QUEK Yeow Hwa for their fantastic help in making this book happen.

Thanks also to my dear friend Caroline Pearce for the help she provided me with the Preface and Introduction.

I would like to thank all the mentors in my career of nanotechnology including my PhD advisor Prof David Neilson, Dr David Kahaner, Dr Gerd Bachmann, Dr Hiroshi Yokoyama and Dr Kazunobu Tanaka for their inspiration and fruitful partnerships; and those who inspired my interest in nanoscience and provided nanostructure research including Dr Robert Clark (Australia), Drs Daijiro Yoshioka (Japan), Alan MacDonald (USA) and Mario Tosi (Italy).

Over the last 10 years, I have the honor to receive help and support from those who drive nanotechnology strategy, research and industry development in Asia, they include Drs Junichi Sone (Japan), Chung Yuan Mou (Taiwan), Maw-Kuen Wu (Taiwan), C.J. Chen (Taiwan), Wiwut Tanthapanichkoon (Thailand), Terry Turney (Australia), Jane Niall (Australia), Khiang-Wee Lim (Singapore), Phan Hong Khoi (Vietnam), Hanjo Lim (Korea), Sang-Hee Suh (Korea), Chunli Bai (China), Lide Zhang (China), Sishen Xie (China), Che Ting Chan (Hong Kong), Winnie Wong (Hong Kong), KaMing Ng (Hong Kong), Paul T. Callaghan (New Zealand), Richard Blaikie (New Zealand) Halimaton Hamdan (Malaysia); and especially the young leaders, Drs T.T. Song (Taiwan), Nick Teerachai (Thailand), Phan Ngoc Minh (Vietnam), HaiXia Zhang (China), and Nurul Taufiqu Rochman (Indonesia).

Lastly, this book would not have been possible without the continuous support and encouragement from my husband Dr Giulio Manzoni.

PREFACE

I have been involved in the world of nanotechnology in Asia for nearly 15 years engaging both in scientific research and in monitoring and analyzing government policy, research and development and industry trends in the field. I am now keen to share some insights into what is currently happening in this exciting part of the world.

I received my introduction to nanoscience in 1992, completing a PhD at the University of New South Wales and subsequently moving to Japan in 1995 to pursue my postdoctoral studies. In 1999 I started concentrating on analysing nanotechnology policy, R&D and industry trends. At this time, nanotechnology was still a relatively small field with few coordinated government funding programs and minimal investment by industry except in the advanced economies such as Japan, Germany, UK and USA. However, after President Clinton announced the US National Nanotechnology Initiative in 2000, the field has undergone profound changes especially in the Asia Pacific region. Public and private funding in nanotechnology has increased more than tenfold and the number of people working in industry and academia has also increased significantly.

Whereas Japan, a developed economy, has long been a leader in nanotechnology R&D and commercialization in this region, since 2003 developing economies such as Thailand and Vietnam have also started to pursue nanotechnology. Even Indonesia has started to become more active in this field and will launch a major nanotechnology initiative in 2009.

Over the last 10 years, I have been conducting research into government policy, R&D and industry trends through site visits, participating in major nanotechnology events and interviewing government policy makers, distinguished scientists, industry leaders and investors. I have been involved in organizing various business events in China, Hong Kong, Indonesia, Japan and Malaysia to promote nanotechnology policy, commercialization and outreach.

Although my main focus has been the Asia Pacific region, I have also been involved in other parts of the world, including the US and Europe. During this time I have had the privilege of working with some of the finest minds and most notable people in the field. One of my earliest mentors was Dr David Kahaner (known, widely, as Mr Asia), who provided me with my first introduction to the world of technology analysis through the Asian Technology Information Program (ATIP) and provided me with advice that has served me well over the years.

Other significant collaborators include the Director of Japan's Atom Technology Program, Dr Kazunobu Tanaka, who has been playing a significant role in developing nanotechnology policy and international strategy in Japan. Also Mr Yuji Tokumasu, who was the nanotechnology planning director during 2001-3 at the Japan Ministry of Economy, Trade and Industry, has provided valuable insights to me in the world of government policy.

Another important figure in nanotechnology in Japan is Dr Hiroshi Yokoyama, with whom I collaborated to create the Asia Pacific Nanotech Weekly (www.nanoworld.jp/apnw) which, since January 2003, has provided a forum for my weekly reports on nanotechnology developments worldwide, mostly covering micro- and nanotechnology policy, research, development and industry trends in the 13 economies reviewed in this book plus USA and Europe.

I have also had the privilege of exchanging information and thoughts on nanotechnology global development with the Chair, Dr M.C. Roco, and Executive Secretary, Dr J.S. Murday, of the Interagency Working Group on Nanoscience, Engineering and Technology (IWGN) at the US government.

I am very fortunate to have the opportunity to work with one of the world's best and respectful nanotechnology foresight experts Dr Gerd Bachmann (VDI, Germany) who always generously shares with me the European nanotechnology development insights and makes me realize there is always so much I don't know.

One of my passions is to bring people together to make a difference. I am extremely grateful to two of the best minds in Asia Dr Hiroshi Yokoyama and Dr Kazunobu Tanaka (one of the contributors to this book), who have shared with me insights into Japan's nanotechnology development and with whom I co-founded the Asia Nano Forum (asia-anf.org) in 2004. This network was established to facilitate collaboration amongst Asia Pacific economies in promoting nanotechnology advancement in the region. Today ANF is a society consisting of a network of members who are the drivers of their own economy's nanotechnology strategy and R&D, and has grown into a successful regional society network especially in supporting education and training of manpower, and standardization and risk management.

Asia is currently the part of the world experiencing the fastest economic growth, with nanotechnology promising to play a significant part in future growth. However, it is not always easy to collaborate successfully in this part of the world due to differences in language and in the culture of disseminating information. This book is intended to fill in some of the information gaps and help advertise the potential provided by nanotechnology in the Asia Pacific region. It summarizes the efforts of thirteen economies in pursuing nanotechnology and their determination to be world leaders in the industry and economic development in the 21st century.

This is the first book to provide a comprehensive view of the Asian capabilities in nanotechnology and should interest government policy makers, academics, industry leaders and investors who are keen to develop opportunities in Asia.

ACKNOWLEDGEMENTS

Let me first thank all the contributors of each chapter for their great support and hard work. I most grateful to my NanoGlobe team and analysts Miss JIANG Jing, Miss Yesie Brama, Dr Doris Ng and Ms ZHAO Hui for their fantastic help in coordination and format editing especially when I was swamped with deadlines and many other commitments.

My special thanks should go to the Singapore Institute of Materials Research and Engineering (IMRE) staff Dr LOW Hong Yee, Eugene LOW, TAN Wei Wei, and Jofelyn LYE as well as ItN Nanovation AG for providing the cover page pictures; and the World Scientific team Rajesh Babu, Jimmy Low, and QUEK Yeow Hwa for their fantastic help in making this book happen.

Thanks also to my dear friend Caroline Pearce for the help she provided me with the Preface and Introduction.

I would like to thank all the mentors in my career of nanotechnology including my PhD advisor Prof David Neilson, Dr David Kahaner, Dr Gerd Bachmann, Dr Hiroshi Yokoyama and Dr Kazunobu Tanaka for their inspiration and fruitful partnerships; and those who inspired my interest in nanoscience and provided nanostructure research including Dr Robert Clark (Australia), Drs Daijiro Yoshioka (Japan), Alan MacDonald (USA) and Mario Tosi (Italy).

Over the last 10 years, I have the honor to receive help and support from those who drive nanotechnology strategy, research and industry development in Asia, they include Drs Junichi Sone (Japan), Chung Yuan Mou (Taiwan), Maw-Kuen Wu (Taiwan), C.J. Chen (Taiwan), Wiwut Tanthapanichkoon (Thailand), Terry Turney (Australia), Jane Niall (Australia), Khiang-Wee Lim (Singapore), Phan Hong Khoi (Vietnam), Hanjo Lim (Korea), Sang-Hee Suh (Korea), Chunli Bai (China), Lide Zhang (China), Sishen Xie (China), Che Ting Chan (Hong Kong), Winnie Wong (Hong Kong), KaMing Ng (Hong Kong), Paul T. Callaghan (New Zealand), Richard Blaikie (New Zealand) Halimaton Hamdan (Malaysia); and especially the young leaders, Drs T.T. Song (Taiwan), Nick Teerachai (Thailand), Phan Ngoc Minh (Vietnam), HaiXia Zhang (China), and Nurul Taufiqu Rochman (Indonesia).

Lastly, this book would not have been possible without the continuous support and encouragement from my husband Dr Giulio Manzoni.

CONTENTS

OVERVIEW

1. Background

The genesis of nanotechnology can be traced back to Richard Feynman's famous lecture "There's Plenty of Room at the Bottom"[1] which he delivered to the American Physical Society in 1959. This lecture has inspired scientists and engineers worldwide to develop technologies to image and manipulate atom and molecules, and to fabricate structures and devices atom by atom, molecule by molecule. Today nanotechnology is commonly defined as the understanding and control of matter at nanoscale dimensions between approximately 1 and 100 nanometers, where unique phenomena enable novel applications. Encompassing nanoscale science, engineering, and technology, nanotechnology involves imaging, measuring, modeling, manipulating and fabricating matter at this scale.

A nanometer is one-billionth of a meter. A DNA molecule is about 2.2 nanometers wide, the typical size of bacteria is in the order of 1000 nanometers (1 micrometer) and the size of human hair is about 100 micrometers. Nanotechnology is going to revolutionize the way we make things and transform the way we live. It is able to transform multiple industries including aerospace, agriculture, automotive, chemical, energy and environment, food, information and communication, medicine and health care, security and transportation. Nanotechnology offers so many possibilities such as providing cheap and clean energy; clean water; lighter and stronger materials; faster, more powerful and energy efficient computers; an exponential increase in information storage capacity; lotus-like self cleaning surfaces; the reduction or elimination of pollution; and early detection and treatment for cancer and other diseases.

The word "Nano-technology" was coined in 1974 by Norio Taniguchi (a professor at the Tokyo Science University in Japan) where he defined the process that consists of the processing of separation, consolidation, and deformation of materials by one atom or one molecule[2]. Manufacturing (in Japanese *Monotsukuri*, meaning making things) has been a focus of the Japanese industry policy. It is known that nanotechnology enables the transformation of advanced manufacturing to make better, cheaper, and greener products. For the last two decades, Japan has been committed to the use of nanotechnology in manufacturing to stay ahead of its competitors.

Since 1999 there have been significant changes in nanotechnology development worldwide. The announcement of the US National Nanotechnology Initiative (NNI) on 21 January 2000 created a strong response from the rest of the world, with a number of countries placing nanotechnology as a priority area in their science and technology policy. Figure 1 shows the timeline of national nanotechnology initiatives and programs in the thirteen economies reviewed in this book with reference to the US NNI[3] and the European Commission (EC)'s 6th Framework Program (FP6) where "Nanotechnologies and Nanosciences, Knowledge-based Multifunctional Materials, New Production Processes and Devices" was included as one of the seven priority thematic areas[4]. In 2001 Japan, China and New Zealand all began major programs focusing on nanomaterials. The following year, Korea, Taiwan, Thailand, Australia, Hong Kong and Vietnam launched national and regional nanotechnology initiatives. Subsequently in 2006, Singapore, Malaysia and Indonesia started nanotechnology programs and are planning to launch coordinated national initiatives in 2009.

As funding has experienced a significant increase over the past decade, nanotechnology is becoming mature for commercialization. The Russian government intends to take a leadership position by launching a ten-year, USD 5 billion nanotechnology commercialization initiative in September 2007 managed by the Russian Corporation of Nanotechnologies (RUSNANO)[5].

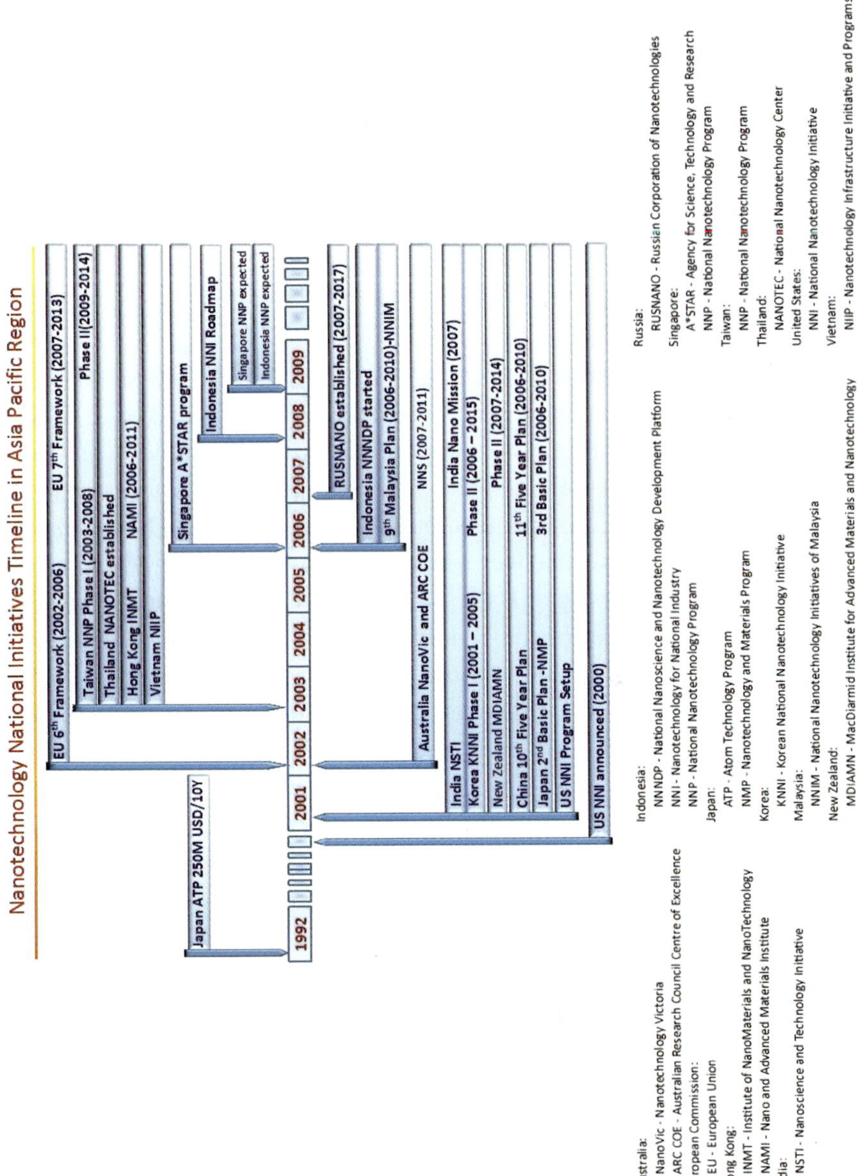

Fig. 1. Government Nanotechnology Major Initiatives Timeline in the Asia Pacific Region.

This book does not intend to provide a review of technical capabilities. Instead it provides a review of those areas that ensure sustainable development of nanotechnology and industry leadership in the thirteen economies in the Asia Pacific region, focusing on government policy and strategy, funding commitment and infrastructure, R&D and commercialization, education and outreach, standardization and risk management.

2. Public Investments

There have been significant changes in the policy of science and technology in AP countries since the announcement of the US NNI. Governments in the AP region started to place nanotechnology as one of the priority areas in their science and technology (S&T) policy and planned dedicated national nanotechnology programs. The NNI was very timely as Japan and China were in the midst of planning their second Science and Technology Basic Plan and tenth Five Year Plan respectively. US government funding in nanotechnology has reached about USD 1.5 billion in 2009[3], more than ten times the amount in 1997[6], This order of magnitude of increase in government funding is typical in economies in Asia, Europe and other part of the world in the last ten years.

Japan was the first in the world to start a major ten year nanotechnology program (the Atom Technology Program) in 1992 with the amount of about USD250 million, and was the largest government investor in nanotechnology R&D until 2003. Japan's nanotechnology development is on par with other world leaders such as the USA and Germany. Figure 2 shows the nanotechnology government funding comparison for Japan, USA, and Germany during 2001-8. Japan started the Nanotechnology and Materials Program (NMP) in 2001 when its second Basic S&T Plan began and nanotechnology was identified as one of the four priority areas (including life science, information technology, environment and nanotechnology). There was a drop in the funding in Japan from 2006 due to a new definition of nanotechnology which excluded some of the university programs[7]. The Japanese government has been investing heavily in nanotechnology and its funding per capita is the highest among the world's top three nanotechnology players. Figure 3 shows the comparison in nanotechnology public investment per capita among Germany, Japan and USA during the period 2001-2008.

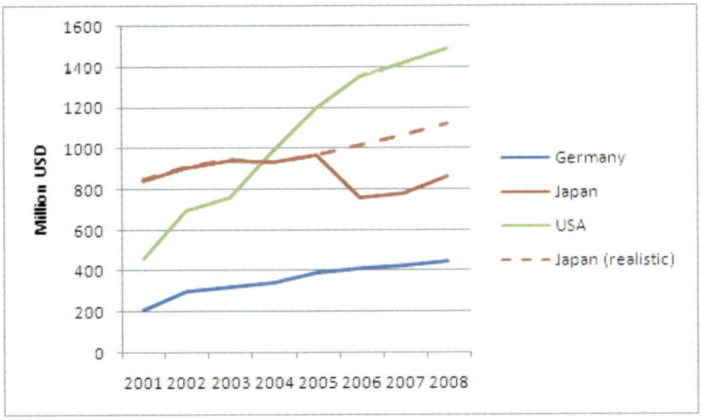

Fig. 2. Nanotechnology Government Funding Comparison among the Three Largest Economies during 2001-8.

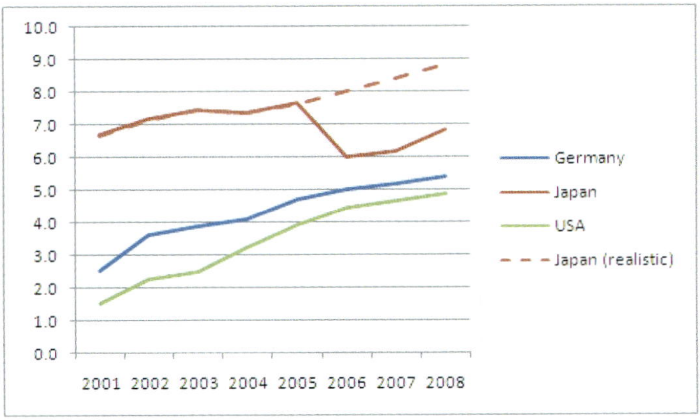

Fig. 3. Nanotechnology Government Funding Per Capita Comparison among the Three Largest Economies during 2001-8.

South Korea started its National Nanotechnology Initiative (KNNI) in 2001 committed 2.391 trillion won (USD2 billion) over the period 2001-10. After Phase 1 (2001-5) of the KNNI, the South Korean government re-launched Phase Two which will continue until 2015. South Korea aims to join the world top three nations in global nanotechnology competitiveness by 2015.

Taiwan is another ambitious player. It launched its first phase of the Taiwanese National Nanotechnology Program (NNP) in 2003 with a total budget of about USD550 million over six years. The second phase of the Taiwanese NNP started in 2009 and continues until 2014 with an allocation of over USD685 million. The aggressive building of infrastructure has been reflected in the launch of major nanotechnology infrastructure building programs for South Korea and Taiwan in 2002 and 2003 respectively.

The world's most populated country China spent RMB 2 to 2.5 billion (USD250-300 million) in the last Five-Year Plan (2001-5) which is a ten fold increase from the 1990s. China's R&D expenditure is reaching 2% GDP in the current eleven Five-Year Plan (2006-10) which is double of that in the previous Five-Year Plan. Nanotechnology is one of the four main basic science research areas which include protein research, quantum manipulation research, and growth and reproduction research.

One of the fastest growing economies, India launched a special Nano Science and Technology Initiative (NSTI) in October 2001 by the Department of Science and Technology (DST) and subsequently the Mission on Nano Science and Technology (Nano Mission) was launched in mid 2007 with an allocation of USD250 million for five years focusing on basic research, infrastructure development, commercialization, education, and international collaboration.

New Zealand's nanotechnology initiative is led by the MacDiarmid Institute for Advanced Materials and Nanotechnology which was granted NZD4.7 million per year over six years from 2001 plus a capital start-up grant of NZD10 million. The institute is joined by the major research institutions in New Zealand and comprises over forty principal investigators, thirty postdoctoral fellows, eighty PhD and twenty Master students.

In Australia in 2002, the Victoria State Government, with three major state universities including Monash University, Swinburne University of Technology and RMIT University, launched an AUD12 million Nanotechnology Victoria initiative (NanoVic). The Australian National Nanotechnology Strategy (NNS) was announced in May 2007 with AUD 21.5 million over four years aiming to focus on initiatives including health, safety and environment, public awareness and engagement, measurement, international engagement and industry activities. The Australian Office of Nanotechnology (AON), based at the Department of Innovation, Industry, Science and Research (DIISR), coordinates the implementation of the NNS.

The Hong Kong Innovation and Technology Commission (ITC) launched two strategic nanotechnology centers in 2003. One is the Institute of NanoMaterials and nanotechnology (INMT) located at the Hong Kong University of Science and Technology (HKUST) and was funded with HKD100 million for four years. The other is the Nanotechnology Center for Functional and Intelligent Textiles and Apparel (NTC) at the Hong Kong Polytechnic University (PolyU) with a total budget of HKD14.7 million for three years. In December 2004, ITC further announced a more aggressive R&D initiative by setting up four strategic R&D centers with nanotechnology selected as one of the focused areas. The Nanotechnology and Advanced Materials Institute (NAMI) was set up in 2006 with a total funding of HKD 400 million for five years. NAMI serves as an open R&D platform for Hong Kong in conducting coordinated and market-driven and demand-led nanotechnology and advanced materials development.

City state Singapore's nanotechnology strategy is led by the Economic Development Board (EDB) in coordination with other sister funding agencies such as the Agency for Science, Technology and Research (A*STAR) and universities. In 2002 the National University of Singapore (NUS) started its nanotechnology initiative called NUS Nanotechnology Science and Nanotechnology Initiative (NUSNNI) aimed at developing research, human capital and long term research capabilities. NUSNNI serves as a platform to coordinate interdisciplinary research activities and set research strategies. The Nanyang Technological University (NTU) started its nanotechnology initiative in 2005 called

Nanocluster which is a network of scientists and facilities to support nano-fabrication, characterization and exploitation of nanotechnology applications. A*STAR started in 2006 Polymer & Molecular Electronics program involving both NTU and NUS in addition to A*STAR research institutes.

Thailand leads the ASEAN countries in setting its National Nanotechnology Center (NANOTEC) in August 2003 committing USD25 million over 2004-8 from the National Science and Technology Development Agency (NSTDA).

The Vietnamese Ministry of Science and Technology launched its national nanotechnology program in December 2003, focusing on building world class research institutions, education and infrastructure as well as strategic research areas.

The Malaysian National Nanotechnology Initiative (NNIM) was officially launched in September 2006 by the then Deputy Prime Minister (who became the Prime Minister in April 2009), with a budget of MYR20 million approved by the cabinet. Malaysia is in the process of setting up its National Nanotechnology Center (NNC).

The world's fourth most populated country, Indonesia, is in the process of setting up its nanotechnology strategy. The Indonesia Society for Nanotechnology just finished the Indonesia nanotechnology recommendation and roadmap for the government. The Indonesian Government plans to launch its nanotechnology program in 2009.

Figure 4 shows the comparison of government funding among the thirteen economies during the years 2003-7. The sum of funding in the thirteen economies amounts to USD7480 million, comparable to that of the entire European Union. Figure 5 shows the comparison in government funding between the Asia Pacific region, the European Union and the USA during the period 2003-7.

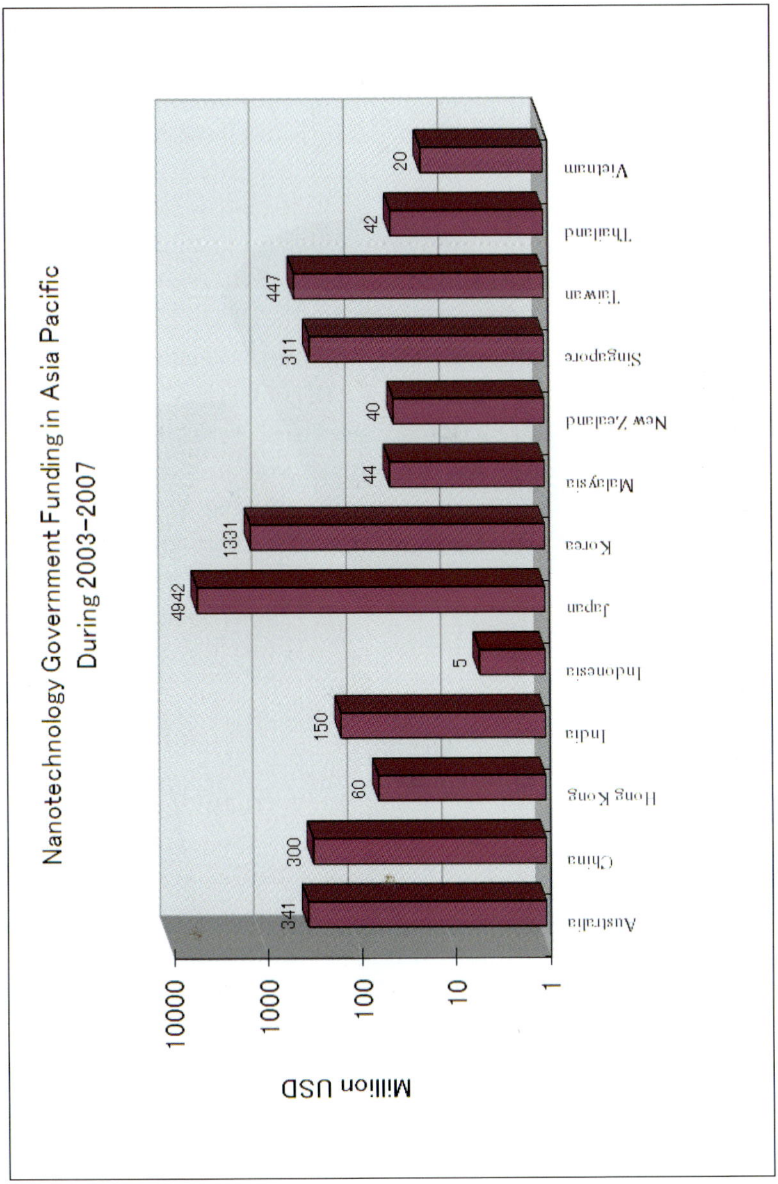

Fig. 4. Comparison of nanotechnology government funding for 2003-7 in logarithmic scale (in million USD) among Asia Pacific economies including Australia, China, Hong Kong, India, Indonesia, Japan, Korea (South), Malaysia, New Zealand, Singapore, Taiwan, Thailand and Vietnam.

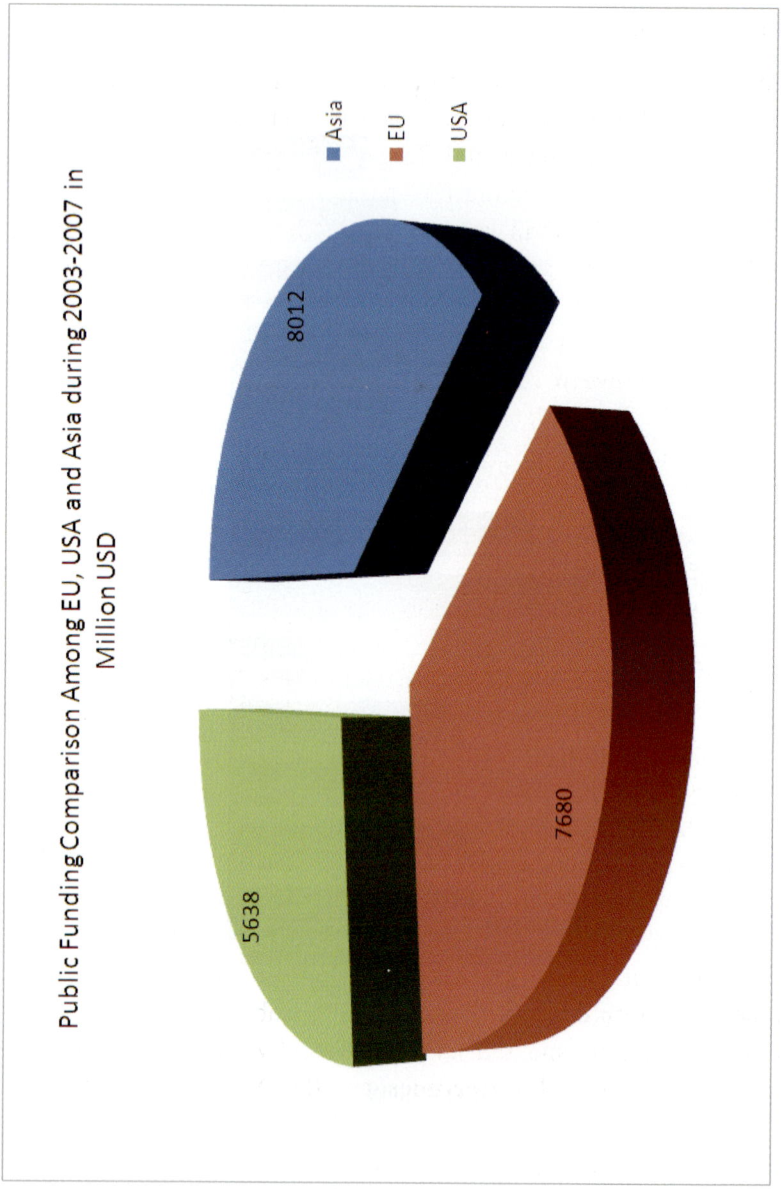

Fig. 5. Government Nanotechnology Funding Comparison for the Asia Pacific Region, Europe and USA during 2003-7.

3. Infrastructure

Japan started to construct comprehensive R&D infrastructure, especially to facilitate collaboration between private and public sector in 1996 when the Japanese first Science and Technology Basic Plan was launched. Dedicated R&D infrastructure for nanotechnology was set up during the second S&T Basic Plan started in 2001 where nanotechnology is identified as a priority area inspired by the US NNI. The Nanotechnology Researchers Network Center of Japan (Nanonet) was set up to provide coordinated facilities and information services and to promote collaboration between researchers domestically and internationally. In the third S&T Basic Plan started in 2006, the Nanonet was expanded, with stronger industry partnership focus, covering 13 centers of excellence consisting of 26 research institutions across Japan. The Japanese Ministry of Economy, Trade and Industry (METI) together with major industry conglomerates launched the Nanotechnology Business Creation Initiative (NBCI) in 2003 and current with over 300 industry and research institution members. Chapter 6, Sections 8 and 9 describe more details of Japan nanotechnology infrastructures and networks.

South Korea and Taiwan have similar structures and goals in their national nanotechnology initiatives emulating the US NNI especially its interagency coordination. Phase One of both economies' nanotechnology initiatives has the common focus of building infrastructure and strategic R&D; while Phase Two will be focusing on accelerating commercialization and creating economic impact.

During Phase One (2001-5) of the KNNI, South Korean has built world class integrated infrastructure for nanotechnology R&D and commercialization especially to promote collaboration among academia, national institutions and industry domestically and internationally. The most notable world class facilities are the National NanoFab Center (NNFC) and Korea Advanced NanoFab Center (KANC). Korea has also established well-organized researchers and business network such as Korean Nanotechnology Researchers Society (KoNTRS) and Korean Nanotechnology Research Association (NTRA). Similarly, the first phase of Taiwan's NNP have also built a major industrial common laboratory located at the Industrial Technology Research Institute (ITRI) and nine other academia core facilities centers across Taiwan. For details of infrastructure and network, please refer to Chapter 7, Section 3.5, and Chapter 11, Section 2.3.

ASEAN and other developing economies including China and India have been especially active in building R&D infrastructures. Thailand government launched its national nanotech center (NANOTEC) in August 2003. One of the key mandates for this center is to establish world class infrastructures which include central core facilities in the NANOTEC Central Research Laboratory, National Network of Centers of Excellence in nanotechnology at eight universities and a network of eight university based centers with focus on textile and cosmeceuticals across Thailand. Most recently a new beam line was installed for nanotechnology R&D at the National Synchrotron Institute. Chapter 12, Sections 3.3 and 3.4 outline more details of Thailand key infrastructure build-up. Vietnam started its nanotechnology infrastructure upgrade program in 2003. See Chapter 13, Section 3 for details. Malaysia and Indonesia both have set up nanotechnology research center of excellence. There are four main recognized nanotechnology research center in Malaysia focusing on zeolites and nanostructured materials, nanoelectronics, advanced materials, catalysts and basic science research with state of the art facilities for characterization, fabrication and processing equipments. Indonesia has budgeted the building of a major nanotechnology infrastructure at Bandung Institute of Technology in 2009 with USD20 million investments. Chapter 5, Section 3 has more information on Indonesia nanotechnology infrastructure.

In 2003, China launched the initiative of building major nanotechnology infrastructure including the National Center for Nanoscience and Technology (NCNST) located in Beijing and the National Engineering Research Center for Nanotechnology (NERCN) located in Shanghai. The NCNST is growing steadily, recruiting overseas Chinese scientists and now with a total of fifty full time faculties supervising a hundred and fifty graduate students plus another a hundred and fifty students from other research institutions who are conducting joint research projects with NCNST. The NCNST provides and manages the national characterization facility and also runs the Key Center for

Biomedical Effects of Nanomaterials and Nanosafety for the Chinese Academy of Sciences (CAS). On the other hand, the NERCN based in Shanghai is currently not in operation. The Shanghai's nanotechnology infrastructure is being coordinated by the Shanghai Nanotechnology Promotion Center (SNPC)[8] which provides not only information, networking, training platform, project funding, but also nanomaterials testing facilities. The national nanotech centers building in 2003 was driven by the National Development and Reform Commission (NDRC) and managed by the Chinese Academy of Sciences (CAS) and Ministry of Science and Technology (MOST). Chapter 2, Section 3 describes more about China nanotechnology infrastructures and key research centers.

Infrastructure development for nano science and technology research is one of the four primary objectives of the India Nano-Mission launched in 2007. There are eleven centers with state of the art R&D facilities across India plus the Center for Computational Materials Science in Bangalore. Chapter 4, Section 2 outlines more details about those centers.

Singapore, the smallest country in the region is well known for its world-class infrastructure for nanotechnology research and development. Singapore's main government research institution A*STAR set up the Science Engineering Research Council (SERC) Nanofabrication and Characterization Facility (SNFC) which serves as a central facility and open to Singapore universities, research institutions, industry as well as international collaborators. In addition, the National University of Singapore (NUS) and Nanyang Technological University (NTU) set up their own coordinated nanotech central facilities under the NUS Nanoscience and Nanotechnology Initiative (NUSNNI) and NTU Nanocluster respectively. Chapter 10, Section 2.1 provides more details.

Australia launched the National Collaborative Research Infrastructure Strategy (NCRIS) in 2005 and allocated AUD550 million to major research infrastructure projects during 2005-11. The total nanotech facilities investment is estimated to be over AUD100 million. Chapter 1, Section 2.3 outlines more about this infrastructure initiative. The MacDiarmid Institute in New Zealand received total of NZD20 million in 2002 and 2007 as capital equipment investment from the government.

Chapter 9, Section 4 describes more details about the institute's capital equipment capacity.

For locations of the main nanotechnology facilities and key research centers of the thirteen economies reviewed in this book, please refer to Fig. 6.

4. R&D and Commercialization

The Asia Pacific (AP) region is advancing to becoming one of the most ambitious and dynamic regions of the world in nanotechnology research and development and commercialization. Japan has already been successful as a world leader in electronics, advanced materials, and precision machines manufacturing. Japan is by far ahead of all other economies in the region in R&D advancement of almost every area and application of nanotechnology. Nanotechnology is considered as a key technology in the 21st century that innovates manufacturing and drives the growth of industries such as electronics, energy, environment, and biotechnology. Notably Japan nanotechnology is already in the market sectors including consumer electronics (FPD, cell phone, digital camera, digital camcorder, and DVD recorder, DRAM, flash memories); water/stain repellent and wrinkle- and shrinkage-free textile; water-resistant foundation cosmetics; DNA chip for analyzing genes; anticancer drug delivery; thin film solar cells; fuel cells; and device manufacturing with finer circuit width. Details of Japan nanotechnology commercialization are outlined in Chapter 6, Part II.

Since 2001 when the South Korea started its National Nanotechnology Initiative, it has made quantum leap in its R&D and commercialization advancement. Three major strategic nanotechnology programs were set up, namely National Program for Tera-Level Nano Devices, the Center for Nanostructured Materials Project and the Center for Nanoscale Mechatronics and Manufacturing. See Chapter 7, Section 4 for more information about these programs. There had been more than three times increase in nanotechnology related Science Citation Index (SCI) publication during 2001-6 and nanotechnology related patent application filed during 2001-5 by South Korea. The major Korean

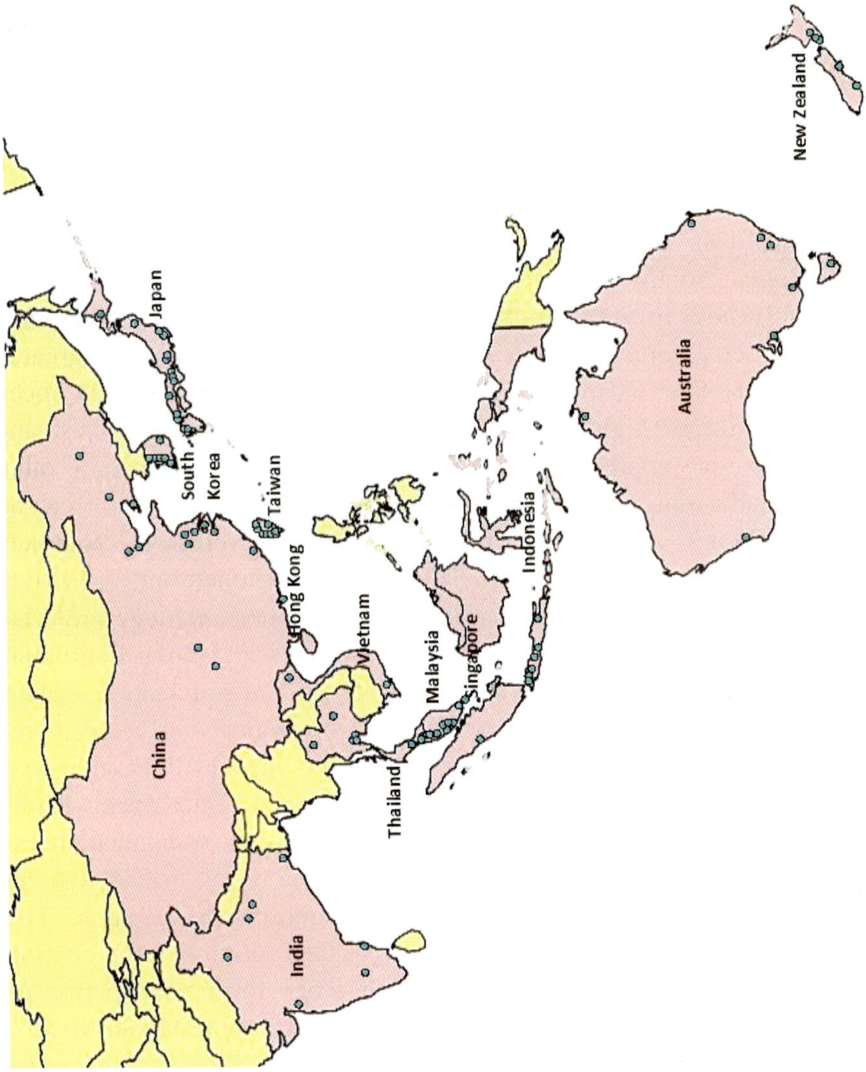

Fig. 6. Location of Nanotechnology Core Facilities and R&D Centers in Asia Pacific Region.

electronic conglomerates dominate the nanoelectronics product market. Samsung Electronics developed the world first 30 nanometer 64 gigabit NAND flash memory. A number of innovative nanotech products are produced by Korea SMEs and start ups including conductive nanosilver ink; antibacterial powders; antiglare coating; color film with nanolayered structure; Carbon Nanotubes (CNT) transparent conductive film; and pharmaceutical products for atopic dermatitis using nano-hybrid technology. South Korea was ranked fourth in nanotechnology competitiveness in 2005. In KNNI Phase Two, Korea is aggressively pursuing commercialization of nanotechnology and international cooperation. See Chapter 7, Section 5 for details of the Korean commercialization highlights.

Taiwan is another bright spot in terms of research and commercialization. Its R&D and commercialization activities in nanotechnology are mainly funded by the government National Nanotechnology Program (NNP). Two thirds of the NNP budget in both Phase One and Two go to industrialization of nanotechnology projects. Two thirds of Taiwan nanotechnology companies are in the traditional industry business[9]. Taiwan nanotechnology related publication during 2003-7 tripled. Nanotechnology innovative products developed by Industry Technology Research Institute (ITRI) include light, thin and flexible seven inch active matrix TFT-LCD panel and electrets-based flexible speaker. Highlights of Taiwan R&D and commercialization are elaborated in Chapter 11, Section 2.5.

China is known for its excellence in nanoscience research. The Chinese Academy of Sciences (CAS) was ranked fourth by total citation in the Topical Citation Report: Nanotechnology during the period of 1992-2002 (behind University of California, Berkeley, IBM and MIT)[10]. Commercialization of nanotechnology in China is still in its infancy. The Micro-Electro-Mechanical System (MEMS) technology is relatively mature and there are over thirty MEMS companies in China today active in MEMS device design, fabrication and foundry services. China Micro-Nano technology R&D and commercialization highlights are shown in Chapter 2, Sections 6 and 7.

Australia stands out in its world class research especially in controlled fabrication of silicon based transistor with nano to atomic scale precision. There are about eighty nanotechnology companies in Australia according to Australian NanoBusiness Forum (ANBF)[11]. Australia nanotechnology products include sunscreen lotion using nanoparticles, supercapacitor using nanostructured materials, bionic ear implants using biocompatible nanomaterials, and drug delivery technology using nanostructured silicon. See Chapter 1, Sections 3 and 4 for more information.

New Zealand is proud to excel in research areas such as nanofabrication and devices, electronics and optical materials, molecular materials, soft materials, inorganic hybrid materials and fusion of nanoscience and biology. In terms of application and commercialization, the MacDiarmid Institute of for Advanced Materials and Nanotechnology shows its strength in biochip for cell imaging; atomic cluster based sensors and interconnects; optoelectronic materials growth (producing world record ZnO diode); and nanoparticles for diagnostics and drug delivery applications. See Chapter 9, Section 3 for details about its research efforts.

Hong Kong nanotechnology places strong focus on commercialization especially in environment; energy; displays and electronic packaging; composites materials; advanced manufacturing; and textiles. Chapter 3, Section 3 provides a summary of the commercialization efforts in Hong Kong.

Singapore intends to build a comprehensive nanotechnology ecosystem to promote nanotechnology application development and the growth of nanotechnology industries. Singapore Economic Development Board (EDB) is the key driver for the nanotechnology strategy in Singapore. Singapore's nanotechnology R&D and commercialization focuses on nano-enabled urban solution (such as cleantech), nanotoxicology, consumer and healthcares. Singapore is the most foreign friendly economy in the region and the government is aggressively attracting multinational corporations and foreign start ups to set up R&D and manufacturing centers in Singapore. Especially since 2008, funding agencies including the National Research Foundation and SPRING launched new funding schemes to stimulate innovation, commercialization and application R&D. The areas of focus in nanotechnology application include biomedical, clean energy, environment and water, and electronics. Each of the areas involve world class R&D infrastructure, researchers, multinationals and start-ups. Details of government grant schemes are available at Chapter 10, Section 1.3, and overview of nanotechnology in Singapore is at Section 2.

India Nano-Mission established seven strategic Centers for Nanotechnology focusing R&D on nanopowders/nanoparticles, nanophosphor materials, drug delivery system, biosensors and nanoeletronics. Nanotechnology application stresses on agriculture and food systems, drinking water, automobiles, energy, biomedical devices, and metrology. More details are in Chapter 4, Section 3.

Thailand has its own priority in food, cosmeceuticals, textile; flexible polymer solar cells, DDS, and sensors for scents.

Vietnam stands out its application R&D in CNT based nanocomposites, CNT based metallic coating and electromagnetic shielding and conductive paint.

Malaysia places its research focus on nanoparticles, CNT, dendrimers, aerogel, OLED, quantum dot, nanomagnetics, single electron transistor, and DDS. Malaysia nanotechnology companies are developing CNT nanocomposites, biosensors, nanocatalysts, and biofertilizers,

Indonesia's research institutions focus its R&D areas in nanostructures, nano-encapsulation, Ag nanoparticles, nanocomposites, and nanocarbon. It envisions its nanotechnology commercialization priority should be nanomaterials manufacturing, processing and product development.

For more information and details about the R&D and commercialization efforts, please read the respective chapters.

5. Nanotechnology Network

The notable regional network is the Asia Nano Forum (ANF) which was founded in May 2004 at its first summit meeting in Phuket co-organized by Thailand and Japan. ANF expanded its network from 11 to 13 in 2005 adding Indonesia and New Zealand, and UAE joined as the 14th network economy in 2008. In 2009, Russia and Iran showed strong interest of becoming a member. ANF's mission is to promote the research and development and industrialization in nanotechnology that educationally, socially, environmentally and economically benefit each economy by fostering the international network and collaboration. ANF was officially registered as a society in Singapore October 2007 and its secretariat is currently being hosted by Institute of Materials and Engineering (IMRE) in Singapore. ANF has had five summit meetings so far hosted by its member economies during 2004-8 where nanotechnology leaders and movers from the invited economies gathered to review nanotechnology development in each participating economy and plan for future collaborations. You may find details of those meetings at the ANF website[12]. The ANF is now a network of fourteen economies in the Asia Pacific region including Australia, China, Hong Kong, India, Indonesia, Japan, Korea, Malaysia, New Zealand, Singapore, Taiwan, Thailand, Vietnam and United Arab Emirates. The thirteen chapters in this book are contributed by the nanotechnology leaders of each economy. Table 1 shows the ANF member organizations and their websites.

For more details about networks in each economies, please refer to each chapter in this book. In addition to the ANF member organizations, the nanotechnology research and industry networks across ANF economies are summarized below:

- Australia - Australian Nano Business Forum (ANBF) and Australia Research Council Nanotechnology Network (ARCNN)
- China - Chinese Society of Micro-Nano Technology (CSMNT)
- Hong Kong - Nanotechnology and Advanced Materials Institute (NAMI)
- Indonesia - Indonesian Society for Nanotechnology(MNI)
- Japan - Nano Business Creation Initiative (NBCI) and Nanotechnology Researchers Network Center of Japan (Nanonet)
- Korea - Korea Nano Technology Research Society (KoNTRS), Korea Nanotechnology Research Association (NTRA)
- Malaysia - National Nanotechnology Center (NNRC), Malaysian Nanotechnology Association (MNA)
- New Zealand - MacDiarmid Institute for Advanced Materials and Nanotechnology (MIAMN)

Table 1. Summary of Asia Nano Forum Network Organizations and Websites

Economy	Network	Network Website
Australia	Australia Nano Business Forum and Australia Research Council Nanotechnology Network	www.anbf.com.au, www.ausnano.net
China	National Center for Nano Science and Technology (NCNST)	www.nanoctr.cn
Hong Kong	Hong Kong University of Science and Technology (HKUST)	www.ust.hk/~inmt, www.nami.org.hk
India	Nano Mission, Department of Science and Technology; International Advanced Research Center for Powder Metallurgy and New Materials (ARCI)	www.dst.gov.in, www.arci.res.in
Indonesia	Indonesian Society for Nanotechnology (MNI)	www.nano-indonesia.org
Korea	Korea Nanotechnology Researchers Society (KoNTRS)	www.kontrs.or.kr
Japan	Nanotechnology Research Institute (NRI), NanoNet	unit.aist.go.jp/nanotech/index_j.html, nanonet.mext.go.jp
Malaysia	Academia Science of Malaysia, National Nanotechnology Center (NNC)	akademisains.gov.my
New Zealand	MacDiarmid Institute for Advanced Materials and Nanotechnology	www.macdiarmid.ac.nz
Singapore	Institute of Materials Research and Engineering (IMRE)	www.imre-astar.edu.sg
Taiwan	National Nanotechnology Program (NNP)	nano-taiwan.sinica.edu.tw

Table 1 (continued)

Economy	Network	Network Website
Thailand	National Nanotechnology Center (NANOTEC)	www.nanotec.or.th/index/php
Vietnam	Institute of Materials Science (IMS)	ims.vast.ac.vn
UAE	Khalifa University of Science, Technology and Research	www.kustar.ac.ae

- Singapore - Singapore Nanotechnology Network (SingNano)
- Taiwan - Nanotechnology Industry Development Association (TANIDA)
- Thailand - National Nanotechnology Center (NANOTEC)
- Vietnam - Institute of Materials Science, Vietnam Academy of Science and Technology (IMS, VAST)
- United Arab Emirates - Khalifa University of Science, Technology and Research (KUSTAR)

6. Education and Outreach

Taiwan stands out in terms of promoting nanotechnology education. A dedicated education program was included in the Taiwan national nanotechnology program to promote life-long learning of nanotechnology, provide higher quality professional education, K-12 education, and to enhance public awareness of nanotechnology. The most notable education materials include Nanotechnology Symphony book used for chemistry, physics and biology education for high school students; the Nano BlasterMan, a 3D comic book created for elementary and middle-school students; and an animated cartoon video Nana and Nono created to help educate younger children. The Taiwan nanotechnology education has inspired the Asia Nano Forum network economies to adopt their education approach and materials in their own nanotechnology education programs. The program has established five regional training centers to train students, school teachers, and professionals across Taiwan. See Chapter 11, Section 2.4 for more details about Taiwan education efforts in nanotechnology.

China is proud to lead education and promote innovation of Nano/Micro technologies among university students across China. The Chinese Society of Micro-Nano Technology (CSMNT) was formed in April 2005 followed by the establishment of an international network called Chinese International Nano Electrical and Mechanical System Network (CINN) in 2006. The CINN launched the first MEMSIC Cup in Spring 2007 which motivated university students in creating innovative Micro/Nano Electrical and Mechanical System (M/NEMS) devices and applications. This contest has attracted great support not only from students but also MEMS industry in China. It recently expanded to International Contest of Application in Nano/Micro Technologies (iCAN). iCAN2009 to be held during January 20-23, 2010. More details about Chinese efforts in Nanotechnology education are at Chapter 2, Section 2.

Korea launched its National Nanotechnology Roadmap in March 2008 with strong focus on long term application development and nurture of scientists.

Japanese media has been actively covering nanotechnology since 2000. The first nanotechnology internet based business magazine called Nikkei Nanotechnology, whose editor-in-chief is Mr Takashi Kurokawa, was launched in November 2001 within Nikkei BP (a prestigious Japanese industry and business publication). Co-currently Nikkei launched another nanotechnology business magazine called "Nikkei Sentan Gijutsu" in November 2001 which was renamed Nikkei Nanobusiness in November 2004 featuring exciting nanotechnology application and products consisting various interviews of nanotechnology scientists, engineers, industrialists in Japan and overseas. This magazine featured articles contributed by Japanese and overseas nanotechnology expert on government policy, academia and industry R&D and commercialization, corporate press releases, patent and market trends. Although the two publications were both terminated in 2005 and 2006 respectively, Mr Takashi Kurokawa is currently in charge of nanotechnology and advanced materials in Nikkei newspaper (the most dominant newspaper in Japan). Japanese public are the most knowledgeable of nanotechnology and its fascinating future opportunities in manufacturing, economy and quality of life thanks to the active media efforts.

Japan took the leadership of hosting the Asia Nano Forum during 2004-7 until its official Secretariat was set up in Singapore October 2007. Dr Hiroshi Yokoyama and Dr Kazunobu Tanaka have been the key drivers of Japan's efforts in international collaboration and network building. Dr Yokoyama heads the ANF Human Resource Working Group and initiated the first Asia Nano Camp 2008 held in Japan brought together 35 top young PhD students and researchers below 30 years old selected from the ANF network economies during 4-21 February 2008. This nanotechnology camp was the event to educate and promote the advancement of nanotechnology and provided a precious opportunity for young scientists from the Asia Pacific region to build friendships. The camp was co-organized by Japan National Institute of Advanced Industrial Science and Technology, National Institute of Materials Science and Tokyo Institute of Technology.

Japan has been the host for the world's largest nanotechnology annual trade show Japan nano tech 200X since 2002 where over 47,000 visitors participated and 600 companies and organizations exhibited from 20 countries at their latest show nanotech 2009[13]. Mr Takahiro Matsui, Secretary General of the Japan nanotech executive committee, initiated and founded this exceptionally successful nanotechnology event. Inspired by this success, Korea started the NanoKorea in 2003. Subsequently, Hong Kong and Taiwan organized their first nanotechnology industry conference and exhibition in 2004. And China started ChinaNano2005 in June 2005 mainly a research conference.

The developing economies such as India, Indonesia, Malaysia, Thailand and Vietnam placed education as a top priority in their nanotechnology and science and technology strategy. Most economies in the Asia Pacific region have set up nanotechnology courses in the MSc and PhD programs at their universities. An interesting trend happening recently is the return of the young generation of overseas educated scientists to their home countries taking up faculty position and program management at research institutions and government funding agencies.

This phenomenon is happening in China, Korea, and Taiwan as well as in the ASEAN countries. For example, the National Center for Nanoscience and Technology of China consistents of 80% "HaiGui" (overseas returned) scientists aged between 30-40 years old. Singapore universities and research institutions are aggressively recruiting renown scientists worldwide both full time and part-time to lead their R&D efforts, and government funding agencies are appointing experienced industry experts to join their advisory boards and funding scheme evaluation panels. The ANF is playing a pivotal role in education in the region through its Education Working Group led by Taiwan and the Human Resource Development Working Group led by Japan. Asia Nano Camp has been experienced to be an excellent platform for the young scientists to learn the international development of nanotechnology and build collaborations at early age.

7. Standardization and Risk Management Efforts

As nanotechnology commercialization accelerates, it has become increasingly important to consider standardization and risk management of nanotechnology products. With no surprise, Japan is the first country in Asia started the societal implication of nanotechnology initiative in August 2004 through an open forum called "Nanotechnology and Society" inspired by the First International Dialogue on Responsible Research and Development of Nanotechnology organized by the USA government June 2004. This open forum became the driving force for both domestic stakeholders and international collaboration in pushing forward nanotechnology risk management, standardization and public acceptance. It stimulated the unique cross ministerial initiative to address the societal implications of nanotechnology in Japan. The outcome of this open forum has led to the inclusion of responsible R&D of nanotechnology as a priority research topic in the recent third Japanese Science and Technology Basic Plan.

Japan is an active member of International Organization for Standardization (ISO) and International Electrotechnical Commission (IEC). Japan provides recommendations and actively involved in nanotechnology standards in nanotechnology technical committee

ISO/TC229 (Nanotechnology) and IEC/TC113 (Nanotechnology standardization for electrical and electronics products and systems). Japan nanotechnology business network NBCI coordinates and facilitates the development of nanotechnology industrial standards together with their over 300 industry members in Japan. Internationally, Japan works closely with USA, Europe and Asia in standardization and risk management. In particular, Japan is a founding member of the ANF which serves as a platform for regional collaboration in different aspects of nanotechnology including standardization and risk management, infrastructure and resources, education and strategic development. Japan's effort in societal implication of nanotechnology implies Japanese nanotechnology industry leadership and commitment for sustainable development of nanotechnology. More details about Japan efforts in societal implication of nanotechnology are outlined in Chapter 6 Part I Sections 3-7.

Korea released its Nanotechnology Standardization Development Roadmap in October 2007 to ensure standards are incorporated even at initial R&D stage and to secure Korea nanotechnology product market share through collaborating with international standards.

Taiwan is a liaison member representing ANF at ISO/TC229 and IEC/TC113. Taiwan pioneered nanotechnology product regulation by setting up nanoMark System. In order to assess risk and systematically manage nanotechnology products in the market, to enhance public trust in nanotechnology development, and protect consumers, Taiwan government set up the nanoMark certification mechanism and guidelines for nanotechnology products. In the Phase 2 of National Nanotechnology Program starting in 2009, it allocated 9% of the total program budget, about USD60 million to tackle health and risk issues. Taiwan has formed the National Nanotechnology Standardization Council in 2009 to centralize planning and communication. Chapter 11, Section 4 provides details about the nanoMark System and list of certified nanotech companies in Taiwan.

Singapore launched the ASTAR Center for Nanometrology Excellence aiming for nanoscale measurement facility set up with SGD10 million committed in 2007. The center is housed in the Singapore Science Park, a hub for R&D and innovation. It provides calibration/measurement services for MEMS, AFMs, LEDs, fiber optics, sensors and piezo-electric devices. The users of the facility include Research Institutes, nano- and biotech firms, semiconductor companies and wafer fabs. Singapore Economic Development Board is currently coordinating the nanotox initiative to be launched in 2009.

Table 2 summarizes the activities in nanotechnology standardization and risk management in 13 economies reformatted from Dr Tsung Tsan SU's presentation given at the Asia Nano Forum Summit 2009 on 16 November 2008.

8. Conclusion

Asia, with more than half of the world's population and including some of the fastest growing economies, has a large and motivated population, and a large and growing market. An increasing number of overseas-educated Asian scientists, engineers and industrialists are taking greater leadership in the advancement of nanotechnology across the region. The region's young leaders have begun to collaborate, facilitated by the Asia Nano Forum. Japan, Korea, and Taiwan are recognized as industry leaders of nanotechnology commercialization, especially in precision instruments, advanced materials and electronics manufacturing. China, India, Australia and New Zealand would be excellent research partners thanks to their excellent research capabilities. Hong Kong is attractive in terms of financing R&D and also in locating R&D headquarters to take advantage of its vicinity to China – a potential manufacturing base. Singapore provides an efficient and strategic location that protects IP and focuses strongly on financing specific nanotechnology applications. It provides excellent funding and business development support from both government and the private sector. Other ASEAN economies, especially Thailand and Vietnam, are catching up quickly.

Over the last ten years (1999-2009) the region has experienced fast and exciting developments in nanotechnology. Japan continues to be the leader in basic research and commercialization with South Korea and Taiwan catching up quickly. Singapore is aggressively building a nanotechnology ecosystem and attracting foreign technology and talent to conduct development and manufacturing. Hong Kong is working in close partnership with mainland China where it taps into its manpower and market while focusing on clean energy, environment and water, textiles and electronics applications. China and India are pumping more funding and looking both long and short term. Progress in Indonesia, Thailand, and Vietnam are driven by young, passionate and foreign educated talent. On the periphery, Australia and New Zealand are both building strong alliances with Asia and the rest of the world.

Lerwen LIU
Managing Director, NanoGlobe Pte Ltd, Singapore
lerwen@nano-globe.biz
April 2009

Table 2. Summary of the activities in nanotechnology standardization and risk management in ANF member economies where information is available

Economy	Technical Standardization Committee	Terminology and Nomenclature	Measurement and Characterization	Health, Safety and Environment	Budget (MUSD)
Australia	NT-001 (2005). Coordinated development of TC229 Business Plan	Very active participation in the work of TC229/WG1 and road mapping	Active participation in the work of TC229/WG2 and road mapping. NMIA have established a nanometrology program	Very active participation in the work of TC229/WG3, road mapping and the development of TR "Health and safety practices in occupational settings relevant to nanotechnologies"	
China	National technical working group for nanomaterials standardization (2003) Committee on Nanotechnology Standardization (SAC/TC279) (2005) Convenor of ISO/TC229/ WG4 NWI	GB/T19619 ~2004 (Basic Standards) GB/T 21511.1-2008	27 Nano-dimensional M&C standards 12 nano materials/ products standards	Established Nanosafety Lab in 2003 at the Institute of High Energy Physics, CAS Nanosafety Join Lab established in National Center for NanoSci. and Tech. (2006, 6.22)	
Hong Kong			Optical traceable AFM for the nanoscale length calibration	Nanotoxicology	
India	National Technical Committee on Nanotechnology 1st meeting (2007.11.23)		Nanometrology at NPL, New Delhi	Major initiative to be launched on toxicology, environmental studies, safety and healthcare and ethical issues	

Table 2 (continued)

Economy	Technical Standardization Committee	Terminology and Nomenclature	Measurement and Characterization	Health, Safety and Environment	Budget (MUSD)
Japan	Nanotech standardization committee (2005), lead Asia collaboration on standardization for photocatalysis materials and products	Working on a draft definition of terms of NanoCarbon	Active in NanoCarbon and Photocatalyst Standardization Effort 5-year Project on "Research and Development of Nanoparticle Characterization Method" (FY2006~2010)	5-year Project on "Evaluating Risks Associated with Manufactured Nanomaterials" (FY2006-2010, 17MUSD)	
S. Korea	Nanotech standardization committee (2006); CNT standard technology committee (2004)	KSL 1622-2006 (Photocatalyst)	Active in nano Ag and CNT; 5 Korean standards (KS D2711- D2715), Active in ISO/TC (contribute 9 items)	3-year Project on Risk Assessment of Nanomaterial	Ÿ CNT: 0.66 MUSD (2006~ 2009) nano-scale: 2.2 MUSD (2006~ 2011)
Malaysia	Technical Committee on Nanotechnology (17 members) (2005)	National TC and WG Meeting (2007~2008)	National TC and WG Meeting (2007~2008)	National TC and WG Meeting (2007~2008)	0.035 MUSD

Table 2 (continued)

Economy	Technical Standardization Committee	Terminology and Nomenclature	Measurement and Characterization	Health, Safety and Environment	Budget (MUSD)
New Zealand				Cosmetics containing nanoparticle → ERMA[(1)]	
Singapore	Working Group for ISO/TC229	Follow ISO's	5-Year metrology roadmap (2004-2008) A new metrology roadmap prepared (2009~2013) Center for nanometrology Excellence (10SGD, 2007)	Health Effects of nanomaterials (limited activities) Nanotox initiative to be launched 2009	10M USD (2005-2007)
Taiwan	Two NanoMark Committees (2004), "Taiwan Nanotechnology Standard Council"	Terminology Standard for Nanomaterials (CNS 14975-2006), Nanotechnology Dictionary (2007. 12)	Consumer Products: 15 Test Protocols, 3 drafts 1 Photocatalyst Standard (CNS15094-2007)	Reviewing Regulatory Framework Active in QD, ZnO	2 MUSD

Table 2 (continued)

Economy	Technical Standardization Committee	Terminology and Nomenclature	Measurement and Characterization	Health, Safety and Environment	Budget (MUSD)
Thailand	Technical Committee on Nanotechnology (2005) lead by TISI(2)	Participation in the work of ISO/TC229/ JWG1 Developing standard terminology for nanomaterials (2009)	Using the existing testing protocols Follow ISO TC229/JWG2 Initiating Inter-laboratory Comparison on Nanoparticle size Measurement Program (2009) Developing material specification of Ag, TiO2, ZnO nano-particles (2009) Preparing Nanometrology for hard disk drive industry Metrology facilities (step height standard calibration)	National Technical Committee on Safety of nanomaterial and nanotechnology Nanomaterials Safety Program (2008~2012) Participation in work of TC229/WG3 and OECD Chemicals Committee Initiating Safe Handling of Nanotechnology (2009) Inter-government forum on chemical safety focusing on manufactured nanomaterials	0.03 MUSD (2007~2008)

Note: (1) ERMA: Environmental Risk Management Authority; (2) Thailand Industrial Standards Institute

References

1. http://www.its.caltech.edu/~feynman/plenty.html
2. N. Taniguchi, "On the Basic Concept of 'Nano-Technology'," Proc. Intl. Conf. Prod. Eng. Tokyo, Part II, Japan Society of Precision Engineering, 1974.
3. National Nanotechnology Initiative www.nano.gov
4. European Commission Nanotechnology http://cordis.europa.eu/nanotechnology
5. Russian Corporation of Nanotechnologies www.rusnano.com
6. Nanostructure Science and Technology, edited by Richard W. Siegel, Evelyn Wu and M.C. Roco (1999)
7. Japan Council for Science and Technology Policy (CSTP) http://www8.cao.go.jp/cstp/english/index.html
8. http://www.snpc.org.cn/english/index.asp
9. Taiwan Nanotechnology Industry Development Association www.tanida.org.tw
10. Science Watch http://archive.sciencewatch.com/july-aug2003/sw_july-aug2003_page2.htm
11. Australian NanoBusiness Forum www.anbf.com.au
12. Asia Nano Forum www.asia-anf.org
13. www.nanotechexpo.jp

Author Biography

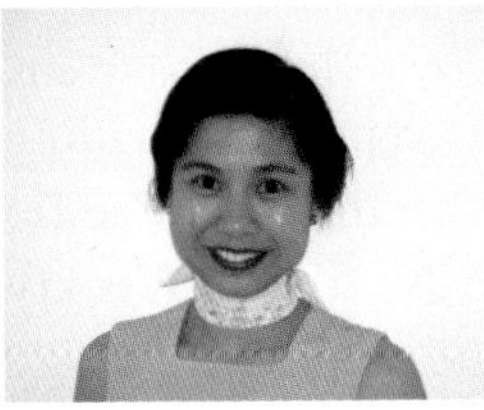

Lerwen LIU is an Asia-based nanotechnology expert specializing in government and corporate strategic services to policy makers and corporate executives. She also specializes in business development and project management for leading nanotechnology companies in their global expansion of R&D, commercialization and market. Since 1999 she has been actively building nanotechnology networks with government agencies, R&D institutions and industries across the world, especially promoting nanotechnology policy and cooperation in the Asia region. She co-founded the Asia Nano Forum (a Singapore-based network organization linking 14 Asia Pacific economies) and founded the SingNano network, with the support of A*STAR, EDB and SPRING in Singapore, which provides a platform for nanotech government, academic and industry leaders to collaborate and accelerate commercialization of R&D. She liaisons closely with government agencies, R&D organizations and industries Asia, North America and Europe and promotes nanotechnology policy, research activities, products and partnerships to the micro/nanotechnology R&D and business community in Asia.

Dr Liu is a director of Zyvex Asia managing its operation in Singapore and its Atomic Precise Manufacturing Program (APM). She is also a strategic advisor and a member of the Board of Directors of Nanostart Asia actively contributing to Nanostart's Asia investment business in Asia Pacific region. She is an invited evaluation panel member for the Proof-of Concept (POC) grant scheme of the National Research Foundation (NRF), Singapore. Dr Liu is the Managing Director and founder of NanoGlobe (www.nano-globe.biz), a consulting company specializes in nanotechnology R&D and business strategy, business development and incubation services.

She is on the advisory board of the first global "Nanotechnology Opportunity Report" published in March 2002, the most comprehensive market report on nanotechnology "Nanotechnology Market and Company Report-Finding Hidden Pearls" written by a team of experts from

Deutsche Bank and University of Ulm in Germany, and the Nanotechnology Expert Committee of the Malaysian Government Researchers' Directory Exchange (REDEX).

Dr Liu has written (during 2003-2006) over 150 reports providing insights on nanotech policy, R&D and business trends in the Asia Pacific region. Her reports are available at the Asia Pacific Nanotech Weekly (APNW, www.nanoworld.jp/apnw) sponsored by the Nanotechnology Research Institute (NRI) of Japan National Institute of Industry Science and Technology (AIST).

Dr Liu has a PhD in physics specializing in many-body effects and transport in semiconductor nanostructures, and has conducted research work in Australia, Japan, USA and Italy.

Dr Liu is an Australian citizen and currently based in Singapore.

Appendix of Glossary

AIST - National Institute of Advanced Industrial Science and Technology (Japan)
ANF - Asia Nano Forum
ARC - Australia Research Council (Australia)
ASEAN - Association of Southeast Asia Nations
CNT - Carbon Nanotubes
DDS - Drug Delivery System
FPD - Flat Panel Display
OLED - Organic Light Emitting Diode
IWGN - Interagency Working Group on Nanoscience, Engineering, and Technology
LED - Light Emitting Diode
MEMS - Micro-Electro-Mechanical System
METI - Ministry of Economy, Trade and Industry (Japan)
NNI - National Nanotechnology Initiative
SME - Small and Medium-sized Enterprise

CHAPTER 1

NANOTECHNOLOGY IN AUSTRALIA

Michelle Y. Simmons

Department of Physics, University of New South Wales
Sydney, NSW 2052, Australia
michelle.simmons@unsw.edu.au

Thomas W. Barlow

Department of Physics, University of New South Wales
Sydney, NSW 2052, Australia
twbarlow@barlowadvisory.com

Nanotechnology research has been growing steadily in Australia in recent years, largely underpinned by the broad support mechanisms that exist in Australia for fundamental research. Current policies suggest that any future expansion in nanotechnology research across Australia will depend upon overall increases in investment for research and development. The indications are, though, that Australia is building a healthy cohort of nanotechnologists, nanotechnology infrastructure, nanotechnology research centres, and nanotechnology companies.

1. Introduction

In relative terms, Australia's greatest strengths in research have lain traditionally in areas reflective of the country's history, economy, and geography. By their share of global publication count, it could be argued that Australian researchers make their greatest contributions in areas such as Astronomy, Plant Sciences, Geosciences, and Education. For many years, though, Australia has also supported excellent research in physics, chemistry and engineering, with over a quarter of all Australian research publications being generated within these fields, a proportion similar to that gained by publications in clinical medicine[1].

With the development of new technologies for interrogating and manipulating nature on the nanoscale, an increasing proportion of research within these latter areas could be regarded as nanotechnology – a fact which has been demonstrated fairly clearly in a recent report on the field[2]. Noting that nanotechnology could include many aspects of modern physics, chemistry, engineering and indeed biology, it is impossible in a review of this length to provide a comprehensive overview of the field in Australia. It is possible, however, to provide a snapshot of some of the research activities that are occurring in this emerging area, and some of the ways in which this research is being supported through government policies.

2. Policy and Infrastructure

A number of Australian Governments have provided support for nanotechnology research and development in recent years. Most of this support, however, has been allocated through normal grant processes and does not represent money specifically ear-marked for nanotechnology.

For example, while the Australian Government's R&D Tax Concession policy has assisted a range of companies involved in nanotechnology research, it has not specifically targeted such companies. Likewise, while there was one year (in 2003) when the Australian Research Council (which supports fundamental research in Australian universities) gave special funding priority to nanomaterials research, this has been a relatively uncommon practice in Australia: for the most part, nanotechnology research initiatives have been funded in open competition with other kinds of research initiatives.

This does not mean, however, that funding for nanotechnology has been slight. A recent estimate suggests that the Australian Research Council (ARC) has spent more than $315 million supporting more than 700 projects in nanotechnology over the past decade[3]. If one factors in additional investment from universities and medical research institutes and from government agencies such as the Commonwealth Scientific and Industrial Research Organisation (CSIRO) and the Australian Nuclear

Science and Technology Organisation (ANSTO), the level of public sector support for nanotechnology would be quite substantial. Indeed, if company investments are included the total investment in nanotechnology research and early-stage commercialization has been estimated at around $100 million per annum in Australia.[1]

Some governments, both state and federal, have announced targeted programs to support the broader national investment. In May 2007, the Australian Government announced funding of $21.5 million over four years to establish a 'National Nantoechnology Strategy' designed to develop regulations and standards, to provide public health assessments of new nanotechnologies and to provide balanced advice to the community about nanotechnology. In 2002, the Victorian Government provided $12 million over three years to Nanotechnology Victoria, a joint venture of three Victorian universities. This funding was provided to coordinate Victorian nanotechnology R&D, with a focus on catalyzing industrial partnerships and commercialization of university nanotechnologies. Predominantly, though, such targeted initiatives have been the exception rather than the rule.

We will look now specifically at some of the networks, centres, facilities, and education programs in nanotechnology currently operating around Australia.

2.1. *National Research Networks*

There exist a range of formal networks, web portals, and emerging community or industry bodies relevant to nanotechnology research in Australia. Many of these are summarized in Table 1. Some of these bodies have sprung up spontaneously from across the Australian research sector; others are the result of specific government programs. Three research networks supported by the Australian Research Council are particularly relevant: the ARC Nanotechnology Network; the Molecular

[1] This is an estimate made by the Australian Government and publicized on the "Invest Australia" website: http://www.investaustralia.gov.au/IndustrySectors/Nanotech/.

Table 1. Selected Australian research networks and portals for nanotechnology

Network or Portal	Description	Website
Australian Nanotechnology Alliance	A membership-based industry body, which seeks to promote nanotechnology with governments and the community.	www.nanotechnology.org.au
Australian Research Council Nanotechnology Network	An ARC research network, which promotes collaboration and intellectual exchange within the Australian nanotechnology research community.	www.ausnano.net
Australian Research Network for Advanced Materials	An ARC research network, which links materials researchers to expertise, facilities, industry partners, and other researchers nationally and internationally.	www.materials.com.au
Future Materials	A portal run by the Australian Materials Technology Network, and partly funded by the Australian Government, to improve Australian companies' access to new materials technologies.	www.future.org.au
Invest Australia	An Australian Government agency that promotes Australia's network of "more than 75 nanotechnology research organisations and over 80 nanotechnology companies".	www.nanotechnology.gov.au

Table 1 (continued)

Network or Portal	Description	Website
Molecular and Materials Structure Research Network	An ARC research network, which links scientists involved in the determination and analysis of molecular structures.	mmsn.chem.usyd.edu.au
NanoVic	A company jointly funded by the Victorian state government and three universities to invest in research infrastructure and to commercialize nanotechnologies.	www.nanovic.com.au

and Materials Structure Research Network; and the Australian Research Network for Advanced Materials.

The ARC Nanotechnology Network seeks to promote collaboration among nanotechnology researchers, to enhance access to nanotechnology research infrastructure in Australia, and to expose nanotechnology researchers to novel ideas and approaches being pursued across all areas of nanotechnology research. It sponsors conferences, workshops, summer and winter schools and short courses in nanotechnology and provides funding support to postgraduate students and early career researchers.

While not exclusively focused on nanotechnology, the Molecular and Materials Structure Research Network unites Australian researchers engaged in the determination and analysis of molecular structures with grid-enabled access to remote instrumentation, visualization tools, structural databases, and computational and analysis facilities. Many of the researchers involved in this network are working on scientific and technological problems at the nano-scale.

The Australian Research Network for Advanced Materials also encompasses a large number of nano-materials research groups. This network seeks to foster greater interdisciplinary interaction and to support early career researchers across four areas of materials science: (i) materials drivers for high-tech information technology, communications and sensor applications; (ii) innovative structural and functional materials for diverse applications; (iii) materials solutions for advanced manufacturing; and (iv) materials for a sustainable Australia.

There are also a number of nanotechnology portals, organized by governments and industry groups, which are providing a growing sense of community and cohesion among nanotechnology researchers and business people around Australia.

2.2. *Key Research Centres*

If we were to list all Australian research nodes with some expertise or research activity relevant to nanotechnology, we would need far more space than is available here. Besides, an effort is already being made to do something along these lines on a regular basis by the Australian Government[3]. For this reason we have chosen to highlight just a few examples of the publicly funded research centres currently performing nanotechnology research in Australia. It should be understood that these centres are indicative only of the diversity of nanotechnology research currently taking place in Australian research organizations. In all cases the centres result from multi-institutional collaborations which are intended to provide a national effort in these research fields.

2.2.1. *ARC Centre of Excellence in Design in Light Metals*

The ARC Centre of Excellence in Design in Light Metals is conducting research into the structural design of alloys and hybrid materials based on aluminium, magnesium and titanium. As part of a comprehensive research program, the centre is developing new alloy nanostructures through controlled nucleation and creating new hybrid materials from nano-scale composites.

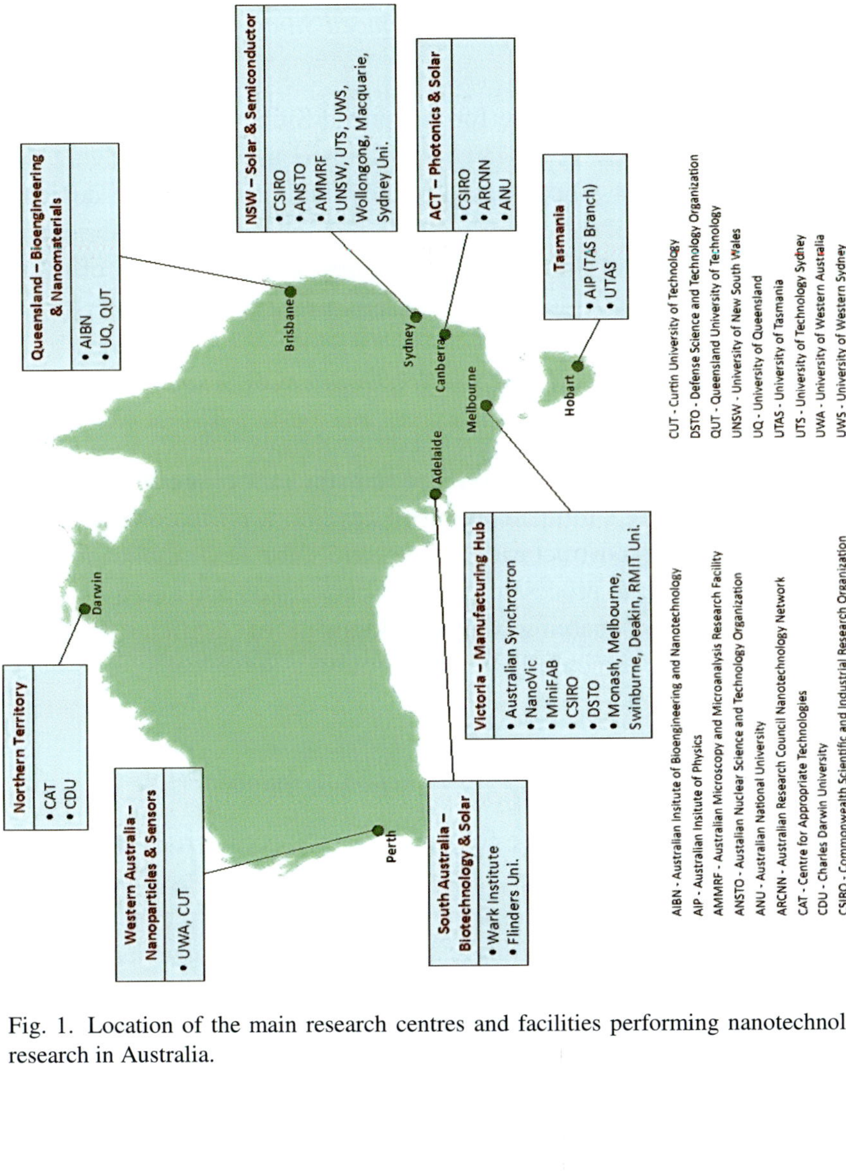

Fig. 1. Location of the main research centres and facilities performing nanotechnology research in Australia.

2.2.2. *ARC Centre of Excellence for Advanced Silicon Photovoltaics and Photonics*

The ARC Centre of Excellence for Advanced Silicon Photovoltaics and Photonics supports a large research program focused on improving the efficiency and cost of silicon-based photovoltaic and photonic devices. An important aspect of the centre's research is focused on the fabrication and characterization of silicon nanostructures to create highly efficient 'third generation' photovoltaic devices.

2.2.3. *ARC Centre of Excellence for Electromaterials Science*

The ARC Centre of Excellence for Electromaterials Science is studying the processes of charge generation and transfer in a range of nanoscale materials. This centre's ultimate aim is to develop ways to control charge transport through nanostructuring of materials for new applications in biomedicine, industrial processing, energy conversion and energy storage – for example, by enabling the development of artificial muscles, synthetic enzymes, electronic textiles, plastic solar cells, and new chemical/bio sensing technologies.

2.2.4. *ARC Centre of Excellence for Functional Nanomaterials*

The ARC Centre of Excellence for Functional Nanomaterials carries out research into the synthesis, characterization and application of nanomaterials including: metal oxide and other inorganic nanoparticles; carbon nanotubes and nanofibres; thin films and molecular sieving membranes; and polymer nanoparticle composites and nanobiomaterials. The centre is especially focused on developing materials for application in clean energy production, in environmental remediation, and in bone and tissue repair.

2.2.5. *ARC Centre of Excellence for Quantum-Atom Optics*

The ARC Centre of Excellence for Quantum-Atom Optics investigates the quantum nature of multiple particle quantum states of atoms and photons, including entangled light and Bose-Einstein condensates (BEC). The Centre combines expertise in quantum optics with atom optics and laser cooling to focus on fundamental research, with the long term goal of underpinning and developing the next generation of quantum technology. Specifically they are able to investigate quantum behaviour from the microscopic or nano world of a few particles up to the macroscopic world.

2.2.6. *ARC Centre of Excellence for Quantum Computer Technology*

The ARC Centre of Excellence for Quantum Computer Technology undertakes research on the fundamental physics and technology of building, at the atomic level, both a solid state quantum computer in silicon and an optical quantum processor using linear optics. The Centre combines expertise in semiconductor device nanofabrication, atom manipulation, epitaxial growth, cryogenic measurement capability and advanced optical circuitry. The research objectives are underpinned by extensive theoretical work in semiconductor device modelling, quantum algorithms, molecular modeling and quantum information.

2.2.7. *CSIRO Niche Manufacturing Flagship*

The Niche Manufacturing Flagship is a substantial research initiative, led by the Australian Government's largest research agency, to support niche industries as well as existing high-value manufacturing businesses in Australia through nanotechnology. The Flagship builds upon CSIRO's research strengths in composite materials, polymer engineering and fabricated devices. The Flagship has also been tasked with research assessing the health, safety and environmental issues raised by nanotechnology.

2.3. *Major National Research Infrastructure*

Australian governments, universities and research agencies support a range of important infrastructure initiatives to underpin Australian research in nanotechnology. There are substantial investments in buildings, laboratories, and equipment right across the research sector. For the most part, these investments are not easy to disaggregate from broader investments in research infrastructure across the physical science and engineering disciplines. In a few instances however, where governments have made targeted commitments to major infrastructure projects, it is possible to point to specific infrastructure that benefits nanotechnology research. Figure 1 shows the locations of main research centers and facilities in Australia.

One example is the Australian Synchrotron, which was recently built in Victoria at a cost of over $200 million. Another example is the OPAL nuclear research reactor at the Australian Nuclear Science and Technology Organization (ANSTO), which was recently built in NSW at a cost of roughly $300 million. Further examples of major infrastructure investment can be found within the National Collaborative Research Infrastructure Strategy (NCRIS), a policy initiative through which the Australian Government is allocating $542 million to major research infrastructure projects between 2005 and 2011. Selected initiatives are summarized below.

2.3.1. *Australian Microscopy and Microanalysis Research Facility*

The Australian Microscopy and Microanalysis Research Facility (AMMRF) is a joint venture of several Australian universities, funded by the Australian Government through NCRIS, which aims to provide nanostructural characterization capabilities and services to the broader Australian research community. With nodes in most Australian states, the facility builds upon existing infrastructure within Australian universities to provide a network of instrumentation including spectroscopic, optical, electron, scanning probes, x-ray and ion-beam tools.

2.3.2. *Australian National Fabrication Facility*

Through NCRIS, the Australian Government has created the Australian National Fabrication Facility to provide nanotechnology fabrication infrastructure to over a hundred research groups distributed around

Australia. The principal node of this facility is Vicfab, a joint venture between the Victorian Government, five Victorian universities, CSIRO and Minifab (a company involved in polymer micro-fabrication). However, there are six other nodes distributed across other states and territories. Collectively this distributed national facility will support infrastructure for a range of purposes, including the nanofabrication of photonic devices, the processing of nano-bio and soft materials, nano-electronics research, optical fibre fabrication, and nano-scale patterning.

2.3.3. *Australian Synchrotron*

No discussion of major national research infrastructure relevant to nanotechnology would be complete without mention of the Australian Synchrotron. This facility, which is based in Victoria, was officially opened in 2007 and will inevitably play an important role in the characterization of materials at the nano-scale across Australia in coming years. At full capacity, the Australian Synchrotron is expected to house more than 30 beamlines. The facility is expected to attract considerable use from nanomaterials researchers right around Australia.

2.3.4. *OPAL – Australian Nuclear Research Reactor*

The newly installed nuclear research reactor (OPAL) at the Australian Nuclear Science and Technology Organization (ANSTO) in NSW provides Australian researchers with significant capability for performing neutron scattering experiments. Recent additional investments have been made through NCRIS to augment the OPAL neutron characterization capability with a deuteration facility. There is a community of researchers based at ANSTO and more widely around Australia who use these facilities for nanostructural analysis.

2.4. *Education Programs*

In recent years, a number of Australian universities have begun to offer degree programs in nanotechnology. Institutions that have offered three-year or four-year Bachelor of Science in Nanotechnology courses (or equivalent) include: the Curtin University of Technology (WA); the Flinders University of South Australia (SA); La Trobe University (Victoria); Murdoch University (WA); the Royal Melbourne Institute of Technology (Victoria); the University of New South Wales (NSW); the University of Technology, Sydney (NSW); the University of Western Australia (WA); the University of Western Sydney (NSW); and the University of Wollongong (NSW). The majority of universities in Australia, however, continue to teach nanotechnology as an element of traditional programs in science and engineering rather than as a degree program in its own right.

It can be noted that a range of vocational education institutions across Australia provide training programs for technicians and tradespersons to develop skills suitable for jobs in advanced manufacturing. These programs may include some nanotechnology components. There are, however, to our knowledge no specific vocational training programs run through the major Australian vocational training institutes that are targeted explicitly at nanotechnology.

3. Research and Development Highlights

There are many examples of outstanding research taking place in nanotechnology across Australia. There are numerous strong and well-known groups led by scientists with international reputations working in areas such as nanobiotechnology, nanoelectronics, nanomaterials, and nanophotonics. Rather than attempt to describe the extensive work undertaken in this area in Australia, the nanotechnology research we are best placed to describe, however, is that which is conducted by one of us (Professor Simmons) in the Department of Physics at the University of NSW. This work provides a good exemplar of directed nano-self-assembly research being developed in Australia in this case for the fabrication of electronic devices in silicon with atomic precision. Here the atomic manipulation capabilities of the scanning probe microscope (STM) are combined with gas phase limited adsorption and atomic precision crystal growth by molecular beam epitaxy (see Fig. 2) to fabricate transistors in which dopants are patterned in silicon with atomic precision accuracy[5] (see Fig. 3).

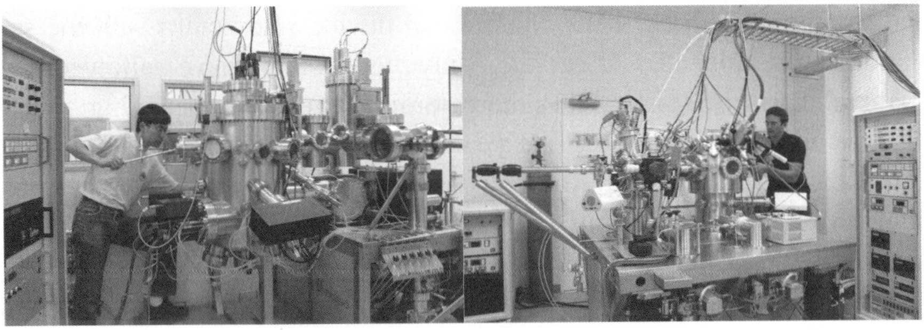

Fig. 2. A unique, customised scanning probe microscope (right) combined with a molecular beam epitaxy growth system (left) within the Atomic Fabrication Facility at the University of New South Wales in Sydney for the controlled fabrication of silicon transistors with nano to atomic-scale precision.

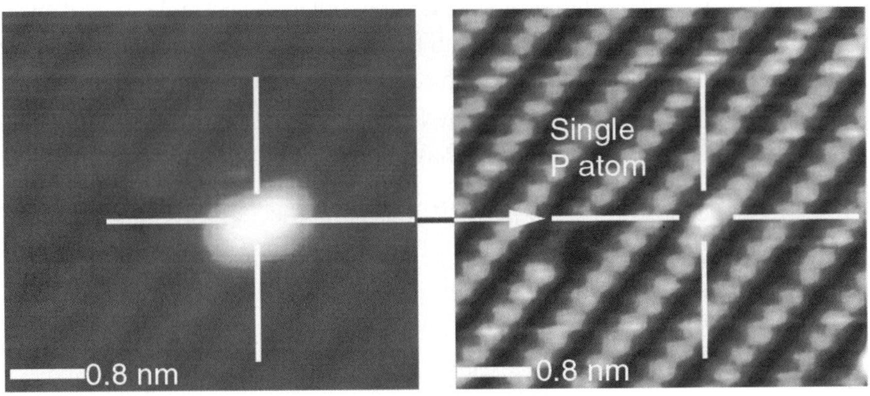

Fig. 3. A scanning probe microscope image of a hydrogen terminated silicon surface where the STM has removed 4 hydrogen atoms (left) before exposure to PH_3 gas and annealing (right) demonstrating the ability to position single P atoms on a silicon surface with atomic precision.

The driving force for the continued expansion of the microelectronics industry is the ability to pack ever more features onto a silicon chip, achieved by continual miniaturisation of the size of the individual components. Over the past three decades this trend, known as Moore's Law has continued with the number of components on a chip doubling roughly every 18 months. However if device miniaturisation continues at the same rate then by 2017 commercial device sizes will reach the sub nanometer scale. To date the only tools that have allowed the manipulation of matter at the atomic level are scanning probe microscopes. Over the past five years our group has developed a unique fabrication strategy for the realisation of nano and atomic-scale transistors in silicon using scanning tunneling lithography for atomic manipulation combined with low temperature silicon molecular beam epitaxy for atomic precision crystal growth.

Whilst the ability to fabricate nanometer and atomic-scale electronic device structures in silicon by STM has long been promised, the realisation of robust devices has been a difficult goal to attain. This is due to many reasons with one of the most significant being the engineering problem of making electrical contact to the STM-patterned buried dopant layer. The problem is that once the sample is removed from the ultra-high vacuum where STM lithography has taken place the actual location of the structure is lost. This is particularly the case when the nanostructured regions are buried under several layers of epitaxially grown silicon. To this end we have developed a technique[6] where

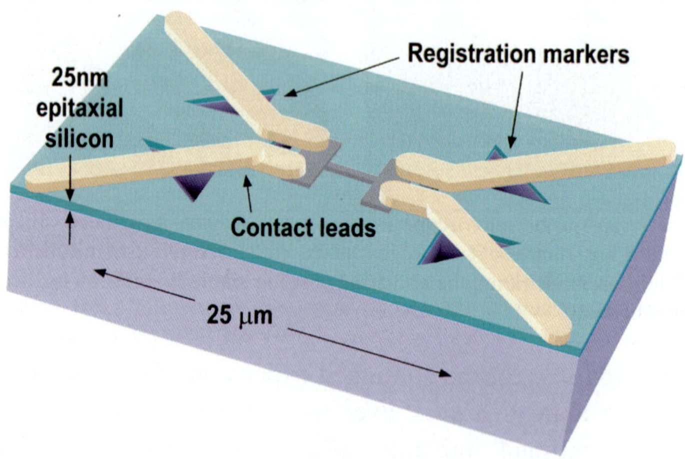

Fig. 4. A 3D schematic of a silicon nanowire patterned by the STM (in grey) with respect to triangular and square etched registration markers in the silicon substrate subsequently used for alignment of surface contact leads (in yellow).

registration markers, defined by conventional optical and electron beam lithography, are etched into the substrate before loading into the vacuum for STM patterning, as shown in Fig. 4. These markers are subsequently used after STM patterning and crystal growth to align electrical contact to the buried, STM-patterned dopants.

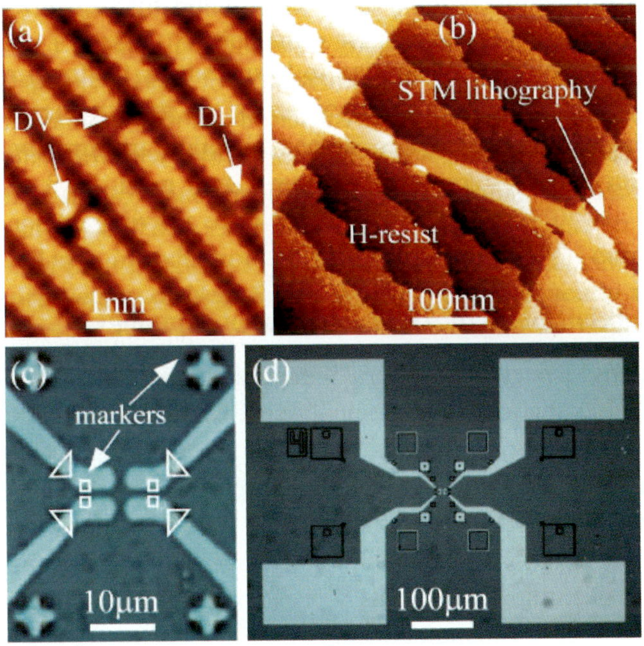

Fig. 5. (a) an atomic resolution image of a hydrogen terminated Si(100) surface (b) the same surface after STM-lithography to pattern a 27×320nm nanowire which is exposed to PH_3 before silicon encapsulation by MBE (c) an SEM image showing the use of the etched registration markers in the silicon substrate to align surface Al contacts once the device is removed from UHV and (d) an optical image of a contacted device.

Figures 5(a) and 5(b) show the STM patterning of a silicon nanowire in a hydrogen terminated silicon surface. The STM patterning of the nanowire is aligned to the etched registration markers shown in Fig. 4. Once exposed to phosphine gas and encapsulated by silicon MBE the sample is removed from the ultra-high vacuum environment of the STM before conventional optical lithography is used to align surface contact leads to the buried dopants, Figs. 5(c) and 5(d). Using such a technology we have been able to realize numerous novel devices, such as the narrowest conducting wires in silicon – down to ~2nm width that still show ohmic conduction, transistors with ordered dopant arrays and nanoscale silicon quantum dots[7].

The fabrication strategy developed allows us to fabricate transistors below the sub 50 nm length scale with controlled dopant profiles and atomically precise interfaces. This capability will enable us to determine what ultimately limits the miniaturization of silicon transistors. It also shows immense promise for the realisation of more sophisticated atomic-scale devices in silicon such as single electron transistors (SETs), quantum cellular automata and a silicon based solid-state quantum computer.

4. Commercialization of Australian Nanotechnology

Clearly many aspects of modern industrial development are dependent upon processes that occur at the nanoscale. Consequently, across a range of Australian industries, there is increasing use of techniques that operate at the nanoscale and a growing range of products being developed that depend upon nanotechnology. This is true in many of Australia's established industries – nanotechnology arguably has an important role to play, for example, in minerals processing, in agribusiness, in defense, and in many areas of manufacturing.

The emerging significance of nanotechnology is also apparent, though, in Australia's emerging biotechnology sector and in other new industries that are highly dependent upon the rapid development of new advanced technologies. A few examples of nanotechnologies that have found commercial application through development by Australian publicly listed companies are provided next.

Selected Examples of Australian Nanotechnology Companies

Advanced Nanotechnology Limited

Advanced Nanotechnology has commercialized intellectual property in mechanochemical nanopowder manufacturing originally developed by researchers at the University of Western Australia. Nanopowders produced using the company's technology are now finding applications in cosmetic products, in sunscreen creams, and in UV-protective paints. The company is developing nanomaterials for additional applications.

CAP-XX Limited

CAP-XX designs and manufactures supercapacitors, predominantly for applications in mobile electronic and communication devices. Founded to commercialize intellectual property developed at the CSIRO, the company is widely recognized for its highly competitive technology. CAP-XX's supercapacitors use nanostructured materials to optimize capacitance per unit volume and to lower resistance across device boundaries. The company produces very high energy and power densities even in very compact devices.

Cochlear Limited

Cochlear is the leading global producer of bionic ear implants. The Cochlear implant was the first cochlea implant to receive approval from the Food and Drug Administration in the USA, and the company now accounts for around 70 per cent of the global market for hearing implants. An important aspect of the company's technology is the material used to interface between the company's device and the nerves in patient's cochlea. Materials used for this purpose continue to be studied at the nanoscale.

Nanosonics Limited

Nanosonics has developed a process to create an aerosol of hydrogen peroxide, with droplets smaller than any bacteria, which can be used to decontaminate an object, chamber, or surface. The company has developed automated systems to deliver this technology into healthcare settings. Its initial products are designed for sterilizing reusable medical instruments.

pSivida Limited

pSivida has commercialized a nanostructured form of silicon, which enables the controlled release of pharmaceutical agents over days, weeks or months. This material was originally developed by QinetiQ, formerly the UK Defence Evaluation and Research Agency. pSivida's initial products have been designed for ophthalmic applications, but the company is developing other therapeutic delivery technologies based around its 'biosilicon' material.

Starpharma Holdings Limited

Starpharma Holdings (and its wholly owned US subsidiary Dendritic Nanotechnologies) is seeking to develop medicinal products based on dendrimer compounds – specialized, complex, nano-sized macromolecules, which can support multiple active sites and which have shown considerable promise as pharmaceutical agents. Starpharma Holding's lead product is a vaginal microbicidal gel that has been developed to prevent transmission of sexually transmitted diseases.

5. Conclusions and Outlook

This article has provided a snapshot of nanotechnology research and policy support as it exists in Australia. It has focused in general on research groups and companies that appear strongly to identify with the nanotechnology concept and, as such, it inevitably overlooks a great deal of the activity related to nanotechnology currently going on in Australia.

There have been no substantial policy initiatives explicitly to support nanotechnology research in Australia, as have been implemented in recent years in some other countries. However, there is a healthy cohort emerging of nanotechnologists, nanotechnology infrastructure, nanotechnology research centres, and nanotechnology companies.

Acknowledgements

MYS acknowledges an Australian Government Federation Fellowship and support from the Australian Research Council.

References

1. Thomson ISI, National Science Indictors database (2006).
2. C. Warris, *Nanotechnology Benchmarking Project* (Australian Academy of Science, 2004).
3. Invest Australia, Nanotechnology – Australian Capability Report (Third Edition, May, 2007).
4. Prime Minister's Science, Engineering and Innovation Council (PMSEIC), *Nanotechnology: Enabling Technologies for Australian Innovative Industries* (11 March, 2005).
5. S.R. Schofield, N.J. Curson, M.Y. Simmons, F.J. Ruess, T. Hallam, L. Oberbeck and R.G. Clark, "*Atomically precise placement of single dopants in silicon*", Phys. Rev. Lett. **91**, 136104 (2003).
6. F.J. Rueß, L. Oberbeck, M.Y. Simmons, K.E.J. Goh, A.R. Hamilton, T. Hallam, N.J. Curson and R.G. Clark, "*Towards atomic-scale device fabrication in silicon using scanning probe microscopy*", Nano Letters **4(10)**, 1969 (2004)
7. F.J. Rueß, W. Pok, T.C.G. Reusch, M.J. Butcher, K.E.J. Goh, G. Scappucci, A.R. Hamilton and M.Y. Simmons, "*Realization of Atomically Controlled Dopant Devices in Silicon*", Small **3(4)**, 563-567 (2007).

Author Biographies

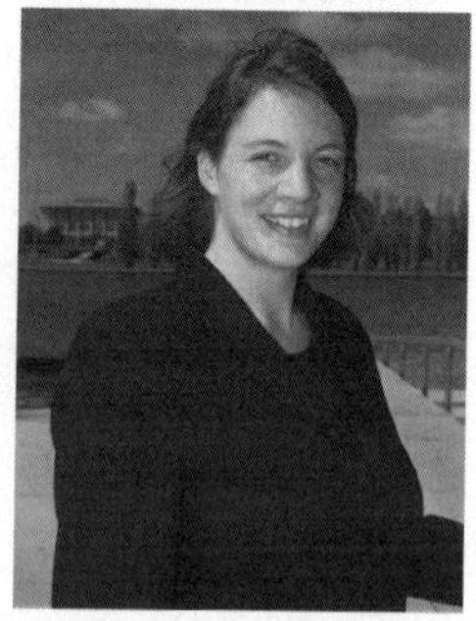

Professor Michelle Y. Simmons is a Federation Fellow and Director of the Atomic Fabrication Facility at the University of New South Wales, Sydney, Australia. In the 1990s, she spent 6 years as a Research Fellow working with Professor Sir Michael Pepper FRS at the Cavendish Laboratory in Cambridge, UK, in quantum electronics. In 1999, she came to Australia as QEII Research Fellow and was a founding member of the Centre of Excellence for Quantum Computer Technology. Her research in nanoelectronics combines molecular beam epitaxy and scanning tunnelling microscopy to develop novel electronic devices at the atomic scale. She has published more than 250 papers in refereed journals (with over 3600 citations), published a book on Nanotechnology, three book chapters on quantum electronics, has filed four patents and has presented over 50 invited and plenary presentations at international conferences. In 2005 she was awarded the Pawsey Medal by the Australian Academy of Science and in 2006 became the one of the youngest elected Fellows of this Academy.

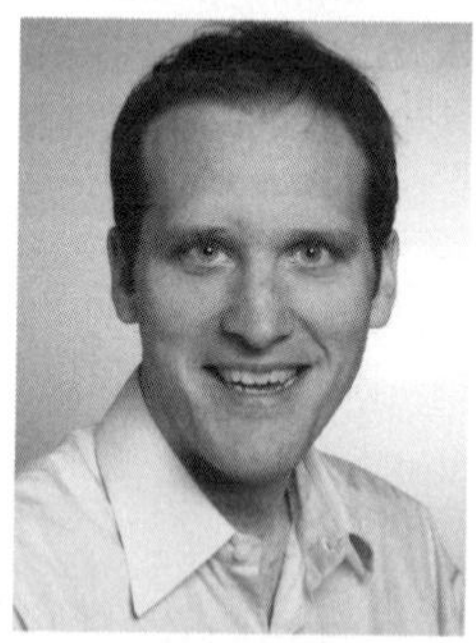

Dr Thomas W. Barlow holds a DPhil in Theoretical Chemistry from Oxford University and a BSc (Hons I) from Sydney University, where he was awarded the University Medal in Chemistry. He has held prestigious research fellowships in chemistry and biomedical science at Oxford University in the UK, and at the Massachusetts Institute of Technology in the USA. He is a leading, independent Australian policy analyst and research strategist and an expert advisor to many of Australia's research-intensive universities, government agencies and technology companies. He regularly publishes major studies on Australian research of significance to decision-makers in research organizations as well as in governments.

His recent highly popular book, *The Australian Miracle* (published by Picador in 2006), has become extremely influential across the Australian research sector. Formerly a research fellow at Balliol College, Oxford, Dr Barlow also worked for a number of years as a columnist with the Financial Times in London, and as a political advisor to a former Australian Minister for Education and Science.

CHAPTER 2

MICRO AND NANO SCIENCE AND TECHNOLOGY IN MAINLAND CHINA

HaiXia (Alice) Zhang

Institute of Microelectronics, Peking University, Beijing, China
zhanghx@ime.pku.edu.cn

ZhaoYing Zhou

Precision Engineering Department, Tsing Hua University, Beijing, China
zhouzy@mail.tsinghua.edu.cn

SiShen Xie

Institute of Physics, Chinese Academy of Science, Beijing, China
ssxie@aphy.iphy.ac.cn

During the last three decades, Micro and Nano Science and Technology has been growing fast as one of the major high-technology fields in China, improving human health, material and energy conservation, and sustainability in the environment. This chapter gives an overview of Micro and Nano Science and Technology in China, including policy and funding strategy from government, the construction of infrastructures and research centers, research highlights and commercialization efforts. As a fast developing country, China is playing more and more important role in international academy events, and its huge market is driving the Research and Development into commercialization gradually, which benefit both citizens and industries.

1. Policy and Funding Strategy

Micro/Nano Science and Technology is emerging as one of the key fields of the 21st century and is expected to enable developments of revolutionary technologies across a wide range of sectors that contribute to improve human health, material and energy conservation, and sustainability in the environment, can benefit citizens and improve industrial competitiveness within twenty years or beyond. The Chinese central government has invested in Micro-Nano science S&T since the 1980s. It became blooming in last 10 years. Several funding agencies, including National Natural Science Foundation of China (NSFC), Ministry of Science and Technology (MOST), Chinese Academy of Sciences (CAS) and National Development and Reform Commission (NDRC), Ministry of Education (MOE) are involved and coordinated by the National Steering Committee for Nanotechnology consisting of 21 distinguished scientists from universities and research institutes.

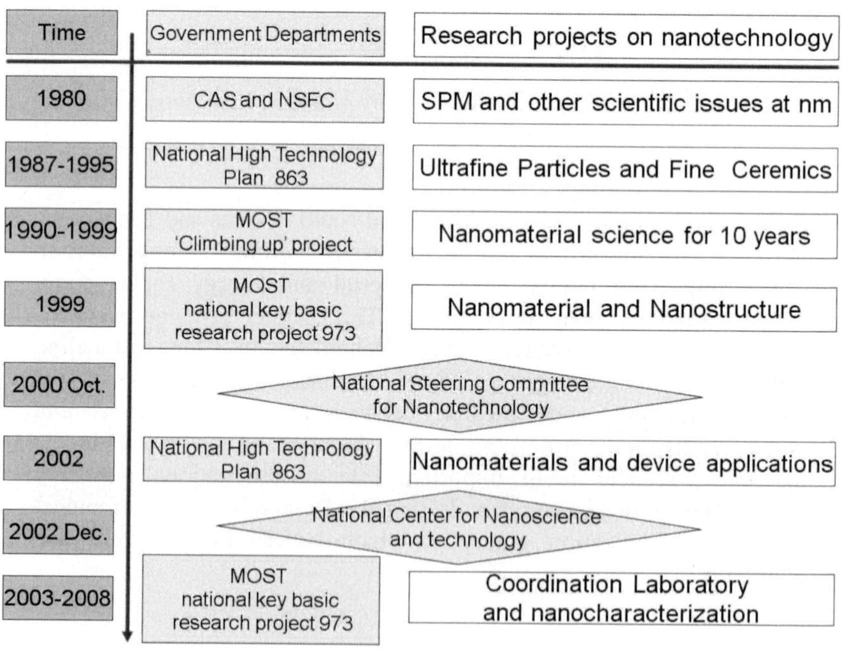

Fig. 1. China major nano science and technology research programs since 1980s.

The Ministry of Science and Technology launched its 1st nanomaterials program in 1990. Figure 1 shows the history of China nanotechnology initiatives.

The National Steering Committee for Nanotechnology recommended in 2001 the launch of China national nanotech programs with the following strategies:

(a) Increasing basic research investment and coordination of R&D to reinforce the industrial exploitation together with scientific excellence and competition
(b) Developing world-class R&D infrastructure and public research platform which match the needs of both industry and research institutions
(c) Promoting the multidisciplinary research/education

The Chinese central government has invested about two billion RMB (about USD230M/5Year) in Micro and Nano science and technology during its 10th Five Year Plan (2001-2005), which almost increases 10 times of the funding in 1990s. From 2006-2007, Funding for Nano is about 400M RMB. The major funding agencies are in charge of difference areas and targets. See Figure 2 for the overall funding structure. The 11th Five Year Plan (2006-2010) launched in 2006 with significant increase in its R&D expenditure up to 2% of China's GDP (this is in contrast to that of the 10th Five Year Plan average 1%). Nanotechnology is one of four main Basic Science Research areas which include Protein Research, Quantum Manipulation Research, and Growth and Reproduction Research. The main research topics of the Nanotechnology area in the 11th Five Year Plan include Nano Processing and Nano Devices; Nanomaterials and Nanostructures; Nano Medicine and Nano Biology, Nanostructure Characterization; Nano Devices and Integration Technologies; Nanosystem Theory and Modelling; and Nano/Micro Scale Bio-mimetics.

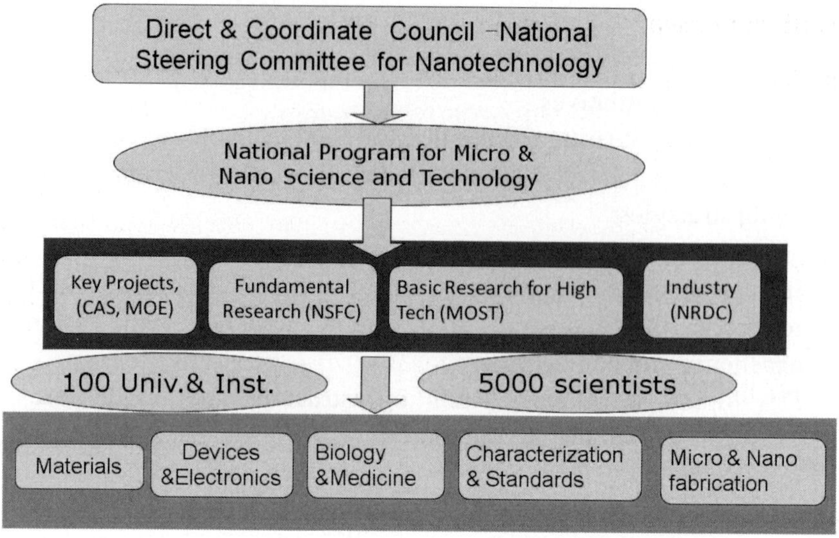

Fig. 2. China policy framework for micro/nano science and technology.

Different funding agencies have their own priorities and focus:

(1) **NSFC** has been a pioneer and key player among funding agencies in fostering developments of micro and nano science and technology in China since 1980s. It supports fundamental creation across all disciplinary principles and technologies.
(2) **MOST** was the key player of 10th Five Year Plan (2001-2005) and The 11th Five Year Plan (2006-2010). Its major target is supporting research in applied science and technology and construction of public research centers.
(3) **MOE** funds infrastructure building and education, and has funded the building up nanoscience and technology R&D laboratories in various universities.
(4) **NDRC** was the former National Planning Agency and has been funding major initiatives providing infrastructure building for nanoscience and technology R&D.

2. National Research Networks and Events

With increasing research activities in micro/nano science and technology, the number of national and international conferences, website portals, researcher societies and networks have been growing. China scientists are playing a visibly important role in international scientific community. A novel contest organized by the Chinese International Nano/Micro Engineered and Molecular Systems (NEMS) Network (CINN) called MEMSIC Cup has been very successful in attracting and motivating young students to create and realize their innovative ideas in Micro/Nano Electrical and Mechanical System (M/NEMS) devices and applications.

2.1. *National Societies and Networks*

Table 1 summarizes the main society and network in China, which are becoming the platform of communication and collaboration among researchers in China. In April 5th 2005, the Chinese Society of Micro-Nano Technology (CSMNT) was established, it is the first national society of this high-technology field, and has more than 100 Institutes and Universities, over 4000 individual members from all over the country. It became one of the key platforms for international and domestic events in Micro and Nano field in China.

An international network was set up since 2006, called Chinese International Nano Electrical and Mechanical System (NEMS) Network (CINN). CINN is a bridge which connects the Chinese researchers all over the world, encouraging sharing information, setting up collaboration and supporting international events. Also, it hosts micro/nano science and technology contest to stimulate interests in the field especially from young scientists. The first contest annual MEMSIC Cup started in Spring 2007 focusing innovative ideas in MEMS.

MEMS and NEMS Society of China (Sub-society of China Instrument and Control Society) is the oldest society of Micro and Nano Science and Technology in China, which is more application focused.

Table 1. Selected China research network

Network	Description	Website
National Center of NanoScience and Technology	Nanocenter is the platform of Nano S&T in China	http://www.nanoctr.cn
International Contest of Applications in Nano/Micro Technologies (iCAN)	iCAN is a global contest for young college students who is interested in Micro-Nano Technology, using Micro-Nano devices from sponsors to create new applications, which should be new driven power in the future marketing.	http://www.ican2009.com
Chinese Society of Micro-Nano Technology (CSMNT)	CSMNT is the national society of micro and nano, has over 4000 members from all over the country	http://www.csmnt.org.cn
Chinese International NEMS Network (CINN)	CINN is a bridge which connects the Chinese researchers all over the world, supporting collaboration and international events	http://www.cinn.cc
MEMS & NMES Society of China	It is a sub-society of China Instrument and Control Society	http://www.micronano.cn/
China Micro Nano Corporation Net	It is a platform of exchanging information for Corporations	http://www.csmnt.com/
Sciencenet	An E-Magazine and information center for Science and Technology	http://www.sciencetimes.com.cn
MEMSIC Cup	MEMSIC Cup is a national annual contest of MEMS/NEMS devices and application since 2007, which sponsored by MEMSIC Ltd.	http://www.cinn.cc/meibei

Micro and Nano fabrication Society of China (Sub-society of China Society of Mechanical Engineering) is a network for fabrication centers and researchers.

2.2. *International Academy Events*

Along with the China blooming economy, the whole country is running fast on its international track. It has been actively participating and organizing many international/domestic conferences and joint seminars. China is also an active member economy of the Asia Nano Forum (ANF). The following are the list of major micro/nano science and technology events held in China recently:

(1) Jan 9-12th, 2007, 1st Integration and commercialization of Micro and nano systems international conference and Exhibition, Sanya, China (Nano2005/2007)
(2) Jan 18-21th, 2006, 1st IEEE International Conference on Nano/Micro Engineered and Molecular Systems (NEMS06), Zhuhai, China
(3) April 24-26th, 2006, MEMS Summit, Beijing, China
(4) June 4-6th, 2007, International Conference on Nanoscience and Technology (ChinaNano2007), Beijing, China
(5) June 9-11, 2005, International Conference on Nanoscience and Technology (ChinaNano2005), Beijing, China (see Fig. 3(a))
(6) Oct. 18-19, 2007, International Workshop on Innovation and Commercialization of Micro and Nanotechnology (ICMAN07), Suzhou
(7) Sept. 19-22, 2007, The 9th domestic conference of China Society of Micro and Nanotechnology (CSMNT), Shanghai, China (see Fig. 3(b))
(8) Oct. 9-12th, 2006, China Japan Joint Seminar 2006, Beijing, China
(9) Mar. 28-Apr. 1st, 2007, China Japan Joint Seminar 2007, Tokyo, Japan
(10) Oct. 22-27th, 2007, Mainland-Taiwan Joint Seminar, Taiwan
(11) Nov. 26-27th, 2004, Asia Nano Forum (ANF Special Workshop on Nanotechnology Societal Impact in the Asia Pacific Region (ANFNSI2004)), Beijing, China

2.3. *International Contest of Applications in Nano/Micro Technologies (iCAN)*

The First International Contest of Applications in Nano-micro Technology (iCAN'2009), is a global contest for young college students who is interested in Micro-Nano Technology, using Micro-Nano devices from sponsors to create new applications, which should be new driven power in the future marketing. This contest is initiated by scientists from China, Japan, USA, Germany, Singapore, Taiwan and Hong Kong. It is supported by Chinese International NEMS Society (CINS), Asian Nano Forum (ANF) and Chinese Society of Micro-Nano Technology (CSMNT) and other societies from each countries and regions. iCAN'2009 is sponsored by following companies, MEMSIC, Nippon Signal, China Star, Nano-Micro Electronics, Chongqing Jinshan, MicroChip, Microsensing, etc.

The Schedule of iCAN'2009

- January-March, 2009: Start the contest in each Country and Region
- April-September, 2009: Team work on Project and domestic contest
- October 15th, 2009: Final list of iCAN'2009 Contest from each Country and Region
- January 20-23, 2010: iCAN'2009 Final contest, IEEE-NEMS2010, Xiamen

Website: http://www.ican2009.com

2.4. *National Contest: MEMSIC Cup for MEMS Devices and Application*

MEMSIC Cup is a national annual contest of MEMS/NEMS devices and application started in 2007 sponsored by MEMSIC Ltd. Located in WuXi (http://www.memsic.com) which is the key manufacturer in consumer MEMS products. The contest targets at application of certain MEMS products, such as accelerometers and magnetic sensors. In 2007, students (include undergraduate and postgraduate students) were invited to submit proposal in spring, 50 proposals were selected in June among 150 submission involved 500 students. The MEMSIC Three-Day summer camp was organized for the selected candidates to share information and establish team work and to visit MEMIC to learn about MEMS manufacturing process. The selected candidates were provided MEMS devices for them to further develop system application based on their proposals. This has become the most popular national contest. The 2008 MEMSIC Cup is currently ongoing (http://www.cinn.cc/meixinbei).

Fig. 3. (a) China Nano 2005, Beijing.

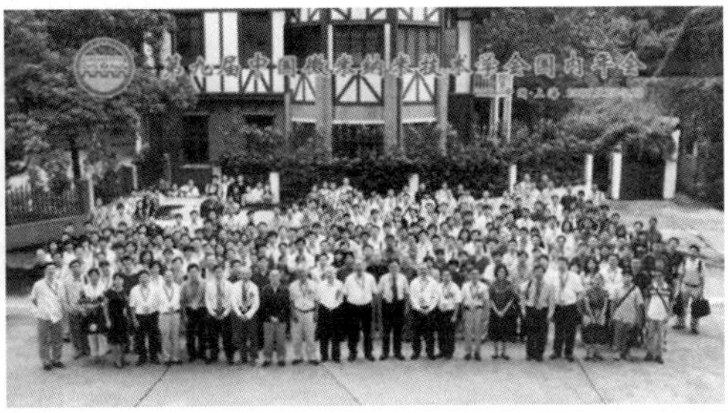

Fig. 3. (b) CSMNT07, Shanghai.

3. Infrastructure and Key Research Centers

China's dedication to developing research centers to facilitate nanotechnology development is an important part of its overall strategy to become a leader in nanotechnology. The Chinese central government is spending between $50 million and $100 million USD annually on nano science and nano technology research (NRDC, MOST and MOE are the major funding sources for construction of Infrastructure). In addition to its usual research and development budget for research at universities, national institutes and companies, Chinese central and regional governments are investing additional resources for the advancement of nanotechnology R&D to build coordinated infrastructures and to accelerate nanotechnology commercialization.

In last 10 years, China set up more than 50 fabrication and research centers, it can be categorized in three levels, national centers, state key laboratories and local centers. Most of them are located in universities and institutes of CAS, as shown in Fig. 4.

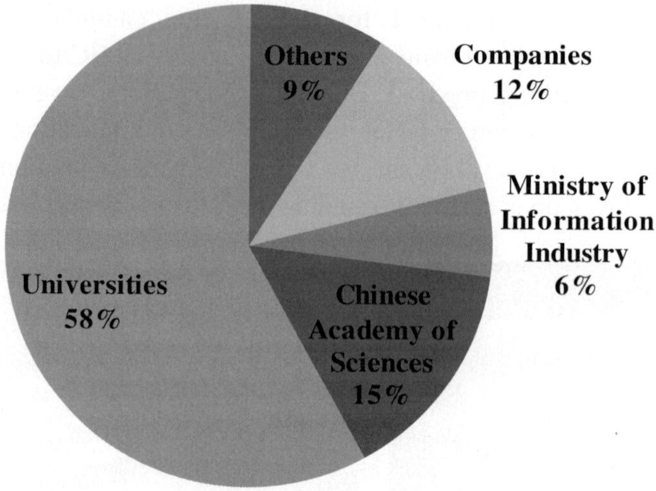

Fig. 4. Micro-Nano S&T centers in China.

It is impossible to go through all of them in this chapter. We selected a number of significant players of to provide a snapshot of recent China micro and nano research and development. (see Fig. 5).

NDRC, the former National Planning Agency and has been funding major initiatives providing infrastructure building for nanoscience and technology R&D, such as the National Center for National Science and Technology (NCNST) in Beijing, and the National Nanotechnology Engineering Research Center (NERC) in Shanghai as well as the Tianjin Nanotech Commercialization Base.

MOST has invested in national/state laboratories in high technology, two of them are very important in Micro-Nano fabrications: The National Key Laboratory of Micro/Nano Fabrication Technology in Peking University and Shanghai Jiaotong University; and the State Key

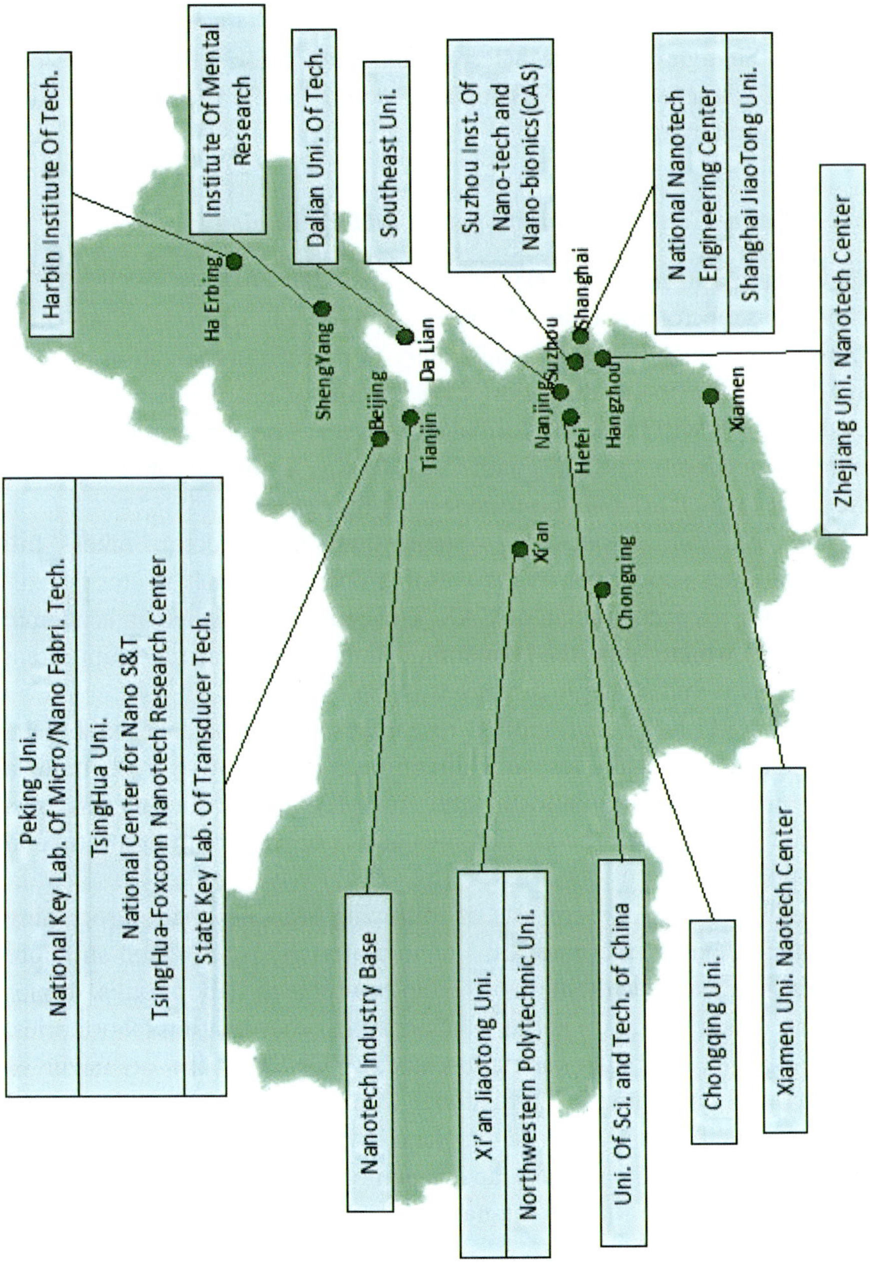

Fig. 5. Main nanotechnology research centers in China.

Laboratory of Transducer Technology shared by Institute Electronics (IE) and Shanghai Institute Microsystems and Information Technology (SIM) of Chinese Academy Science (CAS). Figure 5 shows the location of China's main nanotechnology research centers.

3.1. *National Center for Nano Science and Technology*

The National Center for Nanoscience and Technology (NCNST) of China was co-founded by Chinese Academy of Sciences (CAS) and Ministry of Education. It is a subsidiary non-profit organization of CAS and enjoys full financial support with a status of independent non-profit legal entity. It will have 155 formal employees. The center was officially founded on March 33, 2003, with CAS, Peking University and Tsinghua University which are both co-founders of the center.

The NCNST adopts a system that its director takes full responsibilities supervised by a governing board. Its first director is Prof. BAI Chunli, an academician of CAS. The center established an academic committee which aids the governing board to determine important research areas and strategic directions of NCNST.

In NCNST, basic and applied researches in nanoscience have been positioned as the main research directions. Its objective is to build a public technological platform and research base for nanoscience equipped with state-of-the-art facilities and is open to both domestic and international users.

The NCNST mainly consists of following branches, nano-processing and nano-device laboratories, nano-materials and nano-structure laboratories, nano-medicine and nano-bio-biotech laboratories, nano-structure characterization and test laboratories, coordination laboratories, a website and databases for nanoscience. NCNST is the organizer of the major china biannual International Conference on Nanoscience and Technology 2005 and 2007 (ChinaNano2005/7) which brought together scientists from all over China and a number of distinguished overseas scientists. The center is also the national coordinating body for the study of nanotechnology standardization and risk assessment programs and is an active member of the Asia Nano Forum (ANF).

(http://www.nanoctr.cn)

3.2. *Nanotechnology Industrialization Base of China*

Nanotechnology industrialization base of China (NIBC) was co-founded by NDRC and MOST on December 6, 2000, located in Tianjin. It is a base for commercialization of Micro and Nano technology.

(http://www.nibc.com.cn)

3.3. *National Nanotechnology Engineering Research Center of China*

National Nanotechnology Engineering Research Center of China (NERCN) is funded by NDRC at October 2003 in Shanghai. NERCN funds research projects focusing on application research and development of nanotechnology and connects with the top universities and institutes in Shanghai area.

(http://www.nercn.com.cn)

3.4. *National Key Laboratory of Micro/Nano Fabrication Technology*

National Key Laboratory of Micro/Nano Fabrication Technology (LMNFT) was founded in 1996, and started operating since 1998, located in Peking University focusing silicon-based fabrication and Shanghai Jiao-tong University focusing no-silicon-based fabrication. It is the first academy foundry in mainland China, providing fabrication services to global customers. Meanwhile, a group of professors are working on MEMS and NEMS field, many novel design and devices are coming out every year.

(http://ime.pku.edu.cn)

3.5. *State Key Laboratory of Transducer Technology*

State Key Laboratory of Transducer Technology was established in 1987, and is the first State Key Lab. in the field of sensors and transducers in China. It has two branches, north branch is in Institute of Electronics (Beijing) and south branch is in Shanghai Institute of Microsystem and Information Technology (Shanghai). The laboratory focuses on extensive and various research themes based on MEMS, NEMS, and IC technology, and has made many distinguished accomplishments in the fields such as microsensors and microsystems based on MEMS and IC, theory and techniques of integrated microsystems, biochemical sensors and analysis systems, physical sensors and systems, micro transducers and systems.

(http://www.ie.ac.cn/skl/default.asp)

4. University Nano Center of Excellence

The Ministry Education launched Creating World Class University initiative in May 1998 named 985 Program to develop the selected top nine universities into world class universities. The top nine universities include Peking University, Tsing-Hua University, FuDan University, Shanghai Jiao-Tong University, Xi'An Jiao-Tong University, Nan Jing University, Zhe Jiang University, Harbin Institute of Technology and University of Science and Technology of China. Over the past 10 years, tens of billions of RMB have been invested in this initiative to building infrastructure and facilities.

4.1. *Peking University (PKU)*

(a) MEMS Center in Institute of Microelectronics (IME)

IME PKU was founded in 1996, and started its operation in 1998. It is the first academy foundry in mainland China, not only developing Si-based fabrication technology, also supplying fabrication service to global customers. Meanwhile, a group of professors are working on MEMS and NEMS field. Main research areas of the center include:

- Fabrication Techniques of MEMS and NEMS;
- Standard Process Services: SGADER (Silicon Glass Anodic-bonding Deep Etching Release), Dissolved Wafer Process, Membrane Piezoresistor based Process, Double Side KOH Release Process; Surface Sacrificial Layer Process;
- MEMS/NEMS Design Technology and Tools Development;
- MEMS Devices and Its Applications: Pressure Sensors; Silicon Micromachined Accelerometers and Gyroscope, RF MEMS Sensors, MEMS/NEMS Packaging Technology, Bio M/NEMS System, SiC M/NEMS; Optical M/NEMS.

(http://ime.pku.edu.cn)

(b) College of Chemistry and Molecular Engineering

The Laboratory of Nanochemistry and Nanodevices was established in June 5, 1993, which is affiliated with Institute of Physical Chemistry, College of Chemistry and Molecular Engineering, and Center for Nanoscale Science and Technology (CNST). Its research direction is to develop nanostructured devices for future information technology. The research activities involve nanofabrication and nanostructuring using scanning probe microscopy (SPM), chemical vapor deposition (CVD) growth, chemical assembly and nanoimprint lithography for data storage and logic devices. The lab's main research areas include:

- Logic devices based on single walled carbon nanotubes, covering controlled CVD growth, band structure engineering, Raman spectroscopy and transport measurements,
- SPM memory based on the thermochemical hole burning process of charge transfer complexes,
- Structural biomimetic devices using nanoimprint lithography and biotemplates for sensors.

(http://www.chem.pku.edu.cn/nanochemistry/)

4.2. *Tsinghua University*

(a) Tsinghua-Foxconn Nanotechnology Research Center

Tsinghua-Foxconn Nanotechnology Research Center (TFNRC) was co-founded by Tsinghua University and Mr. Terry Guo, the board chairman of the Foxconn Enterprise Group with a 300 million RMB donation from him. Aimed at promoting both fundamental research and the industrialization of the nano-scale science and technology, the center combines Tsinghua university's scientific resources and Foxconn's industrialization power to support fundamental and applied research, technical innovations and the industrialization of newly developed technologie

The center is located in the campus of the Tsinghua University, Beijing. In Dec. 2003, TFNRC had its inauguration of the 13,000 m^2 building which houses most advanced micro- and nano-scale manufacture and testing facilities. The center provides a top class open platform for academic and industrial researchers both in China and from all over the world. The center's main research areas include:

- Nanomaterials-Carbon Nanotubes,
- Nanotechonology for heat dissipation,
- Nanotechonology for flat panel display,
- Nanoelectronics and Nanophotonics,
- Nano-composite Materials.

(http://www.TFNRC.tsinghua.edu.cn)

(b) Micro-Nano Research Center

Founded in 1996, it consists of several related research groups and laboratories in Department of Precision Instruments and Mechanology, Institute of Microelectronics, and Department of Engineering Mechanics. The research focus of the center is MEMS and NEMS. Its main research areas are:

- Fundamental research in Mirco/Nano technology design, fabrication and characterization,
- Micro/Nano device research: Inertial MEMS, Opto-MEMS, fluidic MEMS, and Bio-MEMS; Nano-scale structures and devices,
- Applications of Micro/Nano technology: Micro-satellite technology; Micro Air Vehicle technology; Biomedical engineering.

Website:
- Department of Precision Instruments and Mechanology (http://www2.pim.tsinghua.edu.cn)
- Division of Micro/Nano Devices and Systems, Institute of Microelectronics (http://dns.ime.tsinghua.edu.cn/english/overview/7.htm)
- Institute of solid mechanics, Department of Engineering Mechanics (http://hy.esinghua.edu.cn/English/about/page4.asp)

4.3. *Shanghai Jiao Tong University (SJTU)*

Research Institute of Micro/Nano Science and Technology (RIMNST) is one of the research institutes directly affiliated to SJTU. Its mission is to perform the basic and applied research in the field of micro/nano science and technology, including micro-electro-mechanical systems (MEMS), nano biomedicine, nano electronics and electronic devices. The institute has world-class research facilities and outstanding researchers in multidisciplinary research. Since 1992, the institute has systematically explored basic and applied research in the field of micro/nano science and technology, such as MEMS, three-dimensional non-silicon microfabrication technology and single electronic devices. The institute has become one of the important foundries for three-dimensional micro fabrication of non-silicon materials. The institute has 4500 m^2, including 800 m^2 clean-room and 2000 m^2 laboratory. Its main research areas are:

- Non-silicon microfabrication technologies and MEMS,
- Nano bio materials and nano-bio engineering,
- Nanofabrication and devices.

(http://mnri.sjtu.edu.cn)

4.4. *Xi'an Jiaotong University*

Institute of Vacuum Microelectronics and Microelectro-mechanical System (VME & MEMS) was established in 1994, the Institute of VME & MEMS is the first institute in the research field of CNT-FED in China. It has become one of the most outstanding institutes on CNT-FED and vacuum microelectronics research in China.

The institute focuses on extensive and various research themes based on field emission and vacuum microelectronics, nanoelectronic materials and devices, MEMS, SAW devices and so on. It has made many distinguished accomplishments in the fields. Its main research areas are:

- Basic theory and key techniques of field emission,
- New method of preparation CNT,
- Structure and characteristics of interface of nano- materials.

(http://www.xanano.org)

4.5. *University of Science and Technology of China (USTC)*

The Laboratory of nanostructure and nanophysics of USTC combines with research team in Hefei National Laboratory for Physical Sciences at Microscale (HFNL) and is one of the most outstanding micro/nano research centers in China. The Laboratory focuses on extensive and various research areas of nanoscale science and technology and has made many distinguished accomplishments in the fields including quantum manipulation on single-molecule level, molecular electronics, assembly of novel molecular architectures, synthesis and characterization of nanomaterials, catalysis of TiO_2 nanostructures, nanoelectronic devices, fuel cell development, and so on. The lab's main research areas include:

- STM study of single molecules with high energy/spatial resolution,
- Single molecule electronics (molecular electronics),

- Quantum manipulation of Single molecules.

(http://nanology.ustc.edu.cn)

4.6. *Southeast University*

The Biological and Medical Nanotechnology Group in the School of Biological Science and Medical Engineering was established in 2000. This group is a significant part of the State Key Laboratory of Bioelectronics. It has been working on the research of basic theory and key techniques of nano-materials, especially magnetic nano-materials, and of the applications in the biomedical field. It has become one of the most outstanding biomedical nano research groups in China. It main research areas are:

- Basic theory and key techniques of Nano-materials and Molecular Assembly,
- Magnetic Nano-materials and its application on diagnosis (MRI) and tumor hyperthermia (MFH, AEH),
- Biomolecule Sensors and single cell optical detection,
- Nano-drug delivery system,
- Biological effects, Biocompatibility and toxicology of nano-materials.

(http://www.Imbe.seu.edu.cn/nano)

4.7. *Chongqing University*

Micro-systems Research Center of Chongqing University is one of the earliest research units engaged in MEMS in China. It is a creative platform of science and technology in Chongqing, and a key platform of the "211 Project" and "985 Project" of Chongqing University.

The center conducts theoretical and experimental research on MEMS/NEMS device and system, micro instruments, along with applied research and technological innovation, particularly in micro-spectrometer and micro biochemical system fields. Also the center has carried out fruitful cooperation in international research projects and academic exchanges with Fraunhofer IZM, University of Technology Chemnitz, Germany, and Santa Anna School of Advanced Studies of Pisa, Italy. It main research areas are:

- Micro-spectrometer,
- Integrated Biochemical Separation and Analysis Chip System,
- Micro Power Generator,
- Grating Light Modulator.

(http://msc.cqu.edu.cn/index.html)

4.8. *Dalian University of Technology*

The Key Laboratory for Micro/Nano Technology and System of Liaoning Province/Key Laboratory for Microsystem and Microfabrication, Education Department of Liaoning Province, was founded by Prof. WANG Liding, academician of CAS (Chinese Academy of Science). The research of the lab involves MEMS, nano technology, and opto-electronic technology. The lab is the leader of the Corporative Union for MEMS Research and Development of Chinese Northeast. It is also one of the key laboratories at Dalian University of Technology supported by "The 211 Project" and "The 985 Project". There are programs for Ph. D candidates in this lab to specialize in MEMS Engineering and Precision Instrument and Mechanics. The lab has been actively developing academic exchange and corporation with the outside, and cooperative relationship has been established with 17 research institutions and enterprises from both abroad and domestic (including Taiwan and Hong Kong). The lab's main research areas are:

- Microstructure design and micro-system simulation,
- Micro processing techniques and testing techniques,
- Bio-MEMS,
- Nano-material and devices,
- Ultra precision gear machining and measurement instrumentation.

(http://mnstlab.dlut.edu.cn)

4.9. *Harbin Institute of Technology*

MEMS Center, Harbin Institute of Technology was established in 1998, and developed based on its research foundation in semiconductor devices, pressure sensors, chemistry sensors and micromachining. It has become one of the most active micro/nano research groups in China. The center's main research areas include:

- Pressure sensors based on polysilicon nanofilms,
- ASIC for integrated accelerometers, gyroscopes and fluxgate sensors,
- Micro direct methanol fuel cell,
- Microchip capillary electrophoresis and micro total analysis systems (μTAS).

(http://www.hit.edu.cn/)

4.10. *Northwestern Polytechnic University*

The MEMS/NEMS laboratory of Northwestern Polytechnic University was established in 1992 to explore science and engineering knowledge related to MEMS and NEMS fabrication, and to develop MEMS and NEMS devices with local high-tech industries viable for commercialization. Interfacing with industries in aviation, aerospace and marine, the laboratory is dedicated to industrial application of miniaturization and system integration. The Lab's main research areas are:

- MEMS Integrated Design
- Micro Smart Structure and Smart Skin
- Micro-system fabrication
- Applied Micro System
- NEMS and Nano-measuring Technology

(http://www.nwpu.edu.cn)

4.11. *Xiamen University*

Since establishment in 2002, Pen-Tung Sah MEMS Research Center of Xiamen University is one of several research centers that own the advanced micro and nano fabrication and test facilities in Mainland China. With the aim of being a world class MEMS Research Center, the center is pursuing to become a public platform for multidisciplinary research, a platform for training technical and creative talents in the fields of MEMS and Nano technology, an experimental base for commercializing and/or industrializing achievements in scientific research. The center's main research areas include:

- Micro and nano fabrication technologies
- MEMS sensors
- POWER MEMS
- Nano Science and Technology
- Semiconductor Devices

(http://memsc.xmu.edu.cn/)

5. Main Research Institutes from Chinese Academy of Sciences

5.1. *Institute of Physics*

The Institute of Physics, Chinese Academy of Sciences, is a comprehensive institute devoted exclusively to fundamental research for the improvement of our knowledge on condensed matter physics, optical physics, atomic and molecular physics, plasma physics and soft-matter physics. The Institute runs a number of Laboratories in physical sciences. It is one of the pioneers in nano science research of china. The institute's main research areas are:

- Nanoscale Physics and Devices Laboratory
- International Center for Quantum Structures

- Laboratory of Microfabrication

(http://www.iphy.ac.cn/)

5.2. *Institute of Chemistry*

As one of the key players in nanotechnology field, Institute of Chemistry of Chinese Academy of Sciences (ICCCAS) hosts National Key Laboratories of CAS and National Center for Chemical Analysis and Testing. Its main research areas are:

- Chemistry of molecular assemblies synthetic,
- Preparation science high-tech (advanced) materials,
- Nanoscience and technology,
- Chemical biology and theoretical chemistry.

(http://www.icas.ac.cn/)

5.3. *Shanghai Institute of Microsystem and Information Technology*

The institute has been focusing its research in developing technologies of making microdevices including sensors and actuators in the past decades. It has developed MEMS high selectivity processes including anisotropic wet etching, sacrificial layer process, and dry etching. Its main research areas are:

- Nano probe
- Nano beam
- Nano Wire

(http://www.sim.ac.cn)

5.4. *Institute of Electronics*

The State Key Laboratory of Transducer Technology Northern Branch was established in 1987. The laboratory focuses on extensive and various research themes based on MEMS, NEMS, and IC technology, and has made many distinguished accomplishments in the fields such as microsensors and microsystems based on MEMS and IC, theory and techniques of integrated microsystems, biochemical sensors and analysis systems, physical sensors and systems, micro transducers and systems. Its main research areas include:

- DNA biochip and micro total analysis systems (μTAS),
- Microarray biosensor detection system of TRAIL receptors,
- Spatial electric field sensor,
- Field programmable gate array (FPGA),
- Wireless integrated network sensors.

(http://www.ie.ac.cn/skl/default.asp)

5.5. *Institute of Process Engineering*

State Key Laboratory of Multi-phase Complex Systems, Group of Material Chemistry and Applied Technology, was established in 1994. It is reputable of preparation/production and application of micro/nano-particles in China. The lab's main research areas are:

- RTV organic-inorganic composition materials for electric power,
- Assembling technique of nanoparticles and nano-device design,
- Standard of nanomaterials and nanometrology,
- Energy-saving materials and related techniques.

(http://www.ipe.ac.cn/mprcas/index.html)

5.6. *Institute of Solid State Physics*

Key Laboratory of Materials Physics, focuses on nano materials research. Its main research areas include:

- Nano powder processing technology and application,
- Enhanced nanometer effects,
- Nanostructure science and technology,

CHAPTER 3

NANOTECHNOLOGY RESEARCH AND COMMERCIALIZATION IN HONG KONG

King Lun Yeung

Department of Chemical Engineering
Hong Kong University of Science and Technology
Clear Water Bay, Kowloon, Hong Kong, HK-SAR, P.R. China
kekyeung@ust.hk

Hong Kong has made significant investment and contributions in nanoscience and nanotechnology research as part of Government's strategy to foster innovation and benefit local industry. Several innovative nano-enhanced and nano-enabled products had been successfully commercialized as a direct result of collaborations and partnerships between government, research institutes, universities and industries.

1. Background

Hong Kong is a city of 7 million peoples situated at the southern coast of China with a per capita GDP of US$ 30,000 in 2008. As in most developed economies, Hong Kong relied heavily on service industries, mainly finance, trade, logistics and tourisms. Hong Kong firms are also heavily committed in light-to-medium manufacturing in China including electronics and electrical appliances, jewelry and watches, textiles and garments, toys and durable goods, and processed foods and beverages. Innovations in products, materials, and manufacturing would therefore benefit industrial productivity. The Government of the Hong Kong Special Administrative Region (HKSAR) supports the manufacturing sector by investing considerable resources in nanotechnology, information technology and biotechnology. Hong Kong's strategy differed considerably from the neighboring economies in being more laissez faire and grass-root oriented enterprise between government and non-government organizations, academic and research institutes and local and international industries.

Hong Kong is active in nanoscience and nanotechnology research since early 1990 with most of the research conducted at the universities and devoted mainly to fundamental science with emphasis on nanoscale phenomena and nanomaterial synthesis. The first organized effort to coordinate nano-related research in Hong Kong and Chinese mainland was spearheaded with the establishment of a Joint Laboratory on Nanostructured Materials and Technology between the Hong Kong University of Science and Technology (HKUST) and the Chinese Academy of Sciences[1] to foster research partnership in areas related to nanomaterials and their applications. This was followed by the HKSAR government investment of more than HK$50 million (US$6.5M) in a single year on basic and applied research in nanoscience and nanotechnology through the Hong Kong Research Grants Council (HK-RGC)[2] to create platform technologies that had broad potential applications.

The Institute of Nano Science and Technology (INST) at HKUST was established in May 2001 with University funding and corporate and private donations to become the first research center in Hong Kong dedicated to nanoscience and nanotechnology[3]. The INST provided start-up research fund to enable young researchers to pursue original and innovative ideas in nano-related fields and to encourage inter- and multi-disciplinary research. The success of the INST provided the government with a template for funding nano-related research and led to the establishment of the Institute of NanoMaterials and nanoTechnology (INMT) and the Nanotechnology Center for Functional and Intelligent Textiles and Apparel (NTC) in 2003.

The INMT was incorporated into a legal business entity, the Nano and Advanced Materials Institute Limited (NAMI) in 2006 and was transformed into a research center and the main funding body for nano-related research in Hong Kong. Besides NAMI, several other funding agencies including the Innovation and Technology Commission (ITC)[4], the Hong Kong Applied Science and Technology Research Institute Company Limited (ASTRI), the Hong Kong Science and Technology

Parks Corporation (HKSTPC), Small Entrepreneur Research Assistance Programme (SERAP)[5] and the Hong Kong Productivity Council (HKPC) provides infrastructure for the development and commercialization of nano-related products and technologies. Table 1 summarizes the nanotechnology centers in Hong Kong.

In addition to the well funded nanotech R&D centres shown in Table 1, there are very competitive nanotech research activities carried out at the City University of Hong Kong (CityU, well-known for its research in nanowires, CNT, and LED), University of Hong Kong (HKU, well-known for its research in OLED, Biomedical and environmental sensing), Chinese University of Hong Kong (CUHK, well-known for its research in metrology, optoelectronics, environment and medical), and Hong Kong Baptist University (HKBU, standing out in nano photonics).

2. Fundamental Research and Platform Technologies

The nanoscience and nanotechnology researches in Hong Kong concentrates in the three major areas of (1) nanomaterials, (2) nano-biotechnology and (3) nano-electronics with strong emphasis on the design, synthesis and fabrication of nanostructured materials of low dimensionality and the associated nanoscale properties that define their functionality and applications. This particular emphasis is well-suited for rapid productization and commercialization.

Hong Kong scientists are among pioneers in the synthesis and study of quantum dots, functional fullerenes, semiconductor nanowires, carbon nanotubes and graphenes. The skill in nanomaterial synthesis is exemplified by the fabrication of the world smallest single-walled carbon nanotubes of 0.4 nm hosted within zeolite channel (Fig. 1a) that exhibit extraordinary large lithium storage capacity and the assembly of ordered macromolecular structure by design to create biomimetic materials that could serve as vehicle for targeted drug delivery (Fig. 1b).

Research in the fundamental design and fabrication of nanodevices provides a platform technology for nanoelectronics and nano-biotechnology. In nano-electronics, the main focus was on developing nanofabrication protocols based new emerging nanomaterials for energy-saving devices such as light-emitting diodes and liquid crystal displays.

Table 1. Nanotechnology R&D centers funded by government since 2003

Location	Funding Amount (HKD)	Funding Period	Research Themes	Website
HKUST	100M	2002-2006	• Energy Storage: Lithium Rechargeable Batteries and Fuel Cells • Nanoelectronics: Nanoscale FETs, Display Technologies • Integrated Manufacturing: Nanoparticles, Fullerenes, Carbon Nanotubes, etc. • Environmental Catalysts: Photocatalytic Air and Water Remediation	www.ust.hk/~inmt
PolyU	14.7M	2003-2006	• Textile and apparel products which exhibit multifunctional properties such as UV resistant, stain and water-repellent and anti-bacterial • Intelligent textiles that could respond to different environment • Nano-structured photonic fibers which could be developed into fabric displays	www.itc.polyu.edu.hk
HKUST	400M	2006-2012	• Nanoparticles and Applications • Energy: Battery and Microfuel Cells • Environment: Nano Sensors and Nano Catalyst • Advanced Composite Materials • Processing Technologies for Nanomaterials • Nanoelectroncs: FET, Display and Lighting	www.nami.org.hk

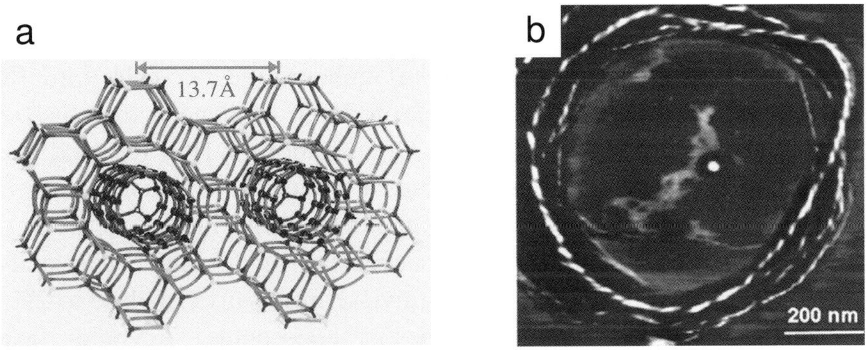

Fig. 1. (a) 3D model of the 0.4nm SWNTs arrayed in the channels of an AFI single crystal and (b) self-assembled macromolecular vesicles.[6]

In nano-biotechnology, research is mainly on nano-encapsulation for targeted drug delivery, nanofluidics and nano-micro-macro interfaces for biological nano-microfluidics and micro-total analysis systems (μ-TAS) devices.

3. Road to Commercialization

The funding structure and industry expectation in Hong Kong demand rapid productization and commercialization of nano-derived products and technologies. Nano-enhanced products are most suitable to the short product development cycle of 18-24 months. This often involves the incorporation of a small quantity of nanomaterials in existing products to create new functionality, or improve existing product performance and life. This has the advantage of existing production facility and customer market. Also, the limited amount of nanomaterials needed allows a rapid transition from laboratory synthesis to manufacturing as the complicated process scale-up and the associated cost could be avoided, at least at the early stage of the commercialization. On the other hand, nano-enabled products are of higher value, but demand larger quantity of high purity nanomaterials and thus require longer development time and greater capital investment.

3.1. *Nano-Derived Products for Hygiene and Environment*

Hong Kong had suffered outbreaks of bubonic plague 1894, bird flu outbreak in 1997 and SARS epidemic in 2003. Hygiene is therefore of outmost concern. The World Health Organization (WHO) reported that one of the most common route for transmission of infectious diseases is by indirect contact with surfaces contaminated with infectious droplets produced by the patient's coughing, sneezing or talking[7,8]. The study carried out by Bellamy et al.[9] found amylase from saliva in close to 29% of the exposed surfaces in domestic households. Reynolds and coworkers[10] extended the study to public places. They detected the presence of hemoglobin (blood marker) in 3%, amylase (saliva marker) in 15%, urea (urine marker) in 6%, and protein (general hygiene marker) in 26% of the 1061 samples taken from the daycare centers, shopping malls, offices, airports, movie theaters, restaurants and gymnasia. They also found an alarming 30% of the surfaces tested positive for biochemical markers were also contaminated with fecal coniforms.

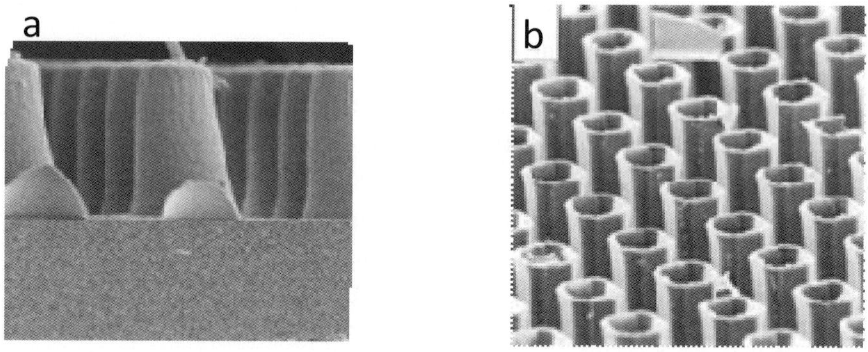

Fig. 2. Scanning electron microscope pictures of (a) closed-tipped and (b) open-tipped microneedles with three microns thick, pure-silica zeolite wall.

The MTR corporation Ltd. in an effort to reduce the risks of infection from casual contacts with contaminated surfaces partnered with local university and industry to develop and use nano silver-titanium dioxide coating on their trains and stations as long-term surface disinfectant[11]. The nanotextured TiO_2 shown in Fig. 2a prepared by using as template the nanometer-sized polymers prepared by microwave-assisted symmetry break-up of polymer thin films display exceptional antimicrobial properties compared to TiO_2 film. The nanotextured surface also exhibit hydrophobic and self-cleaning properties (Fig. 2b). A company sponsored research had successfully developed a new process for preparing nanotextured TiO_2 on large area surfaces for application in sanitary coatings. Entrapping photoactive TiO_2 within the silica network of an aerogel confined and stabilized the nanometer sized TiO_2 crystals. The nano-confinement of the nanometer-sized TiO_2 appeared to increase the photon utilization efficiency of the photocatalyst and increase the formation of free-radical species that reacts and inactivate the microorganisms. Such nanostructured material exhibits exceptional photo-induced, antimicrobial activity capable of log 6 reduction in viable bacteria within 30 min of contact.

A two-level antimicrobial coating with "release-killing" and "anti-adhesion" properties was prepared by nano-encapsulation of ClO_2 in a stable water-in-oil-in-water (w/o/w) double emulsion. A slow sustained release of gaseous ClO_2 at a rate sufficient to inhibit bacteria growth (~1300 μg $ClO_2.g^{-1}.day^{-1}$) was demonstrated for a prolonged period of time (i.e., 28 days). The bacteria adhesion was prevented by the Pluronic polymer used to encapsulate ClO_2, while touch and infectious droplets trigger an increased release of biocide at the sites of contamination resulting in rapid disinfection. This nano-enabled antimicrobial coating is currently being developed for use in hospital and homes with industrial partners.

Hong Kong being a densely populated city with limited land and water resources is very aware of threat of pollution and environmental degradation on the standard of living and quality of life. Several application patents related to photoactive catalyst based on nanostructured TiO_2 had been successfully commercialized as nano-enhance products. The Hong Kong Air Purifier Center, developed several air purifying systems (InnoClean series) using nano-size TiO_2 photocatalyst to remove bacteria, virus, odor, toxic volatile organic compounds in domestic and commercial buildings. The EnvironmentalCare Ltd. commercialized Fotocide photocatalytic oxidation technology based on university research for air and water treatment. The Chiaphua Industry Ltd. created a series of air cleaners and component products based on the patented Nano-Ti Oxide developed at HKUST. The company's customers include North America, Europe and Australia. A local air purifier model (Giabo) is being used by an international hotel chain in Hong Kong and mainland China. The eighteen months it took to bring Nano-Ti Oxide from the laboratory to consumer product is within the government timescale for productization and commercialization of nano products.

A new generation of nanocatalyst for air purification was developed under the sponsorship of the HKSAR Innovation and Technology Fund for INMT. The ambient temperature nanocatalysts were designed to remedy volatile organic compounds (VOCs) and malodor related to nitrogen containing compounds at room temperature without the need of photo-irradiation. The project was launched in January 2003 and in March of the same year, Hong Kong suffered from the outbreak of SARS epidemic. The project goal was expanded to include bioaerosol remediation. Prototyping and on-site testings were accelerated upon the request of the funding agency. The prototype was tested on-location at a Government clinic for elderly and at a children hospice. A significant reduction in indoor bioaerosol was realized with clear decrease in the incidence of upper respiratory illnesses in the staffs and patients. The VOC level and incidence of malodors also decreased during the six months study period. The technology was successfully patented and commercialized through a start-up company, Artenano Ltd. The company is an incubatee of the HKUST entrepreneurship program and the Hong Kong Science and Technology Parks Corporation (HKSTPC). The company had developed over the two years a series of innovative products based on the technology. The products include indoor and in-vehicle air purification systems, and large-scale treatment apparatus for heavily contaminated process air stream from manufacturing sites. Current partnership with a multinational company aim to develop the catalyst to solve malodor problems related to water treatment.

Nanostructured TiO_2 also finds application in water treatment as demonstrated by EnvironmentalCare Ltd. using their Fotocide PCO process. Metal-doped nanostructure TiO_2 catalysts are efficient for decolorization of dye-contaminated water and was tested at Dunwell

(HK) Ltd. Selective adsorbents derived from mesoporous MCM-41 with 3 nm pores could separate different dyes from mixture allowing for their recovery and reuse.[12] Similar MCM-41 derived adsorbent could be prepared for selective separation of precious metals such as palladium, gold and silver from base metals found in mining process and spent electroless plating wastewater.[13-15] The precious metals were recovered at high purity and thus eliminating the needs for expensive downstream purification. Highly dispersible magnetic MCM-41 could be used for on-site treatment of contaminated ground water. The magnetic MCM-41 made of iron oxides and silica is considered environmentally benign and had been recently used to develop highly selective adsorbent for removal of toxic chromium and arsenic oxyanions from surface waters.

Nano-enabled products and technology require strong support from both the government and industry to be successful. This is particularly true for environmental application. The HKUST Chemical Engineering team had developed with partnership with Veolia Environment an efficient nano-enabled technology for the decontamination of water containing refractory organic pollutants that characterizes many endocrine disrupting chemicals including pharmaceuticals, pesticides and herbicides. These pollutants can not be remedied with existing treatment technologies and their presence in drinking water is becoming a common occurrence even in high developed countries in North America and Europe. Minute quantities of these pollutants are known to have a profound impact on health and environment. The new process involves the use of nanoporous membrane, adsorbents and catalysts to enable deep oxidation and effective removal of the pollutants from water. Lab, bench and pilot-scale tests were carried out in Hong Kong and France, and the company is now planning to build a treatment plant based on the technology.

3.2. *Nano-Derived Products for Textiles and Garments*

Hong Kong is one of the five largest exporters of textiles and apparel products in the world and the textile and apparel industry makes great contribution to Hong Kong's economy. The Nanotechnology Centre for Functional and Intelligent Textiles and Apparel at the Hong Kong

Polytechnic University (NTC) was established in 2003 to maintain and enhance the competitiveness of this industry. It received funding from government and industries, most notables are Artex Fashion (Asia) Ltd, Bondex International (HK) Ltd, Cha Textiles Ltd, Glorious Sun Holdings Ltd, Link Dyeing Works Ltd, Sunikorn Knitters Ltd and Wah Tai Piece Goods Ltd.

Nanostructured surface finish that exhibits multi-functional properties including UV blocking, water- and oil-repellant, anti-staining and antimicrobial properties were achieved by controlling polymerization at the fiber surface and through incorporating functional nanomaterials such as nanostructured silver, titanium dioxide and zinc oxides. Prof. Tao's group collaborates with Proctor and Gamble, Saurer and Australia Wool Innovation toward this goal. The group now produces anti-bacterial, water/oil resistant and self-clean fabrics for use in garments, shoes and school bags. Sputter coating method was developed for surface finishing of high sheer apparels including lingerie and negligee. Chitosan nanoparticles finishing was used to improve dyeability of silk, while research into processing of nanometer-sized dye powder is aimed to improve dye solubility and reactivity. Synthesis of nano-pigments is being pursued to improve light intensity and color in the processed textiles and apparel.

The Hong Kong Polytechnic University (HKPU) smart clothes based on smart materials and wearable technologies intended to capitalize on the massive growth in internet connectivity to create a new generation of interactive, intelligent clothing of new functionality ranging from self-cleaning, static-, stain- and impact-resistant and temperature regulation. Flexible E-textile sensors for strain, temperature and humidity are interwoven in the garment based on shape memory foams and textile to provide sensing and response. Radian Fab™ uses photonic fibers to create luminescent fabric display that serves as precursor for more complete display interface for IT and telecommunication devices. The Hong Kong-listed Mascotte group partners with UK-based Eleksen Plc. to produce touch-sensitive interactive textile for wireless fabric keyboards for smart phones and i-pod.

3.3. *Nano-Composites*

High-impact strength polypropylene calcium carbonate and clay nanocomposites are successfully commercialized as structural plastics for use in vehicle, safety equipments and pipings. In addition, organo-clay nanocomposite serves as moisture barrier of high strength, stiffness and temperature resistant with low volatility, creepage and shrinkage, Industrial partnership was critical in optimizing the processing of these polymer-nanomaterials composites. The HKUST and Finetex of Korea established a partnership in the area of nanofibers and nanocomposites research and development at the University. The Chemical Engineering and Mechanical Engineering departments of HKUST improved the ballistic-proof strength of ultra high molecular weight polyethylene fiber by adding carbon nanotubes. The material finds use for light weight, breathable, bullet-proof vest.

3.4. *High Value Nano-Derived Products*

In 2003, the Institute of NanoMaterials and NanoTechnology (INMT) was set up at HKUST with a four years funding of HK$100 million from the government and industry to develop high value, nano-derived products in (1) nano-electronics and displays and (2) energy generation and storage. These represent nano-enabled technologies of high value and impact, but require significantly higher investment and longer development time. The examples given below are still in the early developmental stage and would require at least five years until commercialization.

Research in nanoelectronics focused on component design and fabrication of nanoscale CMOS and non-CMOS transistors with high-K dielectric and power threshold. These includes carbon nanotube field effect transistor, nano Fin field effect transistor (FinFET), wide band-gap AlGaN/GaN HEMT power amplifier and nanocrystals floating gate memory. The research effort in display and lighting is led by Prof. H.S. Kwok with the goal of improving the lighting efficiency of OLEDs by using nanoporous anodic aluminum oxides and establish manufacturing processes for large area displays. Application patents were developed for novel low-cost color LCD, bistable and fast LCD, active and passive matrix driven color sequential LCD and high performance OLED. The HKSAR Government through NAMI had made a massive investment to establish a demonstration Fab-line for the manufacture of high performance displays.

The project on energy generation and storage focused on portable devices. Prof. P. Sheng of HKUST led the project on developing nanocarbon based lithium battery that display reversible capacity substantially higher than the commercial lithium batteries with stable charge-discharge cycle. The micro fuel cell research focused on improvements of fuel cell components and architecture using nanomaterials. Electrocatalysts based on carbon nanotubes are shown to be more active and tolerant to CO poisoning for both hydrogen and methanol fueled micro fuel cells. Polymer electrolyte membrane (PEM) based on Nafion-nanomaterials nanocomposites display lower fuel crossover and improved water management. Nanoporous zeolite membrane was successfully used as PEM in microfuel cell with comparable performance as Nafion membrane for energy generation.

Besides the nano-electronics and energy-related applications, nanomaterials find uses in sensor and diagnostic devices. Latitude Co Ltd. adapted a nanomaterial-based, visible and solar-blind, UV sensor developed by Prof. Sou of HKUST in there products to detect harmful UV radiation. ZnO nanocrystals and nanowire based gas sensors were developed to detect toxic gases, while carbon nanotubes provide rapid humidity sensing. Two mega amplification nanobiolabels were developed based on SuperNova® and SuperSignal® technologies that allowed the design of more sensitive medical diagnostic kits. Microneedles and nanopump anticipate progresses in μ-TAS. The zeolite microneedles are crystalline and have superb strength allowing it to penetrate skin layer without damage (see Fig. 2). The microneedle could be designed to sample body fluid or deliver therapeutic drugs. The porous zeolites allowed for long-term programmed drug delivery. The nanopump device is based on metallized anodic aluminum oxide that is capable of rapid liquid pumping by electroosmosis principle.

3.5. *Manufacturing and Fabrication of Nanomaterials*

The investments on basic nanomaterial research have resulted in successful commercialization of many nano-enabled and nano-enhanced products and processes. Nanomaterials and nanoparticles in particular find uses in biomedicine, biomedical devices, microelectronics, sensors, catalysts and adsorbents, and are becoming common ingredient in paints and pigments, polymers and plastics, textiles and garments, and cosmetics and personal care products. It is therefore becoming

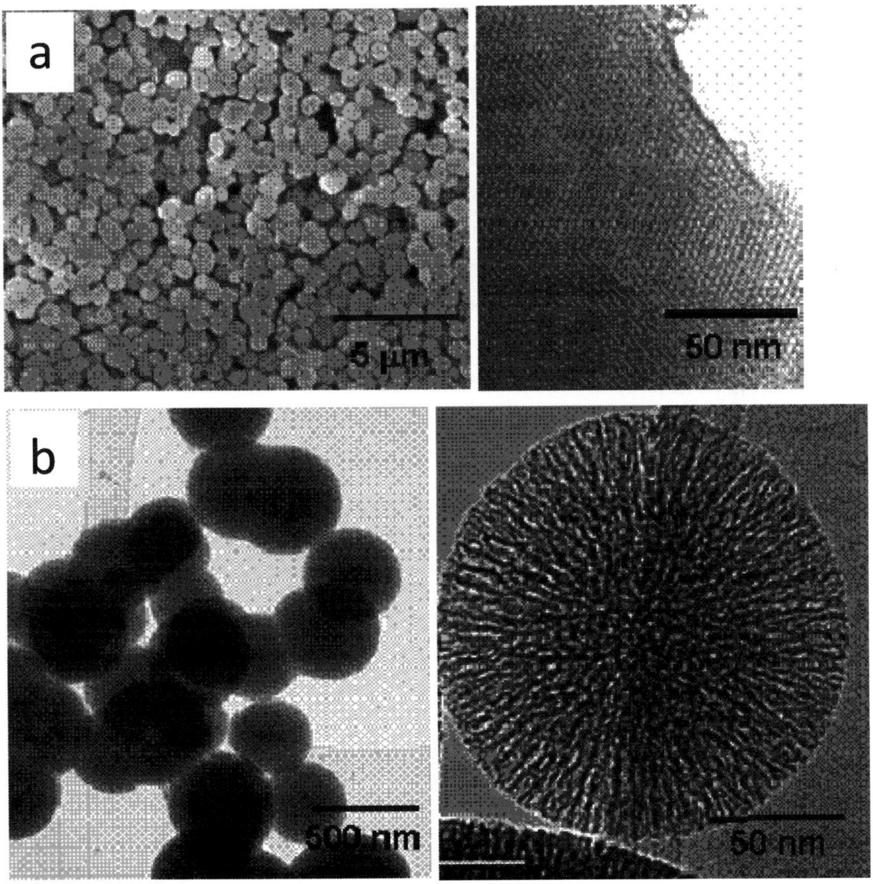

Fig. 3. MCM-41 prepared by (a) batch reactor and (b) microreactor.

imperative to investigate more efficient, cleaner and reliable manufacturing methods for nanomaterials beyond batch synthesis. The group of Prof. K.M. Ng applied basic Chemical Engineering principles to the manufacture and synthesis of nanomaterials. They successfully employed crystallization method for the simultaneous separation of high-purity fullerenes C60 and C70, and used supercritical fluid processing to produce nanocrystals.

Nanomaterial production in microreactors have the advantage of rapid, controlled mixing and fast heat and mass transfer rate ensuring fast production and safe operation. Rapid scale-up by replication shortens the time-to-market, and thus makes innovations more readily available. Figure 3 shows MCM-41 prepared by batch synthesis and microreactor from same synthesis composition and conditions. It is clear that MCM-41 particles prepared from microreactor yield smaller and more uniform particles (i.e., 300 ± 50 nm) compared to MCM-41 produced by batch reactor. The MCM-41 preparation represents two of the most versatile and popular synthesis routes for nanomaterials, the templated synthesis and sol-gel processing. The results would therefore be relevant for synthesis and purification of other nanomaterials.

Fabrication protocols for nanoporous zeolites were patented by HKUST and enabled the incorporation of zeolites as catalyst, membrane, adsorbent and structural material in microsystems. Figure 4 displays the incorporation of zeolite as freestanding membrane for micro fuel cell application. A zeolite film was first grown on the surface of silicon wafer by wet synthesis method. The zeolite film is nonporous with organic template molecules trapped within the zeolite pore channels. The deposited film was micropatterned and the templates were selectively removed from the patterned area to produce permeable zeolite membrane. The silicon beneath the patterned zeolite was removed by etching across the zeolite membrane to create the microchannel shown in Fig. 4a. Gold was deposited on the wall of the silicon channel by electroplating. The electroplating solution diffuses across the nonconducting zeolite membrane by electroosmosis and deposits as a metal layer on the silicon surface (Fig. 4b). The catalyst support precursor was added by a ship-in-a-bottle process (Fig. 4c) and converted by a solvothermal process into 10 nm-diameter hollow silica spheres (Fig. 4d). The Pt catalyst was then added by impregnation. Final deposition of catalyst and electrode on the surface of zeolite membrane completes the membrane electrode assembly.

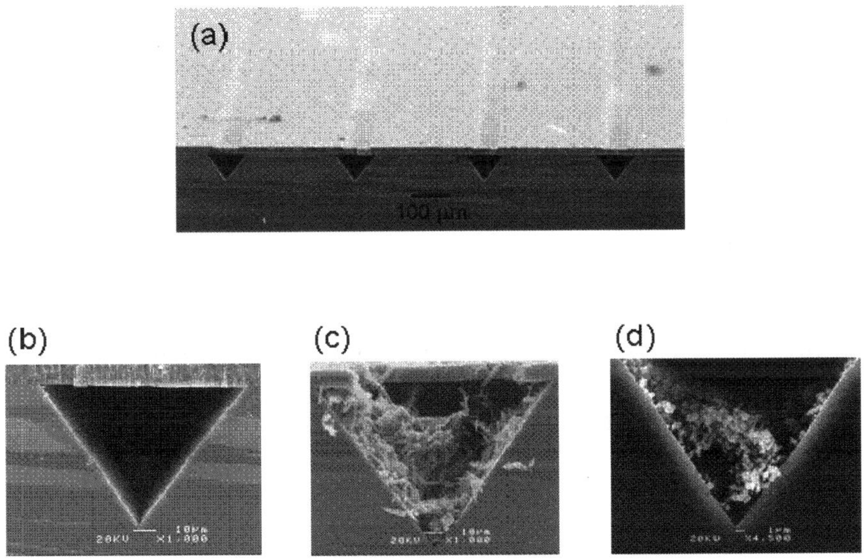

Fig. 4. Scanning electron micrographs of (a) zeolite membrane enclosed microchannels measuring 100 μm wide, 100 μm deep and 10 mm long. Higher magnification, cross-section pictures of (b) zeolite microchunnel, (c) ferrocene loaded microchunnel and (d) microchunnel containing silica nanospheres for catalyst support.

4. Concluding Remarks

The history of nanotechnology research in Hong Kong is less than a decade, but impressive achievements were attained in the last five years in nanomaterials synthesis and fabrication. Aggressive productization and commercialization efforts by industry are critical in bringing innovative nano-derived products to consumer market not only in Hong Kong but also worldwide.

References

1. http://www.ust.hk/en/pa/e_pa010515-96.html
2. http://www.info.gov.hk/gia/general/brandhk/csnano.htm
3. http://instweb.phys.ust.hk/index.php?option=com_content&task=view&id=12&Itemid=41
4. http://www.itf.gov.hk/eng/about.asp
5. http://www.itf.gov.hk/eng/SERAP.asp
6. http://instweb.phys.ust.hk/images/stories/Carbon%20Nanotubes.pdf
7. Hospital Hygiene and Infection Control, Safe management of wastes from health-care activities, World Health Organization.
8. Boone, S. A.; Gerba, C. P. *Appl. Environ. Microbiol.* **2007**, *73*, 1687-1696.
9. Bellamy, K.; Laban, K. L.; Barrett, K. E.; Talbot, D. C. S. *Epidemiol. Infect.* **1998**, *121,* 673-680.
10. Reynolds, K. A.; Watt, P. M.; Boone, S. A.; Gerba, C. P. *Int. J. Environ. Health Res.* **2005**, *15*, 225-234.
11. http://www.mtr.com.hk/eng/sustainability/sustainrpt/2006rpt/sia-2006-ni.html
12. Ho, K. Y.; McKay G.; Yeung, K. L. *Langmuir* **2003**, *19*, 3019-3024.
13. Lam, K. F.; Yeung, K. L.; McKay, G. *Langmuir* **2006**, *22*, 9632-9641.
14. Lam, K. F.; Fong, C. M.; Yeung, K. L. *Gold Bull.* **2007**, *40*, 192-198.
15. Lam, K. F.; Chen, X. Q.; Fong, C. M.; Yeung, K. L. *Chem. Commun.* **2008**, 2034-2036.

- single crystalline nanowire array for nanodevice applications.

(http://www.issp.ac.cn/)

5.7. *Institute of Metal Research (IMR)*

Advanced Nanomaterials are the research topics in IMR. The institute's main research areas include:

- Carbon nanotubes, nanofibers and grapheme,
- One-dimensional nanostructures,
- Nanostructured materials for clean energy and environment applications,
- Nanostructured surface coatings; nanobiomedical materials, etc.

(http://www.imr.ac.cn)

6. The Research and Development Highlights

China is well known for its micro and nano science achievement, especially within the last decade. According to the Topical Citation Report on Nanotechnology during the period from 1992 to 2002, the Chinese Academy of Sciences ("CAS") ranked No. 4th (behind UC Berkeley, IBM and MIT) among research institutes specializing in nanotechnology. This ranking was based on citations and citation impact of the over 100 nanotech papers of CAS published between 1992 and 2002. Additionally, in the Thomson Derwent 2003 review report on the patents of nanoscience and technology from 2000 to 2002, China, holding 12% of world patents, was ranked third, following the U.S. (holding 32% of the world's nanotechnology patents) and Japan (holding 21% of the world's nanotechnology patents). In addition, the APEC 2001 review report ranked China third in the amount of publications concerning nanoscience and technology publications (14.2% of the world total), again following the U.S. (representing 41.6% of the nanoscience publication totals) and Japan (at 15.2% of those totals). In 2004, China got to be No. 1 or No. 2 in Nano-S&T publications, depending on the data sources.

The major achievements of nano technology in China are focused on following areas:

6.1. *Carbon Nanotube and Nano Materials*

China is very competitive in carbon nanotubes (CNT) and nanomaterials research and application, especially general methods for the synthesis of nano-scale materials, and carbon nanotubes materials, Bio-inspired nanomaterials, Fig. 6(a) is research done by University Science and Technology of China (USTC) on well aligned ZnO nanorod array by ultraviolet lasing. Figure 6(b) shows Carbon Nanotubes array done by Tsinghua-Foxconn Nanotechnology Research Center (TFNRC).

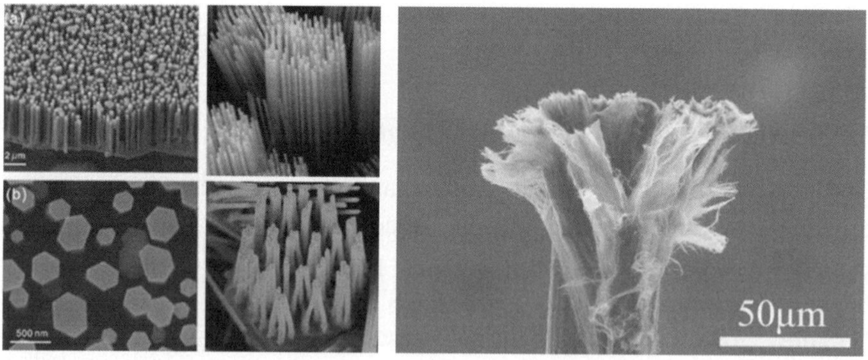

Fig. 6. (a) Ultraviolet lasing of well-aligned ZnO nanorod array from USTC; (b) the carbon nanotube array from TFNR.

6.2. *Micro-Nano Fabrication, Devices and Systems*

MEMS devices and Nano devices have been developing in China in national wide MEMS and Nano fabrication centers. Increasing efforts have been dedicated to application in close with collaboration with industries in china and overseas. Figure 7 demonstrates examples of nanofabrication capabilities. Figure 7(a) shows Nano Coil Array fabricated by Focus Ion Beam (FIB) process and Fig. 7(b) shows Nano Cantilever fabricated by MEMS process.

6.3. *Micro-Nano Biology and Medicine*

Research areas targeting at application areas such as healthcare, implant sensors, drug delivery, bio-mimetics and nanomedicine are getting more and more popular in China. Figure 8(a) shows cicada wing nanostructures fabricated using nanoimprint lithography by Center for Nanoscale Science and Technology, Peking University. Figure 8(b) shows interaction of magnetic nanoparticles and tumor cell from Southeast University.

6.4. *Characterization and Standard*

China has set up the platform of testing and characterization of micro and nano S&T, and the work on standard is moving ahead and gets worldwide recognition.

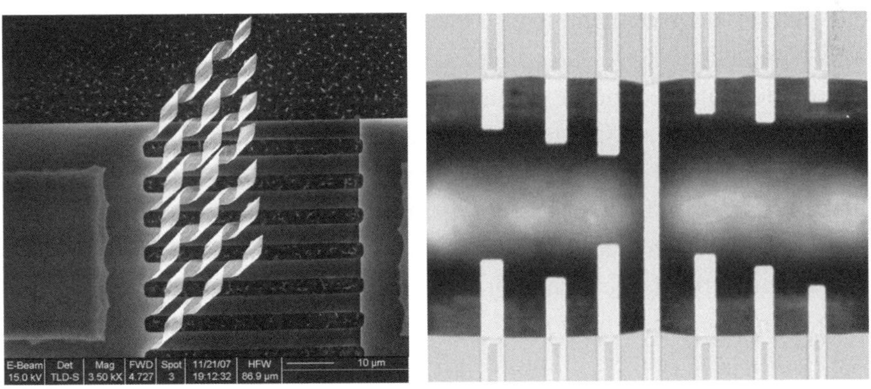

Fig. 7. (a) 3D nano coils array from PKU; (b) nano beams from SIM, CAS.

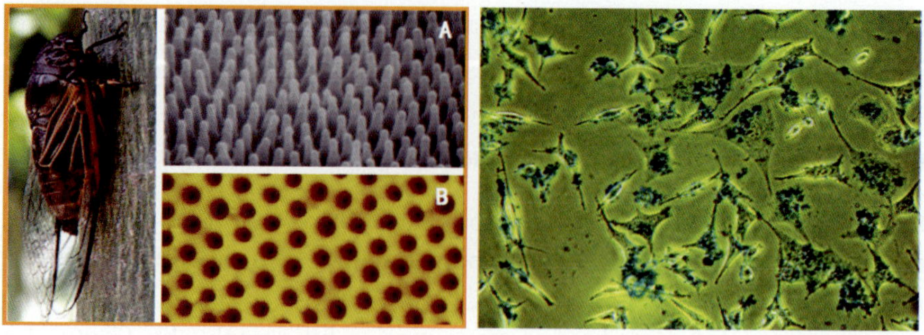

Fig. 8. (a) Cicada wings: Nanoimprint Lithography from PKU; Fig. 7. (b) Cicada wings: Nanoimprint Lithography from PKU.

Besides the above mentioned four fields, Chinese scientists have also achieved notable development in environmental, clean energy and other fields in last 10 years.

7. Commercialization of China Micro-Nano Technology

As a developing country, China is eager to move fast in commercialization and market development. Until the end of 2004, there are more than 600 Nanotechnology enterprises and 100+ institutions involved in Micro and Nano technology. Main Products are MEMS devices, sensors, and nanopowders of Oxides, Metals and their applications in products such as coatings, fibers, papers, ceramics, catalyst and others. Currently there are over 30 MEMS companies in China in MEMS device design, fabrication and foundry services. We expect by 2011, China will become the largest market for MEMS sensors and actuators and other nanodevices. The follow provides brief descriptions on China's key commercial players in micro and nano technology field.

7.1. *MEMSIC*

MEMSIC, Inc. designs, manufactures and markets CMOS Micro-Electro-Mechanical Systems (MEMS) IC products that have on-chip mixed signal processing. MEMSIC is the first and the only company that integrates a MEMS inertial sensor with mixed signal processing circuitry onto a single chip using a standard CMOS IC process. This combination of technology has successfully yielded products at substantially lower cost and higher system performance and functionality than competitive products in the market for sophisticated accelerometers. In addition, this technological approach allows the Company to easily integrate additional functions, or create new sensors, using a standard CMOS IC process to expand into other MEMS application areas beyond accelerometers.

The performance versus price points that MEMSIC's accelerometers can achieve is opening up new applications that until now were not practical. Due to its products are in the extremely high cost performance and dependability, MEMSIC has become a leading supplier of accelerometers in the field of consumer and electronic markets around the world. MEMSIC is well recognized by many world renowned companies such as Lenovo-IBM, SONY, Panasonic, Autoliv, IN Focus, TOMTOM, and MITSUBISHI to name a few. MEMSIC will continue designing and introducing new products to create new markets and opportunities on the basis of continuous improvement and innovation to become a leader in the field of integrated IC and MEMS. (http://www.memsic.com)

7.2. *Capital Bio*

Headquartered in Beijing, CapitalBio is a leading life science Company that develops and commercializes total health-care solutions including a broad range of innovative biochip-related products for genomic and proteomic research, bio-safety testing, clinical applications and to address wider human health needs.

CapitalBio has rapidly evolved from a young innovative biochip developer into a comprehensive life science entity with four affiliates or subsidiaries: AVIVA Biosciences, a San Diego, California based company that develops and markets on-chip patch-clamp technologies for ion-channel studies for drug discovery research; Chipscreen Biosciences, a Biotech company located in Shenzhen, China, for small molecular drug discovery and development; CapitalBio International and CapitalBio Hong Kong, two fully owned subsidiaries of CapitalBio Corporation, based in San Diego, California and Hong Kong, China respectively. CapitalBio's 24,000 square meters of state-of-the-art facilities host more than 400 staff for Administration, research, development and manufacturing. (http://www.capitalbio.com)

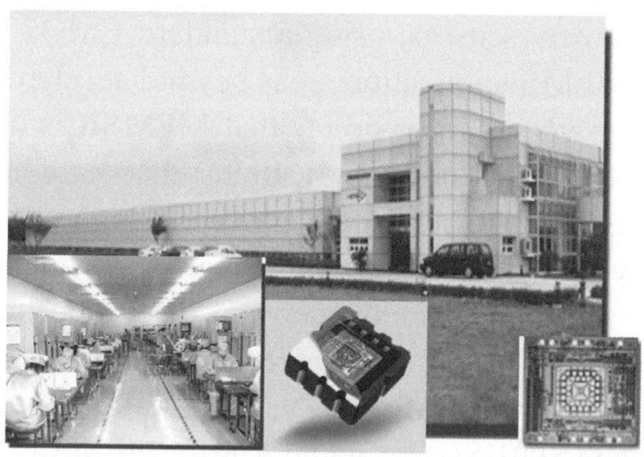

MEMSIC Semiconductor (Wuxi) Co. Ltd.

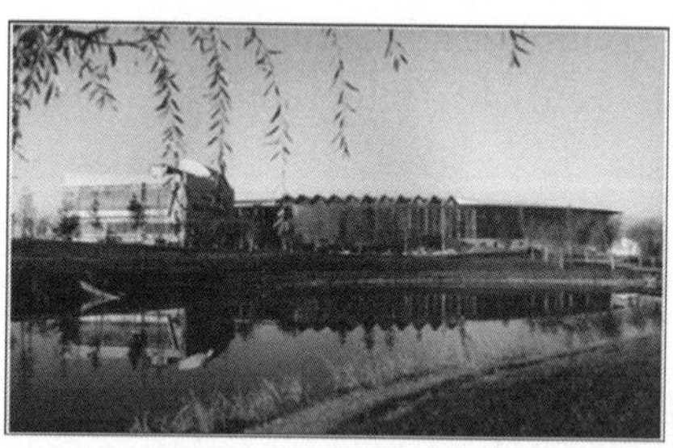

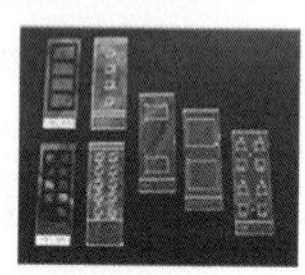

CapitalBio SmartSlide ™-3.0 Multi-sample

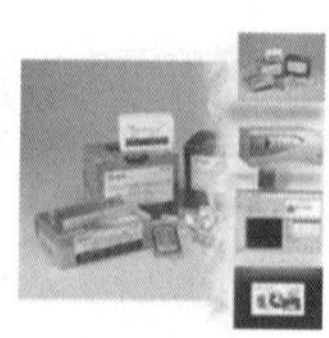

CapitalBio Corporation in Beijing

7.3. *Shenyang Academy of Instrumentation Science*

Shenyang Academy of Instrumentation Science (SAIS) is a national scientific academy. It belongs to the China National Machinery and Equipment Group (CNMEG). The main technologies and products of SAIS include: Transducers, Transmitters, Intelligent Instruments, Optics-Mcchnis-Electrics Integrated Products, Optical Thin-film Products, Optical Filters, Aspherical Lenses, Digital Lenses. SAIS has established science and technology management support for small and medium enterprise. SAIS has fulfilled the requirements of scientific research commercialization, and has built five limited companies, two R&D centers, two research institutes, two industry inspecting and testing centers and published three magazines. (http://www.hb-sais.com)

7.4. *First MEMS*

First MEMS Co., Ltd. is a fast growing high-tech MEMS (Micro-electro-mechanical Systems) company located in Zhongguancun high-tech park, Beijing. First MEMS is the first company that applied the MEMS technology to produce pressure sensors as a batch production in Mainland China. First MEMS is spun off from Institute of Microelectronics, Peking University and National Key Laboratory of Nano/Micro Fabrication Technology, which ensure the strong technology backing of MEMS sensor chips. Among the executive and non-executive directors of First MEMS are a number of highly reputable experts in the specific technologies and processes central to the company's commercial success. (http://www.mems.com.cn)

7.5. *MEMSensing Microsystems Co. Ltd.*

MEMSensing Microsystems Co., Ltd. is an early stage venture focusing on MEMS acoustic devices with proprietary technology, which dramatically improves the performance and reduces the cost of MEMS acoustic devices.

MEMSensing Microsystems Co., Ltd. is an early stage venture company focusing on MEMS acoustic device development. It has developed proprietary technology, including advanced acoustic-mechanical-electronic model, stress insensitive structure, whole silicon microphone fabrication process and the package which is compatible both with high mount and zero-height. The company has six silicon microphone patents. (http://www.memsensing.com)

7.6. *Shanghai Integrated Micro System Technology (SIMST) Co. Ltd.*

Shanghai Integrated Micro System Technology (SIMST) Co., Limited, founded in 2006, is a company focusing on the commercialization of MEMS technology from State Key Laboratory of Transducer Technology in Shanghai Institute of Microsystem and Information technology (SIM). SIM now can provide MEMS pressure sensor chip and the high-g accelerator in high volume. Their applications include, but not limited, on the tire pressure measurement system (TPMS), industrial instruments and consumer electronics (CE), as well as the car assessment system. (http://www.sim.ac.cn)

7.7. *Xi'an Winner Information Measurement and Control Co. Ltd.*

Xi'an Winner Information Measurement and Control Co., Ltd. is located in the science & technology park of Xi'an Jiaotong University (XJTU) engaging in research and development, manufacturing and sales of sensors and instruments. The company has a excellent scientific and technical team with advanced technologies and equipments, and a corporate culture of "Dedication, Credibility, Care and Perfection". It continuously pursues customers' satisfaction with corporate mission towards "keeping ahead, Credibility, Persistence and Innovation". The company has created its own brand products and is making their best efforts to become internationally competitive. (http://www.winner18.com)

7.8. *Suzhou Institute of Nano-Technology and Nano-Bionics, Chinese Academy of Science — A Bio- and Nano-Tech Incubator*

Suzhou Institute of Nano-Technology and Nano-Bionics (SINANO), CAS, located in the well-known Suzhou Industrial Park, is jointly founded by the Chinese Academy of Science (CAS), the government of Jiangsu Province and the government of Suzhou city. The institute intends to build a global platform for technology innovation and incubation services as well as education in nano-technology and nano-bionics areas.

SINANO conducts research and development (R&D) and provides facility services. It has seven R&D divisions including Nanodevices and related materials, Nanomedicine, Nanobionics, Nanosafety, System Integration and IC Design, International Laboratory for Adaptive Bio-nanotechnology (i-LAB) and Interdisciplinary division. Each unit oversees its R&D programs in basic science and strategic technology application, and is responsible for funding application and allocation, facilities management, human resource training and coordination with other divisions. SINANO conducts cutting edge R&D focusing in areas of industry applications and new technologies that enhance the manufacturing competitiveness of local industry. SINANO is equipped with nanofabrication facility, testing and analysis platform (see Fig. 9), computing center, engineering development platform, technology training center, technology transfer center and test and measurement service center for solar cells. The SINANO not only provides infrastructure support for the R&D activities of SINANO and but also serves as an open platform for supporting R&D to other academic institutions and industry nationally and internationally.

SINANO will recruit 450 personnel for research, administration and technical support including 250 permanent positions, 150 project-specific posts, 250 graduate students and 50 visiting scholars and postdoctoral researchers, reaching total of 700 personnel by 2010.

Up to now, SINANO has recruited over 400 personnel, reaching 70% of its target. 90% of the senior researchers are overseas educated, 80% personnel are with higher degrees including PhD. and Masters. The researchers are multidisciplinary with degrees in physics, chemistry, biology, material sciences, computer science, electronic engineering, and mechanical engineering. The average age of personnel is around 31, the youngest research team in Chinese Academy of Science.

SINANO is strongly committed to both domestic and international collaboration. In form of international conference, exchange visits,

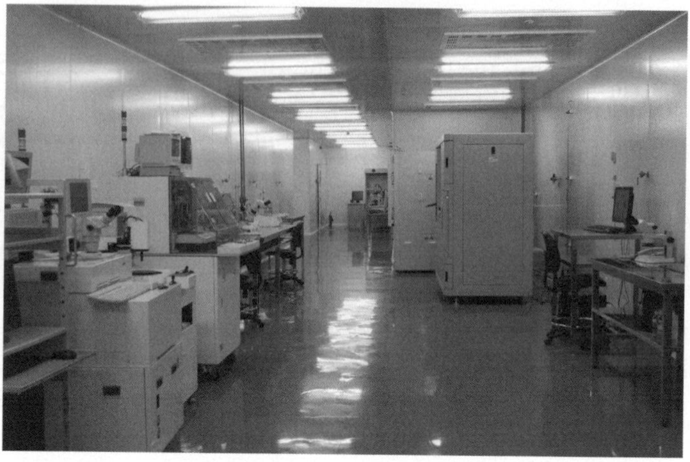

Fig. 9. Platform for characterization and test in Suzhou Institute of Nano-Technology and Nano-Bionics, Chinese Academy of Science.

research cooperation and so on. SINANO has signed agreements with institutes, universities and industries around the world, including US, Canada, Japan, Finland, Germany, and Hong Kong. Currently SINANO is working on establishing collaboration with institutes and universities in Russia, Israel, England, Singapore and other countries of European Union.

8. Summary and Outlook

In China, the 11th Five Year Plan (2006-2010) is ongoing right now. The government is investing 2% GDP in R&D, and nanotechnology is a research field of national focus.

China has hosted its 1st Olympic games in Beijing in 2008. It has been another big milestone for this huge and historic country. After 30 years amazing development, the facility and infrastructure have been well constructed. The central government continues to increase its investment in R&D and this will provide increasing opportunities to grow micro and nano science and technology in China, not only in research, also in commercialization. China is emerging not only as a world economic power, but also a leading nation in nanotechnology.

Author Biographies

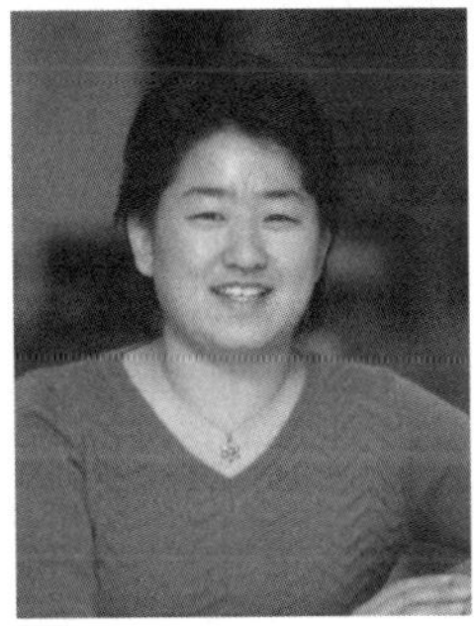

Dr Haixia Zhang is a professor at the Institute of Microelectronics, Peking University. She received her PhD. from Huazhong University of Science and Technology in 1998. She was a visiting professor in UC Davis and Case Western Reserve University in USA from 2004 to 2006. Her research field include, MEMS Design and Fabrication Technology; MEMS Sensors and Actuators and Nano Electromechanical System (NEMS). She published more than 50 papers and received National Invention Award in Science and Technology in 2006 for Si MEMS Technology and Application.

She is a co-founder of the Chinese International NEMS Network and the Deputy Secretary General of the Chinese Society of Micro-Nano Technology. She has served as conference chair of a number MEMS/NEMS international conferences including IEEE NEMS2009, ICSICT2008, ASME MNC 2006, and others. She also serves as an organization chair of Transducer 2011.

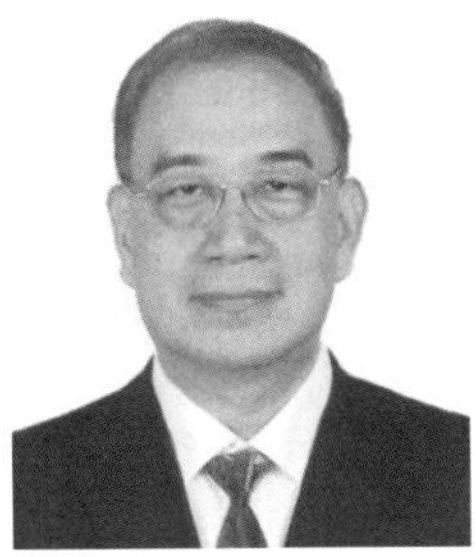

Prof. Zhaoying Zhou is a professor at Micro and Nano Research Center, Tsing-hua University. His research interests include MEMS, measurement and control system technologies, bio-medical devices. He has published 8 books, over 300 academic papers indexed by SCI, EI and ISTP. 600 more papers cited by SCIC and CSCD. Prof. Zhou has obtained more than 50 patents including foreign patents. He was awarded the National and Ministry Awards for Science and Technology Invention over 10 times. He is the Vice President of China Instrument Society, President of Micro/Nano Devices and System Branch Society, Honor Vice President of China Micro/Nano technology, Editor of the international Journal of Micromechanics, Sensors and Actuators A Physical, and Chief delegate of the World Micromachine/MEMS Summit.

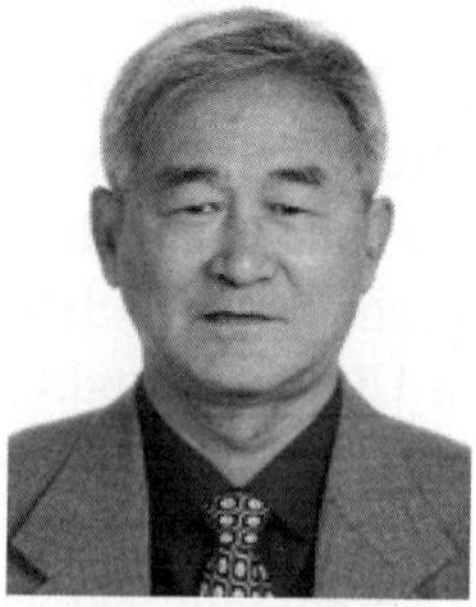

Dr SiShen Xie is a profressor at Insitute of Physics, China Academy of Sciences (CAS). He is a Deputy Director of Chinese National Center for Nano Science and Nano Technology. Prof. Xie is the Director of Center for Nanoscience and nanotechnology of CAS, and Director of Vacuum Physics Laboratory of CAS. He was the Chief Scientist of the National Key Basic Research Project (973) 'nanomaterials and nanostructure'. Prof. Xie has worked in the fields of High Tc superconductors and his recent research focus are in the fields of nano-scale materials fabrication, structure and properties, including nanotube and nanowires of oxides and silicon.

He was awarded by the Natural Science and Technology First Prize and Third Prize by the National Science Foundation China in 1989 and 1990 respectively. He was awarded the First Prize by CAS in 2001, the Qiao-Kou-Long-Ji Prize of Chinese Materials Researchers Society in 1999 among other awards. Prof. Xie has over 120 scientific papers published in prestigious international journals, including Nature, Science, PRL, PRB, APL, CPL and others. He became an Academician of the Chinese Academy of Sciences in 2003

Prof. Xie graduated from Department of Physics, Peking University in 1965. He received his Ph.D. from Institute of Physics, CAS, Beijing in 1983. He was a Post-Doctoral Scholar in Department of Electrical Engineering and Computer Science at the University of Colorado, USA during 1984-1986.

Author Biography

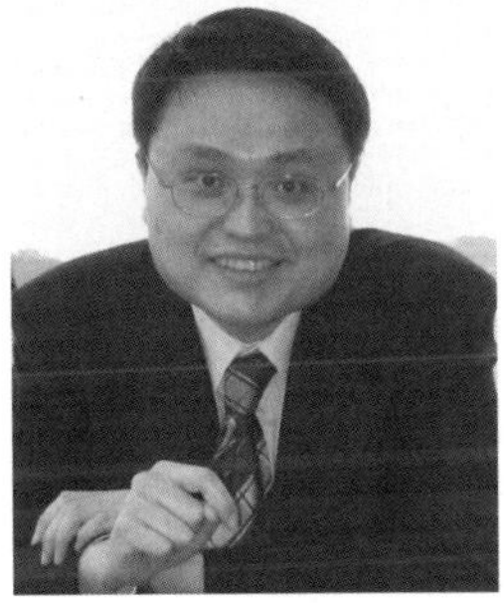

Dr King Lun Yeung is an Associate Professor of Chemical Engineering at the Hong Kong University of Science and Technology. He obtained his Ph.D. in Chemical Engineering at the University of Notre Dame, U.S.A. in the area of surface science and catalysis. His current research focuses on the rational design and molecular-level engineering of functional nanoporous and nanostructured materials with chemical, environmental and bio-related applications.

CHAPTER 4

NANO SCIENCE AND TECHNOLOGY — THE INDIAN ODYSSEY HAS BEGUN

Venkatesh Rao Aiyagari

Department of Science and Technology, Technology Bhavan, New Mehrauli Rd.
New Delhi, India, 110 016
venktesh@nic.in

Praveen Kumar Somasundaram

Department of Science and Technology, Technology Bhawan
New Delhi, India, 110 016
praveen@nic.in

An overview of nanoscience and nanotechnology research in India is given, with emphasis on the R&D activities of the Department of Science and Technology, Ministry of Science and Technology, government of India.

1. Introduction

Indian Science and Technology have always been a part of the civilization which many may find difficult to unravel the mystery of drifting of the profound traditional knowledge of the ancient society in mathematics, metallurgy, astronomy, medicine and several other areas. The world witnessed a renaissance in the first half of the 20th century. Indian S&T also responded with vigor. Traditionally, Indian scientific research has focused on generating relevant innovative technologies; preserving, protecting and adding value to indigenous resources; adopting an appropriate mix of traditional, conventional and modern technologies; developing and nurturing human resource; strengthening basic research in the areas of frontline science and engineering. The government is committed to making S&T an integral part of the socio-economic development of the country. Unlike developed countries, almost 80% of the resources for R&D come directly or indirectly from the Government. The S&T infrastructure in the country accounts for about 1% of the GNP. India has made significant progress in the areas of atomic energy, space, defence, agriculture, industry; and certain Indian institutions are well-known for academic excellence. The new India has been looked upon as one of the most advanced countries in nuclear technology including production of source materials. The country is self-reliant and has mastered the expertise covering the complete nuclear cycle from exploration and mining to power generation and waste management. Accelerators and research and power reactors are now designed and built indigenously. Space research and developing remote-sensing satellites have endorsed the scientific strength of the country. Yes, S&T in India is entering a new frontier.

'Science and Technology Policy 2003' envisages: "Every effort will be made to achieve synergy between industry and scientific research. Autonomous Technology Transfer Organizations will be created to associate organizations of universities and national laboratories to facilitate transfer of the know-how generated to industry. Increased encouragement will be given, and flexible mechanisms will be evolved to help, scientists and technologists to transfer the know-how generated by them to the industry and be a partner in receiving the financial returns. Industry will be encouraged to financially adopt or support educational and research institutions, fund courses of interest to them, create professional chairs etc. to help direct S&T endeavours towards tangible industrial goals. There has to be increased investments by industry in R&D in its own interest to achieve global competitiveness to be efficient and relevant. Efforts by industry to carry out R&D, either in-house or through outsourcing, will be supported by fiscal and other measures. To increase their investments in R&D, innovative mechanisms will be evolved." It's obvious, Nano Science and Technology in the country has more to work around this policy statement.

The technology spur in around 'bottom up' and 'top down' have significant impact across the world which otherwise been entangled with technological revolutions in the last two decades in sectors of increased human interactions. The promises and expectations in nano scale technology, with due consideration of magnanimous funding by federal governments, are still running high in spite of only a few technological breakthroughs and commercial products in the market. India also responded to the urgency of initiating programmes of diversified contours for overall development of nanoscience and technology in the country. The Indian S&T system consists of a vast network of S&T institutions/laboratories; universities and other scientific, educational institutions. India has always had vibrant and stabilized basic research groups scattered around the country. Indeed more recently multinational companies and foreign institutions are locating or relocating their R&D units in India because of the availability of broader spectrum of intellectual capability within the country. India has tremendous opportunities in the field of nanoscience and technology. It is aknowledge-intensive area and India, with its long tradition of R&D can be a major player.

A journey through the post independent development of Indian science, its vibrant research agenda, reveal the profound influence of Professor C.N.R Rao, the Chairman-Scientific Advisory Council to the Prime Minister in the S&T structure of the country. The Nano Science and Technology Programmes are also steered under his intellectual stewardship. Professor Rao himself being an active researcher has encouraged a large number of researchers from various Institutions across India with expertise in different disciplines - Chemistry, Physics, Materials Science, Biotechnology etc.

This chapter gives an overview of nanoscience and technology research in India with particular emphasis on the R&D activities of the Department of Science and Technology, Ministry of Science and Technology, Government of India. We do not claim to present a complete picture of India nanotechnology research, indeed there are many players primarily Government agencies including Central and State Governments.

2. Indian Initiatives in Nano Science and Technology

Various Ministries/Departments of Government of India such as the Department of Science and Technology (DST), Department of Information Technology (DIT), Defence Research and Development

Organization (DRDO), Council of Scientific and Industrial Research (CSIR) and Department of Biotechnology (DBT) have been supporting R&D in Nano Science and Technology.

The Nanotechnology Development Programme of Department of Information Technology, Government of India, with a plan to create infrastructure for research in nanoelectronics and nanometrology at national level and also to fund small and medium level research projects in specific areas such as nanomaterials, nanodevices, carbon nano tubes (CNT), nanosystems, nanometrology, etc., was started in the year 2004. Under this programme, eight small and medium R&D projects and two major projects namely Nanoelectronics Centres a joint project at Indian Institute of Science (IISc) Bangalore and Indian Institute of Technology (IIT) Bombay - and Generic Development of Nanometrology for Nanotechnology at National Physical Laboratory (NPL), New Delhi have been initiated.

DST launched a special Nano Science and Technology Initiative (NSTI) in October 2001. The NSTI focused on research and development in nanoscience and technology in a comprehensive manner so that India can become a significant player in the area and contribute to the development of new technologies besides carrying out basic research at the frontiers of knowledge. The programme supported R&D projects, strengthening of characterization and infrastructural facilities, creation of centres of excellence, generation of trained manpower, joint projects between educational institutions and industry for technology development etc.

As a result of this initiative, about 150 R&D projects have been granted to individual researchers in 60-70 institutions across the country on synthesis and assembly of ceramic nanoparticles, nanotubes, nanowires, nanoporous solids, DNA chips, nanostructured alloys, etc. The main focus has been on adopting chemical methods for synthesis of these materials.

Pieces of specialized equipment have been granted to facilitate characterization of the synthesized materials. Less expensive piece of equipment like ordinary STM/AFM, light scattering set-up, etc., have been made available to individual researchers. And, major pieces of equipment like Field Emission TEM with CCD, Nanomanipulator with

SPM, AFM/STM/TEM, Optical Tweezers, etc., have been established as common facilities.

Projects with potential application in the field of nanolithography, nanoelectronics, drug/gene targeting, DNA chips, nanotubes, nanostructured high strength materials, quantum structures, etc., have been awarded to prospective researchers.

On the manpower development front, Advanced Schools and International Symposia and Training Workshops for Research Scholars/students in this promising area have been held.

In a coordinated move, the Government of India has established eleven Centres for Nano Science for quality research. These Centres are centred around established scientists of known caliber and spearheads the nano research activities in the country. The location and focus of the Centres for Nano Science are given below (see Fig. 1 for their geographic location).

2.1. *Jawaharlal Nehru Centre for Advanced Scientific Research, Bangalore*

The Centre focuses synthesis and characterization of a variety of new nanomaterials - nanocrystals, inorganic nanorods, nanocomposites and other structures, employing innovative procedures. Significant achievements have been made in device applications such as carbon nanotubes (see Fig. 2) based supercapacitors, oxide nanorods based gas sensors etc. Nanoscale patterning of metals and inorganic materials employing electron beam lithography and AFM based techniques have been worked out.

2.2. *Indian Institute of Science, Bangalore*

The Centre primarily centred on soft lithography aspects of nanoscience. This includes biomolecules, organic thin films, polymers etc. Bile acid-based organo- and hydro-gels with nanoparticles have also been investigated. Liquid crystal-gold composites have been investigated for their electrical conductivity and dielectric behaviour. In the area of biomolecules based soft films, a DNA monolayer on a Langmuir film of

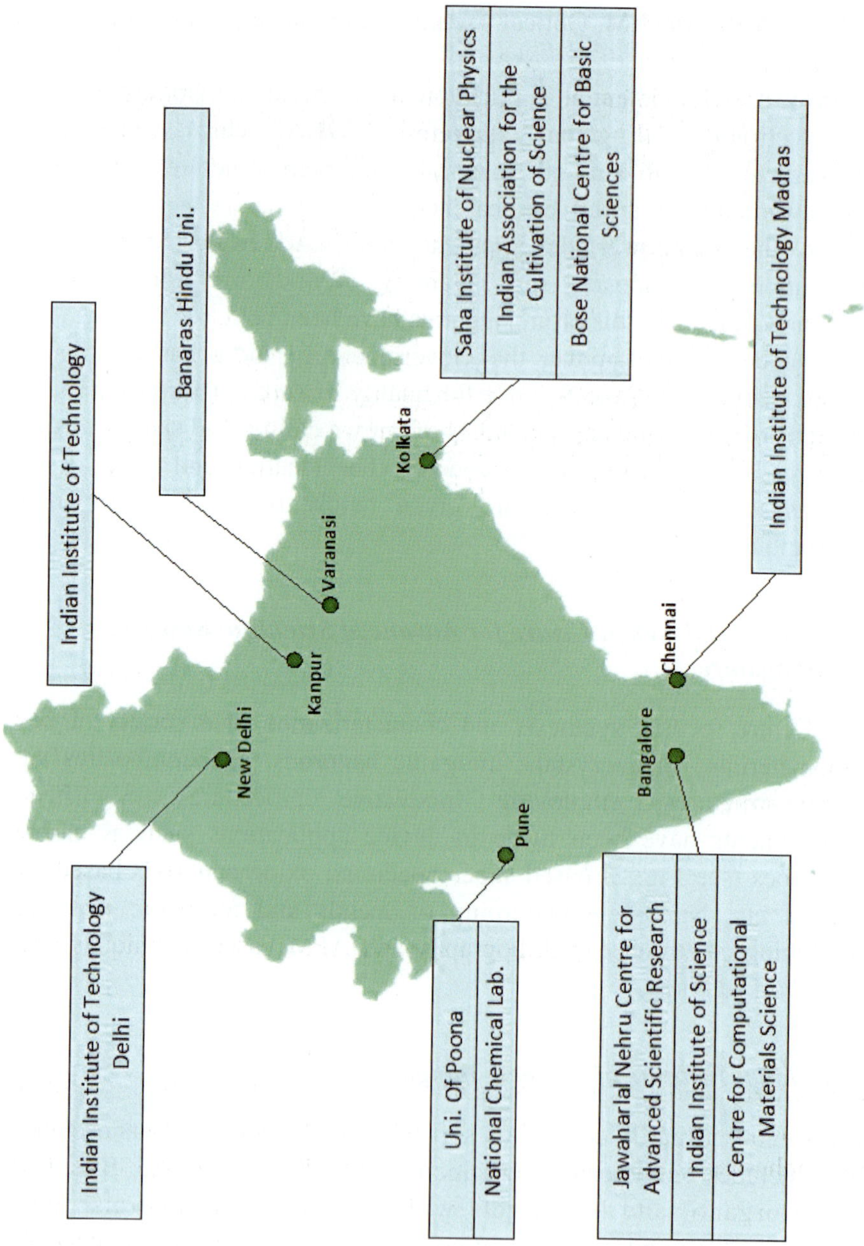

Fig. 1. Main nanotechnology research centers in India.

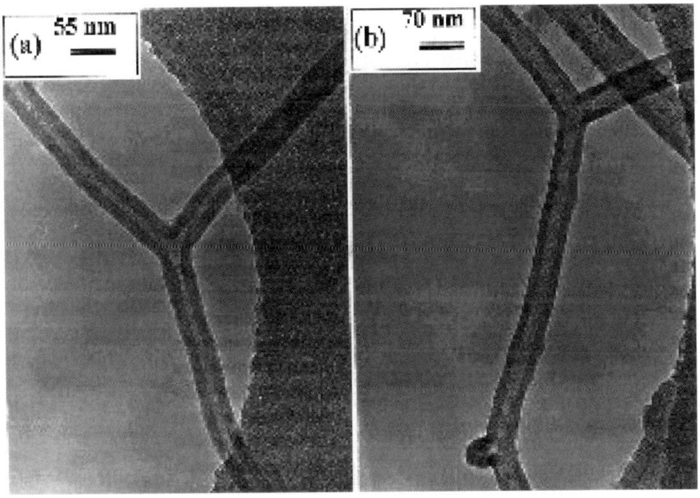

Fig. 2. Single walled Carbon Nanotubes.
Source: Jawaharlal Nehru Centre for Advanced Scientific Research, Bangalore.

metal arachidate has been formed to follow interactions of specific DNA-protein complexes. Well oriented, compact self assembled monolayers have been formed on metallic substrates and used to understand their interfacial properties with various solvents. The formation of droplets during pulsed laser deposition is of concern in the formation of thin films. The system containing $La_{0.5}Sr_{0.5}CoO_3$ (LSCO) on $LaAlO_3$ (LAO) substrate has been studied using atomic force microscopy.

2.3. *Indian Institute of Technology, Kanpur*

The emphasis of the Centre is on novel techniques that not only have rich underlying science, but are also suitable for commercialization in large area, rapid and inexpensive patterning required for the bulk-nano applications such as the optical coatings, structured colors, super-hydrophobicity and smart adhesives. These ideas also have potential applications in engineering of self-organized meso-scale patterns in the soft materials for opto-electronics, MEMS, lab-on-a-chip devices and in nanoscale understanding of interfacial phenomena such as adhesion, debonding, wetting and friction at soft surfaces. Magnetic nanoparticles have attracted considerable attention in recent years because of their technological applications in areas such as high density data storage, ferrofluid mechanics and biomedical/drug delivery systems. These applications are driven by the unique magnetic behavior of the individual magnetic nanoparticles. Since a large fraction of atoms in a nanoparticle is on the surface, a different physio-chemical environment of these atoms influences magnetic properties of magnetic nanoparticles in a nontrivial manner. The magnetic alloys of particular interest in nanoparticle form are the ordered and disordered phases of CoPt and FePt. Near the equiatomic composition, CoPt and FePt form a magnetically soft disordered face centered cubic (fcc) structure which can be transformed to an ordered face centered tetragonal (fct) structure on annealing at $T >$ 700°C and 550°C respectively. Preliminary studies of electron transport in rectangular lattice of nanosize CoPt squares on NbN superconducting films reveal fascinating consequences of the antagonism between superconductivity and magnetism.

2.4. *Indian Association for the Cultivation of Science, Kolkata*

The Centre focuses on synthesis and characterization of a series potential nanomaterials and nanocomposites. Composites containing elements like Ti, Cr, Fe, Co, Sn etc inside a mesoporous silica framework were synthesized by hydrothermal technique. They exhibit interesting surface properties. Nanoporous inorganic and organic-inorganic hybrid materials were also prepared using different ionic surfactants. Nanostructured $CuInS_2$ thin films were fabricated by ion layer gas reaction method. The particles were found to have radii in the range 6 to 21 nm. Ga_2O_3-SiO_2 nanocomposites were prepared by a sol-gel method. Ga_2O_3 particles had diameters in the range 2 to 5 nm. ZnO nanocrystals with morphologics like nanorod arrays and flower-like assemblies were grown. Surface confined living radical polymerization (SCLRP) technique was used to prepare coreshell particles comprising of metal (Au) or metal oxide (ZnO, TiO_2) core and polymer shells of controlled thickness. Colloidal metal nanoparticles-peptide conjugate were synthesized using newly designed tripeptides. Spongy gold nanocrystals with pronounced catalytic activity were made by modified-citrate reduction technique. Poly (3-hexyl thiophenc) (P3HT) – organically modified Montmorillonite (om-MMT) polymer nanocomposites were prepared and characterized. In situ Ag nanoparticles were produced by reduction of Ag+ with DMF in the presence of poly (Vinylidene fluoride) (PVF_2). Analysis of data by Avrami equation showed that two dimensional nucleation was operative with linear or diffusion controlled growth. Silver molybdate glasses were used as templates to grow Ag_2MoO_4 nanoparticles, $Ag_6Mo_{10}O_{33}$ nanowires, $Ag_2Mo_2O_7$ nanorods and α-AgI nanocrystals. Silver nanowires of length 150 nm and diameter 30 nm were grown in Ag_2S doped tellurite glasses by electrodeposition technique. GaN nanowires of diameter 15 to 40 nm and lengths of a few microns were grown from the edges of platelets at a temperature of 1050°C. By annealing at 800°C dendrite-like GaN nanostructures were obtained. GaN nanorods of diameters in the range 40-100 nm were grown in ZnO/quartz substrates. Nanometer-sized silicon powders were formed by 13.56 MHz radio-frequency capacitive glow discharge plasma of silane-argon gas mixture.

2.5. *Bose National Centre for Basic Sciences, Kolkata*

The focus of the Centre is on the synthesis of nanomaterials and their arrays, investigations on science (physical and chemical properties) of size reduction and confinement, science of tools for nanosciences research like the physics of Scanning Probe Microscope, Computational materials science of nanoscale materials and science of confined biomolecules and biomolecular recognition. In addition to bulk and time averaged measurements the activities at the Centre include both spatially resolved (Scanning Probe Microscope based) as well as temporally resolved measurements that include noise spectroscopy (low frequency) as well as ultra fast optical spectroscopy down to few tens of picosecond. Significant results have been obtained in synthesis of metal nanowires, nanotubes and complex functional structures using a novel electrodeposition technique and investigation of their transport properties from 3K - 700K to elucidate changes in transport properties on size reduction; tuning of ground state properties of perovskite oxide nanomaterials (films, particles and wires) synthesized by chemical routes; synthesis of ZnO nanocrystals, nanowires and films whose optical properties including band gap and Photoluminescence can be tuned by doping with Cd and Mg; growth of ordered arrays of nanoparticles and wires using chemical and electric forces using a combination of nanoimprint lithography, optical lithography and e-beam lithography; Physics of dynamic force microscopy and force spectroscopy; investigation of processes in biomolecules and confined biomolecules using picosecond spectroscopy to investigate effect of confinement on biomolecules, understanding the energy transfer between biomolecule and science of micelles and reverse micelles. Interesting results have been reported on Fluorescent Quantum Dots.

2.6. *National Chemical Laboratory, Pune*

This Centre focuses on the synthesis of nanomaterials by novel methods utilizing the micro-organisms, foams as templates and their subsequent assembly using biological templates like DNA etc.; study the applications of nanomaterials and nanomaterial-polymer composites as drug delivery systems and medical implants. It has been found that synthesis of Co and Ni nanoparticles in presence of an anionic surfactant sodium dodecyl sulphate (SDS) and oleic acid under specific concentration conditions lead to the formation of highly ordered assemblies of monodisperse spherical aggregates. These particles show large magnetocaloric effect in the vicinity of a low temperature magnetic transition in Co and $Co_{core}Ag_{shell}$ systems. It has been observed that surface modification of polyimide membranes by NH_3 plasma treatment and the subsequent layer-by-layer assembly of Au nanoparticles and aminoacid lysine could lead to the gradual lowering of the surface contact angle which is regarded as an essential step for the utilization of polymer membranes for bio-implants and cell growth.

2.7. *University of Poona, Pune*

The Centre primarily focuses on Novel quantum confined structures: Synthesis, analysis, applications and theory; Magnetic and ferroelectric nanomaterials; Bio-nanomaterials and Lithography for nanodevices. Synthesis of novel CdS nanoparticles using DNA has been reported. In addition, using chemical method of synthesis a variety of nanoparticles of CdSe, ZnO, SnO_2, $TbMnO_3$, ZnSe etc. have been developed. The group has made substantive study on systems like Mn doped ZnO, II-VI semiconductor clusters, titanate nanotubes and composites etc.

2.8. *Indian Institute of Technology Madras, Chennai*

The Centre is largely concerned with the chemical interactions of nanoparticles, nanorods and sub-nano particles in solution and in the form of thin films. Such chemical interactions result in novel chemistry and new physical phenomena. The chemical properties such as the degradation of pesticides on nanoparticles have been commercialized. The flow induced potential on nanoparticle surfaces has been extended to anisotropic nanostructures such as nanorods. Chemical interactions with carbon nanotubes lead to the distortion of the nanotube structure, which in turn results in visible emission from the nanotubes. Noble metal nanoparticles interact with proteins strongly and the chemistry of the proteins is preserved in some cases. Using nanoparticle chemistry, drugs have been delivered to cells. The chemistry of anisotropic nanostructures such as nanorods is similar to spherical particles in several cases, but there are also significant differences. A new model has been developed to understand the temperature dependent rotational dynamics at nanoparticle surfaces. The chemistry to subnano dimensions with the recently discovered sub-nano clusters of gold was explored.

2.9. *Indian Institute of Technology Delhi, New Delhi*

The Centre focuses on the development of nanomaterial based sensors. Hydrogen induced changes in the electronic and optical properties of Pd and rare earth metals make these materials highly suitable for sensor and switching applications. In a study on the hydrogen sensing properties of Pd nanoparticles, it has been shown by controlling nanoparticle size, surface conditions and thickness, the opposing and entangled effects due to electronic and geometric changes can be controlled and thus separated. Due to the presence of interparticle gaps and complete conversion to PdH_2, the initial response is followed by a substantial decrease in resistance due to closure of conducting paths. This study thus sets the foundation for fabricating a novel gas sensor devices having pulse-like response. A new methodology of improving the hydrogenation properties of solid state materials by using ion-induced nanotracks as two-way transport routes has been successfully implemented. A large enhancement in the hydrogen stoichiometry value (H/Pr) from 7.2% to 17.8% is observed with increase in the ion doze. The role of nanotracks in hydrogen removal during deloading is even more remarkable. It is observed that about 31% H can be removed during deloading in ion-irradiated samples (maximum possible is 33.3%, corresponding to PrH3 to PrH2 transformation) in comparison to only 12% in case of un-irradiated layers.

2.10. *Banaras Hindu University, Varanasi*

Freestanding monolithic uniform macroscopic hollow cylinders having radially aligned carbon nanotubes with diameters and lengths up to several centimeters were fabricated and used as filters to eliminate heavy hydrocarbons from petroleum and to filter bacterial contaminations from water. In another work, structural, electrical and mechanical properties of multi-walled carbon tubes (MWNT) – polyethylene oxide (PEO) composites were investigated. Composites with different wt% (0 to 50 wt%) of MWNTs, were prepared and characterized. The conductivity measurements on the MWNT (~50 wt%)-PEO composite films showed an increase of eight orders (~7.5×10^{-8} to 6.52 S cm^{-1}) of magnitude in conductivity to that of a PEO film. Formation and microstructural characterization of coaxial carbon cylinders consisting of aligned CNT stacks were also investigated. Some other investigations involving CNT relates to admixing CNT to $NaAlH_4$ to increase the hydrogen desorption kinetics and admixing of CNT to MgB_2 superconductor to increase the critical current density. Efforts have also been made to synthesize gallium nitride nanocrystals on aligned carbon nanotubes. The cuprous oxide (Cu_2O) nanostructures have been synthesized by anodic oxidation of copper through a simple electrolysis process employing plain water as an electrolyte. Two different types of Cu_2O nanostructures have been found. One type got delaminated from copper anode and collected from the bottom of the electrochemical cell and the other was located on the copper anode itself. The obtained nanostructures are nanothreads, nanorods and nanocubes. Similar growth of nanostructures of SnO_2 has also been obtained. For ZnO, vapour phase growth involving reaction of Zn vapour with oxygen has led to the formation of various nanostructures, e.g. nanotetrapores, nanorods, nanocombs etc. Correlation of nanostructural features with optical and field emission studies have been carried out. Other nanomaterials synthesized correspond to silicon carbide and titanium carbide nanosttructured films. Yet another variety of nanomaterials studied are the quasicrystalline materials: ((Co, Ni)Al_2O_4).

2.11. *Saha Institute of Nuclear Physics, Kolkata*

R&D on the compound semiconductor based nano-electronic device materials have been initiated. This is a unique effort to develop and establish core technology for the futuristic devices based on epitaxial materials. Novel fabrication techniques that permit precise control of the structure, properties, size and position of nanoscale elements are the most important aspects of this activity. The Centre focuses on two-types of materials, namely Si-Ge and III-V systems. Molecular Beam Epitaxy (MBE) facility for growing Si-Ge materials and a metal-organic vapor phase epitaxy (MOVPE) system for growing III-V materials will be realized. The focus of the MOVPE system will be the growth of InP nanowires, doping, nanowire heterostructures and exploit their potential application in futuristic quantum devices. Study of the electronic structures of nanomaterials and nanostructures are also very important since they are responsible for the physical, chemical and other novel properties. Angle resolved photoemission spectroscopy (ARPES) is the most direct momentum-resolved technique that is employed for the investigation of electronic structure of crystalline materials. The focus of the studies will be the low-dimensional semiconductor nanostructures grown by MOVPE-MBE systems.

2.12. *Centre for Computational Materials Science, JNCASR, Bangalore*

The Centre spearheads the study of theoretical aspects of computational materials science like phase transitions, structure, and dynamics of several metastable systems including supercooled liquids, gels etc. Interesting results were reported on ab initio molecular dynamics simulations of nanoclusters of room temperature ionic liquids, first principle calculations on ferroelectricity in epitaxial thin films of $BaTiO_3$ and $BaTiO_3/SrTiO_3$ superlattices, quantum chemical calculations on stable metal clusters as well as on the origin of magnetism in geometrically frustrated lattices. Scaling relations for the dependence on size, material and structure of the elastic and vibrational properties of nanoclusters and periodic systems were obtained using density functional theory.

3. Mission on Nano Science and Technology (Nano Mission)

While very good and internationally competitive fundamental studies are being carried out on nanoscale systems, India has to catch up in the area of Nanotechnology. The Government has launched, in mid 2007 a Mission to encourage research work in the field of Nano Science and Technology with an allocation of US $ 250 M for five years. The primary objectives of the Nano-Mission are: (i) Basic Research Promotion – Development of fundamental understanding of matter that enables control and manipulation at the nanoscale. Multidisciplinary research will be especially encouraged. Shall provide support to individual investigators and interdisciplinary groups of investigators. Creation of Centres of Excellence and major research facilities at various locations where knowledge-base exists are also envisaged; (ii) Infrastructure Development for Nano Science and Technology Research – Investigations on the nano scale require expensive equipments like Transmission

Electron Microscope (TEM), Scanning Probe Microscope (SPM), Optical Tweezer etc. For optimum use of resources, it is proposed to establish a chain of shared facilities across the country; (iii) Public Private Partnerships and Nano Applications and Technology Development Centres – A strong linkage between academic institutions, R&D institutions and industries is called for. We need to evolve a system involving co-operative, multidisciplinary, public private ventures, with adequate access to risk capital and funding. To catalyze Applications and Technology Development Programmes leading to products and devices, the Mission proposes to institute activities like Public Private R&D Projects, Nano Applications and Technology Development Centres, Nano-Technology Business Incubators etc.; (iv) Human Resource Development – The Mission shall focus on providing effective education and training to researchers and professionals in diversified fields so that a genuine interdisciplinary culture for nanoscale science, engineering and technology can emerge and (v) International Collaborations – Research and Development in Nano-Science and Technology has been identified as a priority area in our S&T co-operation agreements with many countries. Apart from exploratory visits of scientists, organization of joint workshops and conferences and joint research projects, co-ordinated access to sophisticated research facilities is an important component of such agreements. Academia-Industry partnerships will also be nurtured under these programmes.

3.1. *Research Themes — Nano Mission*

We have done an extensive project study on defining the priorities of research. In choosing the national priorities the following guiding principles were taken into consideration:

(1) Areas of national interest and priorities like agriculture, energy, health care, drinking water, etc;
(2) Areas where national strength and expertise in R&D exist;
(3) Areas having application and commercial potential including that for export.

Based on these considerations, the following emerge as the research themes of the thrust areas for the Nano Mission.

3.1.1. *Agriculture and Food Systems*

Majority of Indian population still depends on agriculture and the economy predominantly an agrarian. Large scale agriculture in the country also makes the country's environment and eco-systems dependent on agricultural practices. Nanotechnology offer enormous possibilities to increase and value-add to our agricultural produce, reduce dependence on fertilizers and pesticides and help in mitigation of environmental degradation due to agricultural practices.

Agricultural and food systems security, disease treatment, delivery systems, new tools for molecular and cellular biology, new materials for pathogen detection and protection of the environment are examples of the important links of Nanotechnology to the science and engineering of agriculture and food systems. The exciting possibility of combining biology and nanoscale technology into sensors holds the potential of increased sensitivity and therefore a significantly reduced response-time to sense potential problems. It is possible today to envision a nanoscaled surveillance system for the safety and security of today's agricultural products with capabilities of identity preservation and tracking.

3.1.2. *Drinking Water*

Nanomaterials have great potential to solve water purification problems. With increasing population and increase in standard of leaving, the potable water requirement in our country is bound to increase. This has already become an urgent and basic need. Nanotechnologies are expected to offer several solutions to this problem and our scientists have the necessary competence achieve to such breakthroughs. Significant lead results and commercialization of nanoparticle based filters have been reported.

3.1.3. *Nanotechnologies for the Automotive Sector*

India is the biggest market for 2- and 3-wheelers in the world. World's biggest companies are Indian. Interestingly, India has also emerged as the manufacturing hub for auto components for all the major companies of the world. Nanotechnology is expected to play important roles in anti reflective and hydrophic surface coatings, catalytic nanoparticles as a fuel additive, nanoporous filters to minimize emission, nanoparticles as filler in tyres, self-healing coatings etc.

3.1.4. *Nanotechnologies and the Energy Sector*

Given our huge population, our energy requirements are large. Development of new and novel energy producing, energy saving and energy storage devices is a challenge that has to be met by our scientists. Storage, and fuel cells, batteries etc. are the immediate deliverables. Our scientists are already active in these areas. Imparting a commercial direction to their R&D efforts will bring in considerable benefits for the country.

3.1.5. *Nanoparticles as Drug Delivery Systems*

This area is of immense importance to India. Any R&D in the area of drugs and pharmaceuticals has tremendous societal relevance. India has been traditionally strong in chemistry and the chemical, pharmaceuticals and biotech industry have been extremely forward looking and vibrant in our country. Faced with the challenges of the new IPR-regime, they are putting considerable effort and resources in R&D leading to development of new drugs, diagnostics and delivery mechanisms. Nanotechnologies offer enormous possibilities for drug delivery mechanism. Given our strengths, this is an area where India can be an internationally competitive player.

3.1.6. *Nanobiosensors*

India has a vibrant biotechnology sector. We also have a number of excellent institutions pursuing cutting-edge research in Life Sciences. A vibrant nanobiosensor programme will greatly enhance the capabilities of these groups leading to new understanding of life processes.

3.1.7. *Nanocrystalline Phosphors*

Nanophosphors have a wide range of applications from energy saving in devices to displays to their use as and biological luminescent tags. It is an area where intensive materials science research is required and India has a large number of scientists who can fruitfully contribute to this area.

3.1.8. *Other Nanomaterials*

Indian scientists have contributed a great deal to the science of carbon nanotubes. Also coating and paint industry is fairly robust in our country. It is realistic to expect that commercial production of nanopowders, nanocomposites, CNT's etc. will give further push to this industrial sector. We can also fruitfully aim at value addition of materials locally available in plenty.

3.1.9. *Nanometrology*

Nanotechnology is looking forward to major developments and breakthroughs in future. Also, developments in this area are crucial for establishment of industrial and regulatory benchmarks. India has considerable strength in scanning probe/microscopies and related instrumentation. Indian scientists can play significant role in this area with considerable knowledge-gaps.

3.2. *Centres for Nano Technology*

The R&D in this emerging field in the country is anchored by Centres for Nano Technology located at the following Institutes.

(1) *Indian Institute of Science, Bangalore*: The Centre focuses on the development of nanodevices based on thin film technologies (self-assembly, LB films), synthesis of molecules (surfactants used for

preparing various assemblies) and polymers. In addition, it also focuses on technology development in the areas of nanocomposites and nanobiosensors. It promises to develop biosensors capable of detecting carbohydrates mediated pathogenic infections. In the area of nanodevice, the main focus would be on development of high density nonvolatile memory elements for random access memories and nanotube based integrated circuits and devices. Energy storage devices with large capacity e.g. super-capacitors, would be fabricated using CVD-produced manganese oxide nanocomposite coatings.

(2) *Indian Institute of Technology Bombay, Mumbai*: This Centre has state-of-the-art equipments for carrying out research in the broad areas of nanobiotechnology, nanoelectronics and nanomaterials. A few selected research activities that would be undertaken in this centre include the development of engineered/hybrid nanomaterials for healthcare especially magnetic nanoparticles for hyperthermia treatment of cancer and MRI contrast agent; liposome based, magnetic nanoparticles attached with drugs for targeted delivery. The Centre also aimed at the development of novel surfactants nanoparticles for treatment of respiratory diseases, nanocomposites for dental and orthodontics use, micro-devices for cardiac use nano conducting polymers as building blocks for IC, actuators as sensors, valves and pumps, organic LEDs etc.

(3) *National Centre for Biological Sciences (TIFR), Bangalore*: This Centre would focus on the study of the nano-scale in biology. The main thrust of the Centre would be to develop versatile platforms to allow modular implementation of high-content, high-throughput screening devices based on nanoscale biosensors for a variety of cellular processes. Development of biosensors using unusual nucleic acid structures and their applications to probe cellular processes are envisaged.

(4) *SN Bose National Centre for Basic Sciences, Kolkata*: The Centre would develop technology platforms, devices and products in MEMS and NEMS. Specifically, they would focus on development of technology and technique for sub-micron MEMS/NEMS to fabricate Si wafer arrays of Si/Si_3N_4 structures; technology to grow nanowires and nanoparticles or arrays of them of particular size at a

predetermined location on the MEMS/NEMS and technology to attach nanoparticles to fabrics and natural fibres and related surface treatments.

(5) *Indian Institute of Technology, Kanpur*: The Centre falls in the area of printable electronics and nano-patterning for novel manufacturing paradigms. It offers nanotechnology based device development with specific focus on printable electronics using soft materials such as molecular solids and polymers and their heterostructures with inorganic systems. Development of organic semiconductor based RFID tags, sensors etc. are envisaged.

(6) *Indian Association for the Cultivation of Science, Kolkata*: This Centre focuses on design and synthesis of materials suitable for applications in photovoltaics and sensors. Humidity and other gas sensors employing nanocomposites based on metal-metal oxide interfaces would be developed. In addition, the researchers at the Centre would look at design and synthesis of core-shell and quantum well structures based on group II-VI semiconductors with high efficiency for photovoltaic applications.

(7) *Amrita Institute of Medical Sciences, Kochi, Kerala*: The Centre is dedicated to implants, tissue engineering and stem cell research. The R&D at the Centre include biomimetic nanoparticles for use in scaffolds, cytotoxicity and cell-surface interactions of nanoparticles and nanocomposites on bacterial and human cells, creation of nanocomposites and synthetic extracellular matrices, mechanical behavior and high resolution microstructural characterizations. Three alternative synthetic routes for nanobiomaterials are explored, namely, chemical and biochemical, electrostatic and thermomechanical. Chemical routes are explored for the processing of biomimetic nano hydroxyapatite crystals for incorporation in polymer matrices and on surfaces and within nanofibers. Electrostatic methods are employed in the electrospinning of nanofibers of biodegradable polymers and blends of biodegradable polymers to control biodegradability and biocompatibility as well as mechanical properties. Finally, thermomechanical approaches are used to process nanocomposites of biopolymers and biopolymer

blends containing nanohydroxyapatite crystals for implant applications.

3.3. *Human Resource Development*

An intensified HRD programme is structured under the Nano Mission. A postdoctoral fellowship in Nano Science and Technology which was initiated in 2005 will be expanded in terms of number of fellowships. A postgraduate teaching programme (M.Sc and M.Tech) is under finalization.

3.4. *Public-Private-Partnerships*

We have some success stories in public-private-partnerships in R&D in areas like Drug and Pharma, Biotechnology etc. We aim to expand these partnerships to other areas too. Obviously, the first choice is nanotechnology. We have initiated some activities in this direction.

Development of High performance rubber nanocomposites for tyre engineering: This project being implemented at M.G University, Kottayam envisages the development of novel technologies in tyre engineering band on nanosize fillers in collaboration with Apollo Tyres. The *Nano Functional Materials Technology Centre (NFMTC) at the Institute of Technology Madras, Chennai*: The Centre addresses cost effective method for the production of oxide ceramic powders of nano size; consolidation and sintering of Nanocrystalline Oxide Powders for the production of bulk ceramics; Nanocrystalline diamond (NCD) films/coatings on die-inserts and plugs to increase wear-resistance and durability; cost effective production of large scale and highly pure random and aligned Carbon Nanotubes (CNT); nanostructured multi-drug-delivery system for hard tissue applications and CNTs for laser based treatment of cancer by photodynamic therapy. The participating industries are Murugappa Chettiar Group and Orchid Chemicals and Pharmaceuticals. Yet another research program on *Smart and Innovative Textiles (SMITA)* at the Indian Institute of Technology Delhi aims at fundamental understanding of generation of novel materials such as nanofibres, nanofinishes, and encapsulated phase change materials with desired characteristics; investigation of novel methods that are suitable for integrating above materials to textile substrates; fundamental understanding of the effect of the above materials and methods on functionalization of textile structures for developing smart textile; development of technology for upscaling the above processes for industrial benefit and creating new products for high value addition in the textile sector and creating comprehensive expertise and competence within the country by man-power training and enhancement of technical knowledge base. The participating industries are Resil Chemicals, Bangalore; Purolator India Limited, Gurgaon; and Pluss Polymers Pvt. Ltd., Delhi.

There has been some commercial spin-out from these activities.

4. Drinking Water — Indian Efforts

The technology developed at ARCI allows coating of ceramic candles with nanosilver throughout the surface of the candle, with silver loading ranging from 0.4-0.6 g/candle (see Fig. 3). Laboratory tests conducted on the nanosilver-coated ceramic candle filters have confirmed their excellent antibacterial action. The method of synthesis being simple and less expensive, the above technology has high potential for commercialization. The fact that these gravity filters do not require any

> **Nano-technology has been successfully employed by Indian scientists to develop appropriate technologies for filtering contaminated water for drinking purposes. Water filters using nanosilver and nanogold *particles have been developed by scientists at International Advanced Research Centre for Powder Metallurgy and New Materials* (ARCI), Hyderabad and Indian Institute of Technology Madras (IIT-M), Chennai. These filters have been field tested and are commercially available in the country.**

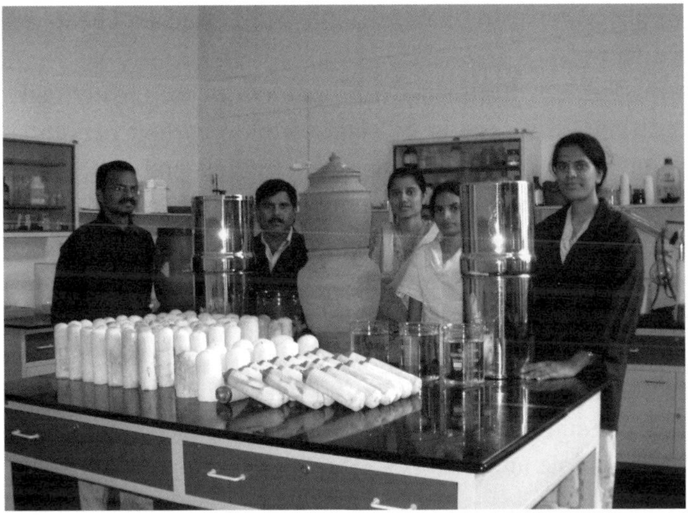

Fig. 3. A batch of nanosilver coated candles (ARCI) ready to be installed in filter housings.

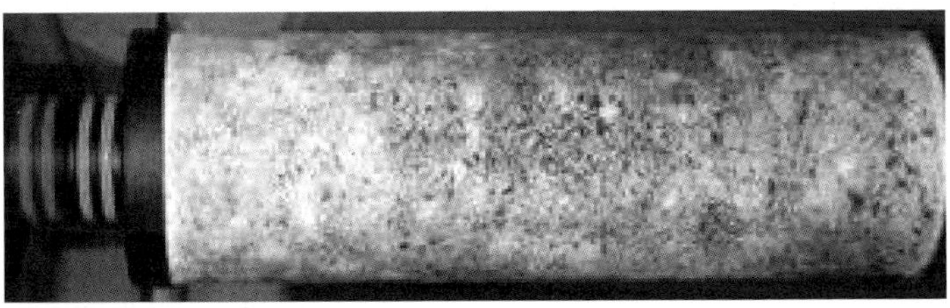

Fig. 4. Cartridge developed at IIT Madras for pesticide removal.

electrical power to exhibit anti-bacterial function is particularly relevant for places where no electricity and tap water are available.

The research group at IIT Madras discovered the use of nanotechnology for the removal of pesticides (see Fig. 4) of relevance to India from drinking water and a new product based on this nanotechnology has been commercialized. This product was reported to be world's first nanomaterial based water filter. Similarly, the same group has found cheap mechanism for the removal of micro-organisms from drinking water using nanoparticles supported on cheap and locally available materials.

A remarkable discovery of induction of electrical signals (voltage/current) in a sample of single-walled carbon nanotube bundles along the direction of flow of various liquids and gases over it was reported in a project being implemented the at Indian Institute of Science (see Fig. 5). The electrical response generated by the flow of liquids is found to be logarithmic in the flow speed over a wide range. In contrast, voltage generated by the flow of gas is quadratically dependent on the gas flow velocity. For the liquid, the Coulombic interaction between the ions in the liquid and the charge carriers in the nanotube plays a key role while electrical signal generation due to gas flow is due to interplay of Bernoulli's principle and Seebeck effect.

5. Conclusions

In measures of quality and quantity India occupies a significant share in nanoscience. Translating the science into technology is our traditional

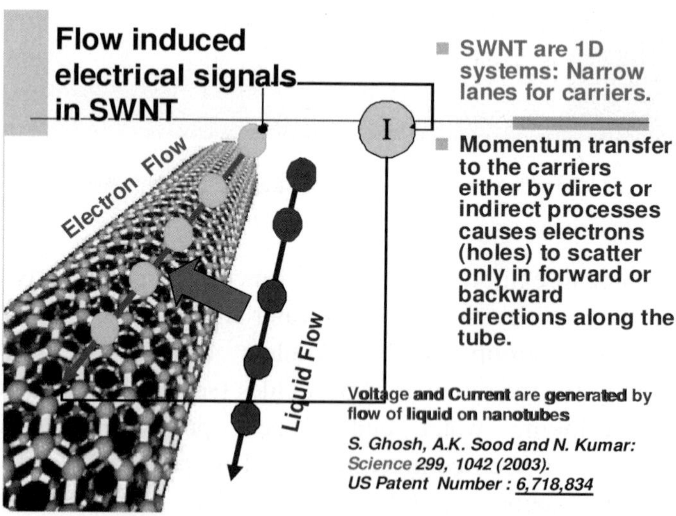

Fig. 5. Flow induced electrical signals in SWNT (Indian Institute of Science).

weakness. Clever ways of translation need to be put in place with key involvement of all stake holders including the government and industry.

Let us also join the question forum. How much nanotechnology is advisable? If our citizens have an average intake of 20 nanoparticles a minute, what would be the result? Experience tells us that most of them do no harm in an environment with natural nanoscaled components. We, Indians, do indeed bother too much about ethical issues. Naturally nanotechnology offers more concerns.

Green technology is the need of the hour. It is certain, the nature forces us to mutate, we need to change. Nanotechnology is yet another driver. *The Indian Odyssey has begun.*

References

1. Science Policy 2003, Ministry of Science and Technology, Government of India
2. Report on National Review and Coordination Meeting on Nano Science and Nanotechnology, Department of Science and Technology, Government of India
3. http://www.dst.gov.in
4. http://www.mit.gov.in
5. http://www.jncasr.ac.in
6. Mid-Term Appraisal of 10th Five Year Plan (2002-2007), Planning
7. Commission, Government of India
8. A little risky business, The Economist November 24, 2007.

Author Biographies

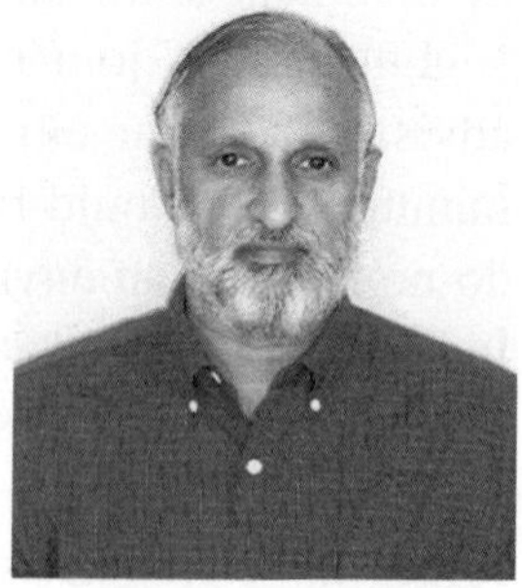

Dr Venkatesh Rao Aiyagari has been working in the India Department of Science and Technology for since 1972 and has made significant and outstanding contributions to the various activities related to promotion of R&D in Science and Engineering and also to various issues related to formulation of policy statements on science and technology. He has greatly contributed to the Review of the Scientific Policies and in the preparation of the Technology Policy statement in 1983 and more recently the new Science and Technology Policy in 2003. He has greatly contributed to the formulation of the science and technology plans in the country and also has considerable international exposure to the issues connected with science policy and management of R&D. In the Department of Science and Technology, he is handling one of the most important programs titled "Science and Engineering Research Council (SERC)" which is responsible for promoting R&D in newly emerging and frontier areas of science and engineering. The SERC has been making increasing efforts to promote basic research particularly in the universities and academic institutions and encouraging young scientists and engineers. SERC has done a yeoman service to the scientific community over the years and has helped to achieve outstanding research results. This has resulted in establishment of major National Programs and Research Centres of Excellence in areas such as Smart Materials for Application, Carbon and Nanomaterials Technology, Computational Fluid Dynamics, Laser Processing of Materials, Application of Lasers in Medical Science, Robotics and Manufacturing Science, Fuel Cell Technology, Bio-Medical Resonance, and Display Technologies.

Dr Aiyagari has greatly contributed to the success of the SERC program which involves intricate management and coordination skills involving scientists and scientific institutions coming from various academic institutions, national laboratories and industry. Some of the new initiatives include (a) A fund for strengthening of research infrastructure in Universities and other academic institutions; (b) A national program of Fellowship to attract students to Science and Engineering as a Research Career; (c) Swarnajayanti Fellowships in Basic Research for young scientists below the age of 40 years for pursuing World Class Level Science; (d) A program to encourage women scientists to pursue research, and more recently; and (e) Nano-Science and Technology Mission. He has outstanding leadership qualities and has been leading a dedicated group of young scientific officers committed to their tasks. He is very well known in the scientific community and has established an excellent rapport with them. Dr Aiyagari has been actively involved with the Technology Management Program for the UNDP-Government of India Country Cooperation Framework. He has also made significant contributions to the programs of the United Nations System in the field of Science and Technology.

Dr Aiyagari received his PhD from Sri Krishna Devaraya University (Ananthapur), Masters degree in Systems Engineering and Management and Operations Research from Institute of Industrial Administration, Union College (New York), and Bachelors degree in Mechanical Engineering specializing in Industrial Engineering, Ranchi University.

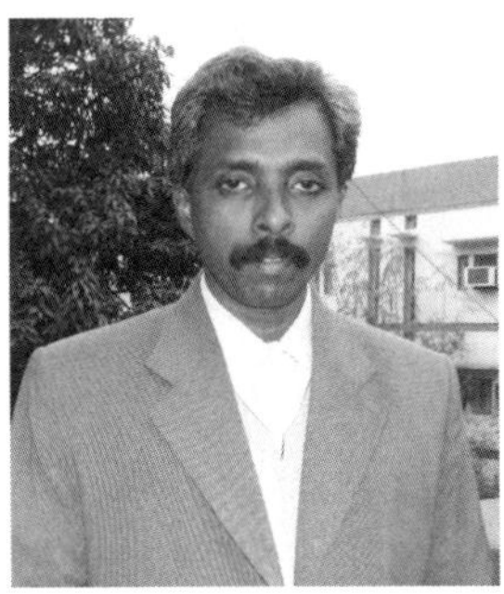

Mr Praveen Kumar Somasundaram holds a Masters Degree in Physics with specialization in Applied Electronics and has been associated with one of the most important programmes titled "Science and Engineering Research Council (SERC)" of the Department of Science and Technology (DST), Government of India. SERC is responsible for promoting R&D in newly emerging and frontier areas of science and engineering in India. Right from the inception of the 'Nano Science and Technology Initiative' (NSTI) of the Department, he was actively involved in the implementation of this programme. The NSTI has been responsible for initiation of several R&D programmes including establishment of Centres of Excellence in Nano Science, Centres for Nano Technology, HRD programmes in Nano Science and Technology etc. He was also worked as an Examiner of Patents and Designs in Patent Office, Government of India.

CHAPTER 5

INDONESIA NANOTECHNOLOGY DEVELOPMENT: CURRENT STATUS OVERVIEW

Nurul Taufiqu Rochman

Research Center for Physics, Indonesian Institute of Sciences
Kawasan PUSPIPTEK Serpong, Tangerang 15314, Indonesia
nuru006@lipi.go.id

Yesie L. Brama

NanoGlobe Pte Ltd, Singapore
yesie@nano-globe.biz

Indonesia, a country with large natural and human resources has to take advantage of nanotechnology for its development. This requires an appropriate strategy with regards to Indonesia's potential and capability in technology advancement. This chapter outlines identified milestones in Indonesia nanotechnology evolution, our proposed strategy and roadmap for nanotechnology development to the Indonesia government and summary of nanotechnology players and recent activities in Indonesia. Researchers who were educated abroad started nanoscience research at public research institutions in early 2000. In 2004, the Mochter Riady Center for Nanotechnology and Bioengineering was established. This private center inspired further formation of nanotechnology policy and R&D efforts in Indonesia. Platform for national nanoscience and nanotechnology development was proposed in 2006 and nanotechnology development roadmap for strengthening the national industry was laid down in 2008. About 43% increase in the number of institutions conducting nanotechnology R&D was observed from 2005 to 2008. In 2009 Nanocenter, which has been budgeted for US$ 20 million, will be initiated during 2009–2010 in Bandung Institute of Technology (ITB). These examples indicate how serious Indonesia now in developing and implementing nanotechnology for enhancing its national competitiveness worldwide.

1. Background

Indonesia is the world's fourth and Asia's third most populated country with ±230 million people and the largest archipelagic country with 17,508 islands and land area of 1,919,440 km^2, which consists of wide variety of flora and fauna. Indonesia is also blessed with abundant natural resources primarily minerals, oil and timber, which has made Indonesia a great supplier of raw materials. Nevertheless, the exploitation of these natural resources is currently not maximized as it goes without further value-add process thus resulting in lower commercial values.

Therefore in order to enhance Indonesia's industry competitiveness, Indonesia should develop its technology and manufacturing capability in manipulating its natural resources to produce value-added nanomaterials and the related products.

2. History, Strategy and Roadmap

Research and development of nanotechnology in Indonesia was started by researchers who just returned from abroad and worked at several research institutions (LIPI, BATAN, BPPT, LAPAN, MRC, etc) or universities (ITB, UI, ITS, Unand, UGM, etc). However, they conducted research in nanotechnology independently as nanotechnology was not yet a priority in Indonesia for the following reasons:

- Poor and uncoordinated infrastructures that are located in separate locations
- Short of skilled human resources
- Lack of research funding
- Prioritized research areas that are beneficial to Indonesia have not been established in nanotechnology

Nevertheless, today nanotechnology development has gotten much more attention either from research institutions or government funding agencies. The history of nanotechnology development is summarized as follow:

- ~2004: Several research centers and universities started nanotechnology research that was mainly conducted by young researchers who just came back from abroad.
- May 2004: Mochtar Riady Centre for Nanotechnology and Bioengineering (Mochtar Riady is the founder of Lippo Group, a billion dollar conglomerate in Indonesia) was established.
- 2004~2005: The State Ministry of Research and Technology started to identify research on nanotechnology.
- 2005~2006: The State Ministry of Research and Technology gave financial support for nanotechnology R&D with total funding about US$ 200,000.
- April 2005: Indonesian Society for Nano or Masyarakat Nano Indonesia (MNI) was established. MNI acts as international focal point on nanotechnology issues. In order to accelerate the development of nanotechnology R&D in Indonesia, MNI has done several activities namely organizing a number of workshops, national and international conferences, and nanotechnology roadshow, establishing MOU with universities to educate students and enhance its research collaboration, creating information web portal for nanotechnology, and constructing the nanoscience and nanotechnology alliance of Indonesia that will integrate and coordinate interests from academics, industry and government.
- 2006: Research on nanotechnology has become an important topic at research centers and universities in Indonesia. In addition, a platform of national nanoscience and nanotechnology development, as shown in Fig. 1, was proposed by the Ministry of Research and Technology.
- December 2008: a nanotechnology roadmap for strengthening the national industry, which was budgeted at US$0.1 million, was finally laid down by the Ministry of Industry, assisted by MNI.

The roadmap is constructed to provide some directions in efforts to increase national technology cluster capabilities with nanotechnology implementation.

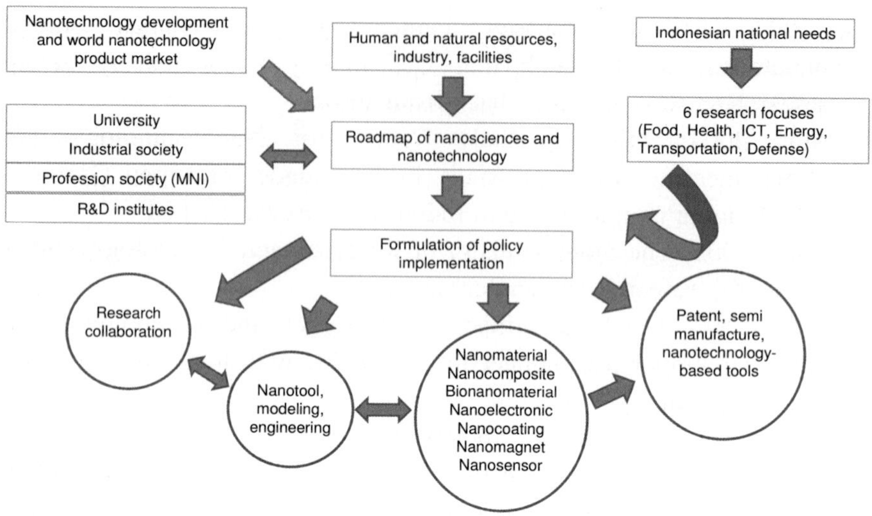

Fig. 1. Platform of National Nanoscience and Nanotechnology Development (source: KNRT).

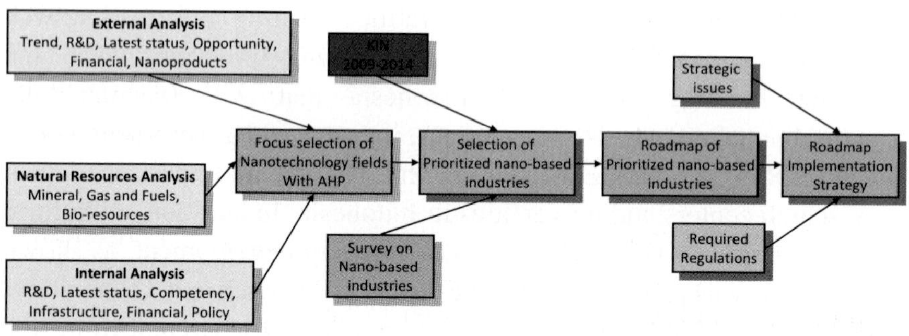

Fig. 2. Research design of nanotechnology development roadmap (source: The Ministry of Industry, RI).

Several activities were involved in order to construct this roadmap such as coordination meeting, participation in the 3rd International Dialogue on Nanotechnology in Brussel (2008), forum group discussion, visit to research institutes, industry survey, and priority allocation for

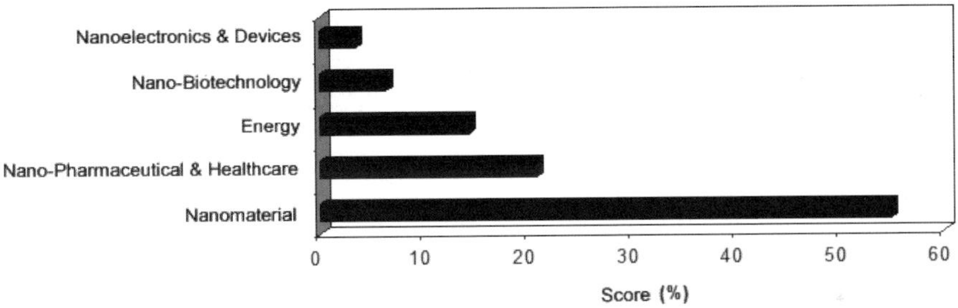

Fig. 4. AHP analysis result for focus selection of nanotechnology fields (source: The Ministry of Industry, RI).

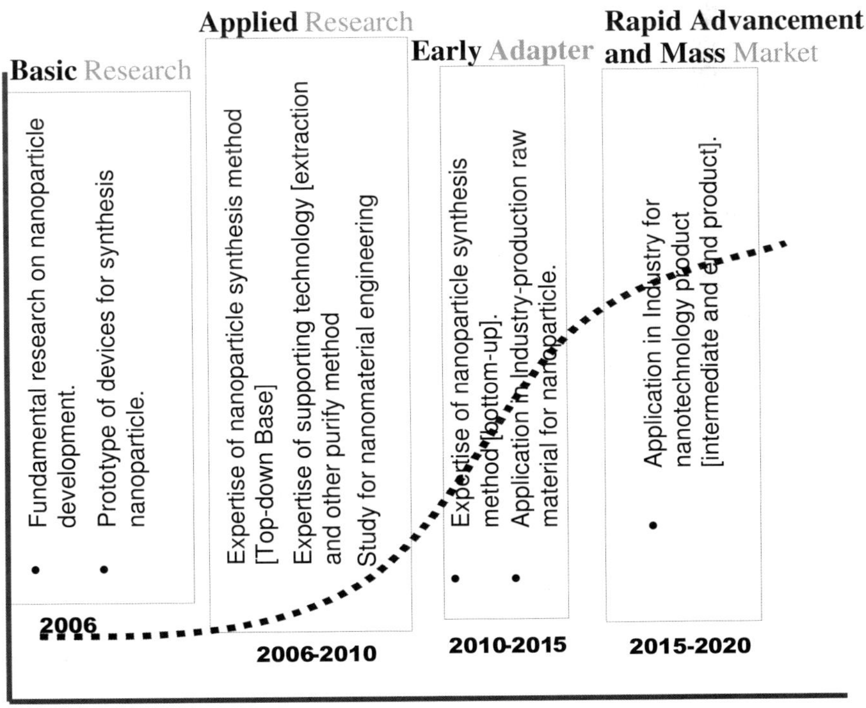

Fig. 5. Nanomaterial and nanotechnology product based on local resources development roadmap.

3. Infrastructure and R&D Players

Several nanotechnology facilities available in Indonesia include:

(1) Gas-phase processes
- Gas condensation with thermal evaporation
- Vacuum evaporation on running liquids (VERL)
- Thermal plasma synthesis
- Combustion synthesis

(2) Wet chemical processes
- Chemical precipitation
- Hydrothermal processing
- Sol-gel processing

(3) Sol-gel synthesis flow chart
- Thermochemical synthesis
- Sonochemical synthesis
- Hydrodynamic cavitation

(4) Solid-state processes
- High energy milling
- Mechanochemical synthesis

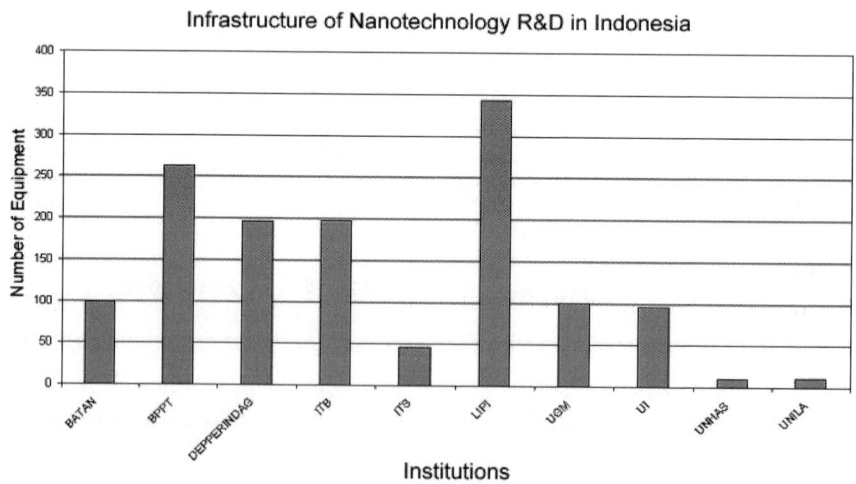

Fig. 6. Available infrastructure for nanotechnology development in Indonesia.

In 2008 several tools for nano-characterizations such as particle size analyzer, atomic force microscope (AFM), high resolution scanning electron microscope (HR-SEM), surface area analyzer, etc were added as new facilities. During 2009–2010, Bandung Institute of Technology (ITB) will build Nanocenter, which has been budgeted for US$20 million.

The number of equipment available to support nanotechnology development in Indonesia is summarized in Fig. 6.

Encouraging progress in nanotechnology R&D in Indonesia can be seen in Fig. 7. We see increasing number of institutions conducting R&D in nanotechnology from 70 institutions in 2005 to 100 institutions in 2008.

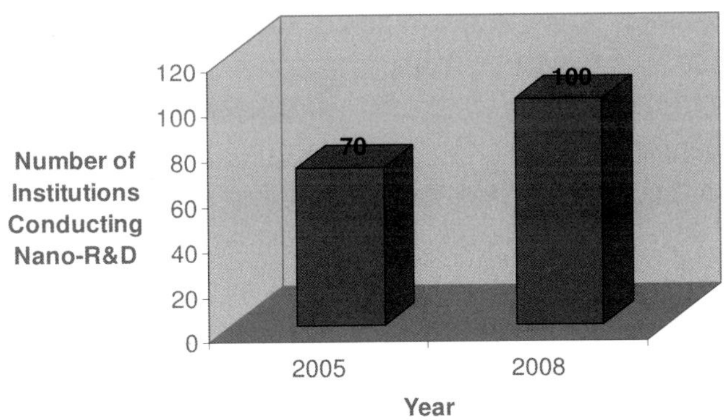

Fig. 7. Number of institutions conducting R&D in nanotechnology.

Nanotechnology R&D in Indonesia can mainly be grouped into five main areas, as follows:

(1) Nanomaterial

The research has mainly utilized Indonesian natural resources such as local clay, minerals, iron sand, etc as raw materials. Prototype tools for attaining nanoparticles in the simplest way such as high-energy ball mill have been developed. Several research topics have been conducted to solve technological problems such as research on thin film, coating etc. In addition, the development of new material with excellent properties has also been done such as research on nanocomposites, nanosilica for high strength concretes, nanosteels etc.

(2) Nanobiotechnology
Since Indonesia has abundant beauty of nature with many varieties of species in flora and fauna, research and development on this field has become another priority in developing the country. Therefore, many key players have been involved very actively in organizing workshops, local and international seminars and discussions.

(3) Nanodevices
Research and development in this field generally has strong relationship and collaboration with partners abroad. Fabrication of silicon quantum dot was done by different methods.

(4) Nanochemistry
Research in this field has been focused on the field of catalyst along with membrane development for fuel cell application.

(5) Nanoscience and Education
We recognized that nano-education is the key issue to disseminate nanotechnology information to the public. Therefore, MNI members have developed a tool kit named Nano-Edu to educate young generation about nanoscience and nanotechnology from young age.

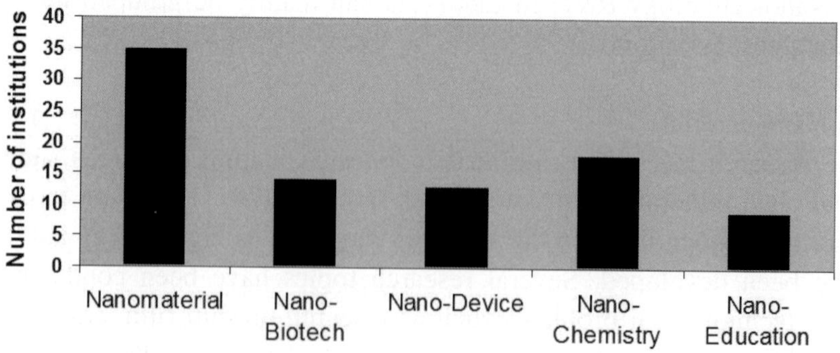

Fig. 8. Number of institutions involved in the main areas of Nanotechnology R&D.

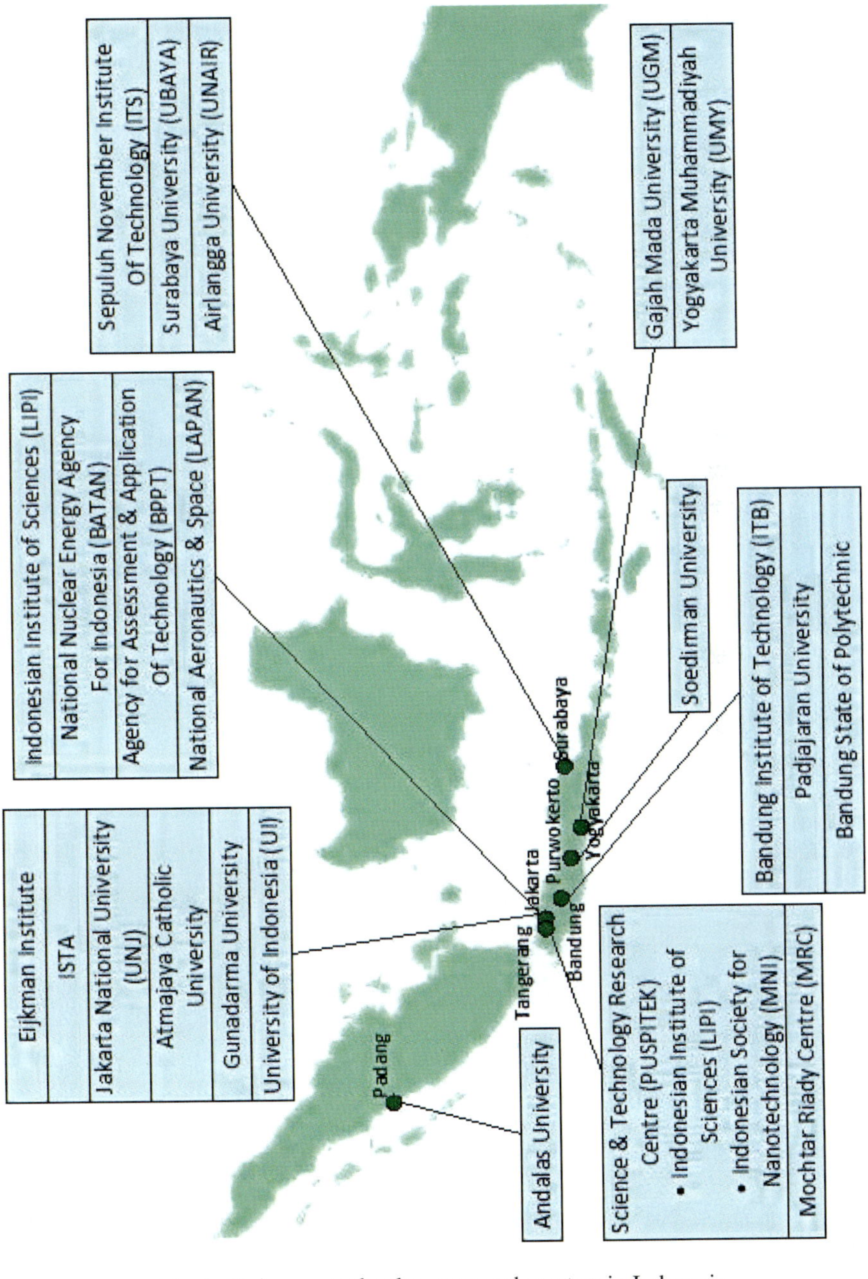

Fig. 9. Main nanotechnology research centers in Indonesia.

Table 1. Summary of current status nanotechnology R&D in Indonesia

Name of institute	Group	Research area				
		NM	NB	ND	NC	NE
Indonesian Institute of Sciences (LIPI)	Research Center for Physics	X		X	X	X
	Research Center for Chemistry	X			X	
	Research Center for Calibration, Instrumentation and Metrology	X		X		
	Research Center for Electronics and Telecommunication	X		X		
	Research Center for Metallurgy	X	X			
	Research Center for Biotechnology		X			
	Research Center for Electrical Power and Mechatronics	X		X		
National Nuclear Energy Agency for Indonesia (BATAN)	Nuclear Fuel and Recycle Technology R&D Center	X				
	Center for Isotopes Application and Radiation Technology	X	X			
	Technology Centre for Nuclear Industrial Materials	X		X		
Agency for Assessment and Application of Technology (BPPT)	Center of Pharmaceutical and Medical Technology	X	X		X	
	Center of Polymer Technology	X			X	
	Center of Materials Technology	X		X	X	
LAPAN	National Aeronautics and Space			X		

Table 1 (continued)

Name of institute	Group	Research area				
		NM	NB	ND	NC	NE
University of Indonesia (UI)	Department of Physics	X		X		X
	Department of Metallurgy and Material Engineering	X				
	Department of Electronical Engineering			X		
	Institute of Human Virology and Cancer Biology		X			
	Faculty of Public Health		X			
	Department of Chemical Engineering				X	
Bandung Institute of Technology (ITB)	Department of Physics	X		X	X	X
	Department of Physical Engineering	X		X		X
	Department of Chemistry	X			X	
Gajah Mada University (UGM)	Department of Chemistry	X			X	
	Department of Chemical Engineering	X			X	
	Department of Industrial and Mechanical Engineering	X	X			
	Faculty of Pharmacy	X	X			
	Department of Physics	X		X		X
Sepuluh November Institute of Technology (ITS)	Department of Physics	X		X	X	X
	Department of Material Engineering	X			X	
Andalas University (Unand)	Department of Chemistry	X			X	X

Table 1 (continued)

Name of institute	Group	Research area				
		NM	NB	ND	NC	NE
Surabaya University (UBAYA)			X			
Airlangga University (UNAIR)			X			
Yogyakarta Muhammadiyah University (UMY)	Department of Chemical Engineering	X			X	
Atmajaya Catholic University			X			
ISTA	A. Yani Institute of Science and Technology				X	
Soedirman University	Department of Physics		X			X
Padjadjaran University	Department of Physics	X			X	
Bandung State of Polytechnic	Chemical Engineering Department				X	X
Jakarta National University (UNJ)		X				
Gunadharma University		X				
Islamic National University (UIN)		X				

Table 1 (continued)

Name of institute	Group	Research area				
		NM	NB	ND	NC	NE
Muchtar Riady Center (MRC)		X	X			
Eijkman			X			

Note: NM: Nanomaterials, NB: Nanobiotechnology, ND: Nanodevice, NC: Nanochemistry, NE: Nanoscience and Education. Data is mainly from Indonesian Nanoletter, Vol. 1, No. 1, published by MNI.

Figure 8 illustrates the number of institutions involved in the R&D of the five main areas of nanotechnology, while Table 1 lists all involved institutions in details. It is apparent that currently nanomaterial is the area that gains most attention and interest from our researchers. Also, Fig. 9 shows the location of the main nanotechnology research institutes.

Progress of nanotechnology R&D in Indonesia can also be seen from the number of publications made in 2005–2008, as shown in Fig. 10. The research was conducted in several nanotechnology applications namely environment, food, defense, transportation, health and medicine, ICT and energy. LIPI and BATAN are currently the institutions that produce the highest number of publications in nanotechnology.

Table 2 tabulates several fascinating research activities and achievements in nanotechnology conducted by Indonesian researchers and their partners.

Table 2. List of Several Nanotechnology Researches and Their Achievement

Research Title	Institution	Achievement
Bi-Mn-O Nanobelts	Research Center for Physics, LIPI	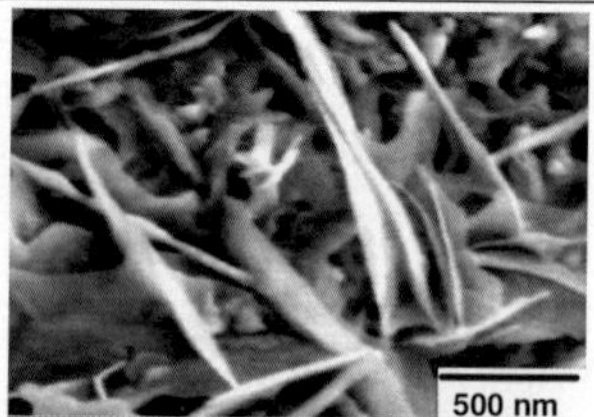
Nanocomposite Fe_x-C_{1-x}	Tech. Center for Nuclear Industrial Materials, BATAN	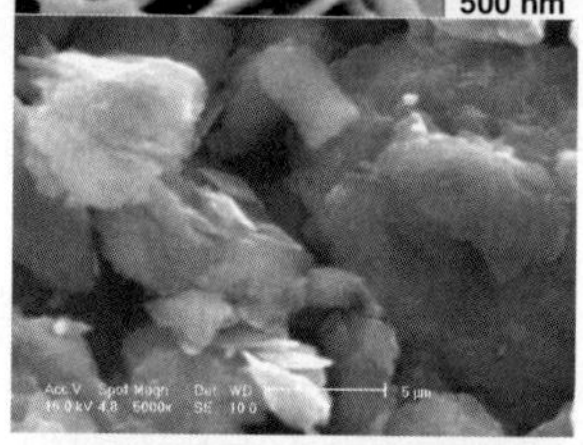
Production of Hydrogen and Nanocarbon	Department of Chemical Engineering, University of Indonesia (UI)	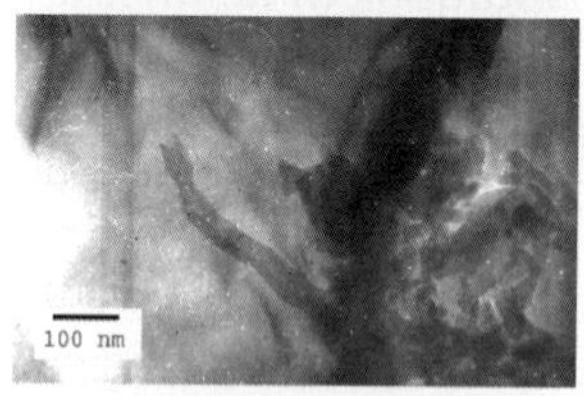
Modeling and Simulation in Nanostructured Materials Design: Polyethylene Oxide Nanocomposite	Bandung Institute of Technology (ITB)	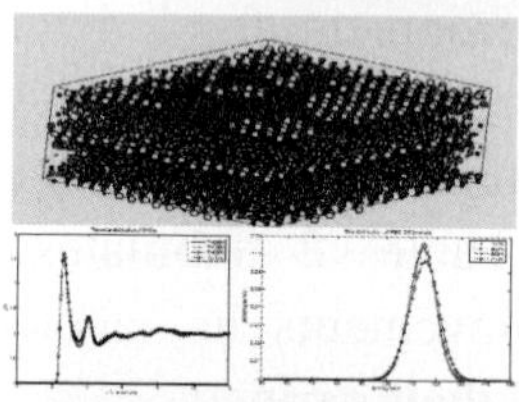
Synthesis of Ag Nanoparticles from Photography Liquid Waste	University of Sebelas Maret	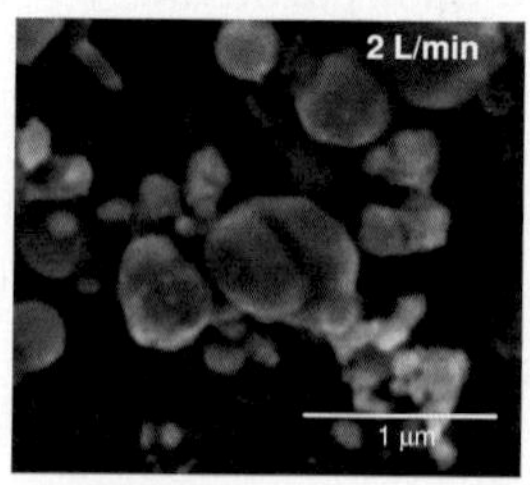

Table 2 (continued)

Research Title	Institution	Achievement
Development of Nano-encapsulation	Center of Pharmaceutical and Medical Technology, BPPT	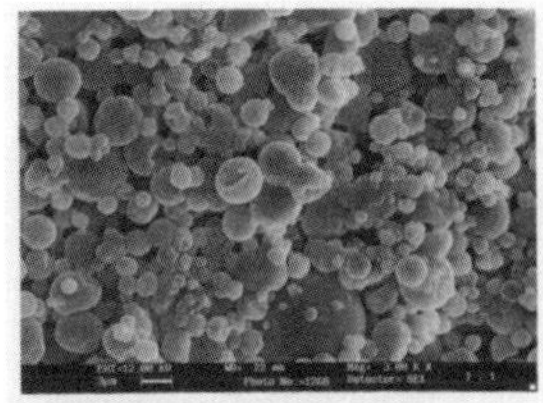
Development of High Energy Milling 3D	Research Center for Physics, LIPI	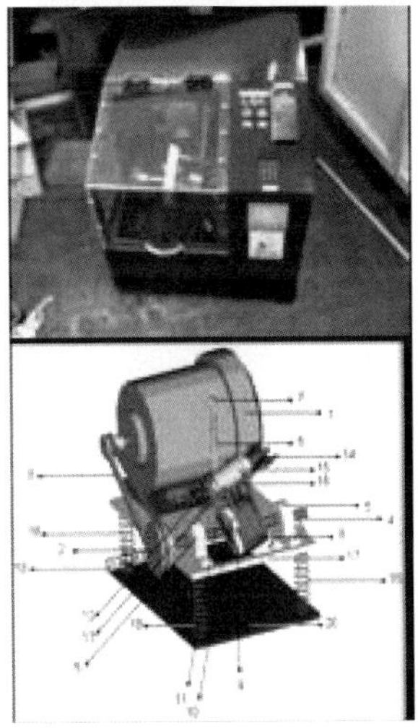

4. Nanotechnology Outreach and Recent Activities

New nanotechnology paradigm has changed the way of perceiving science and technology, so it has to be educated to all levels of society. This is to ensure an improvement in quality and quantity of human resources to face nanotechnology challenge in the future.

Figure 11 illustrates nanotechnology outreach strategy involving all stakeholders on Nanotechnology development in Indonesia. Government research institutions coordinated by state ministry of research and technology are the main players and source of information on

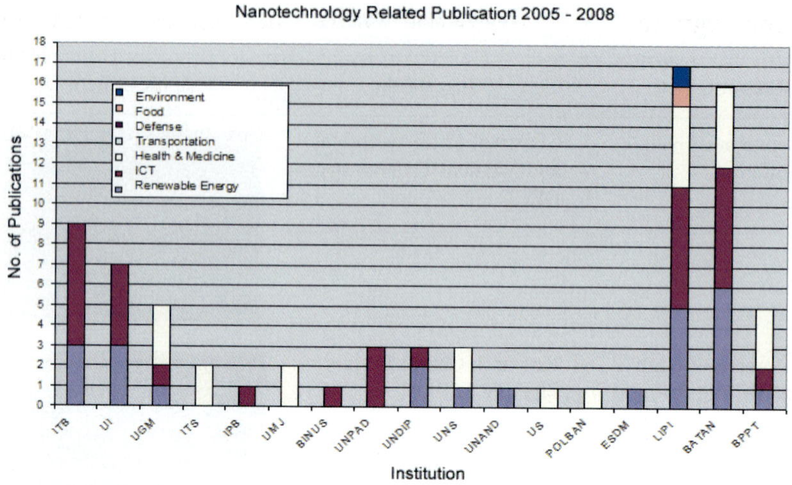

Fig. 10. Number of nanotechnology related publications in 2005–2008.

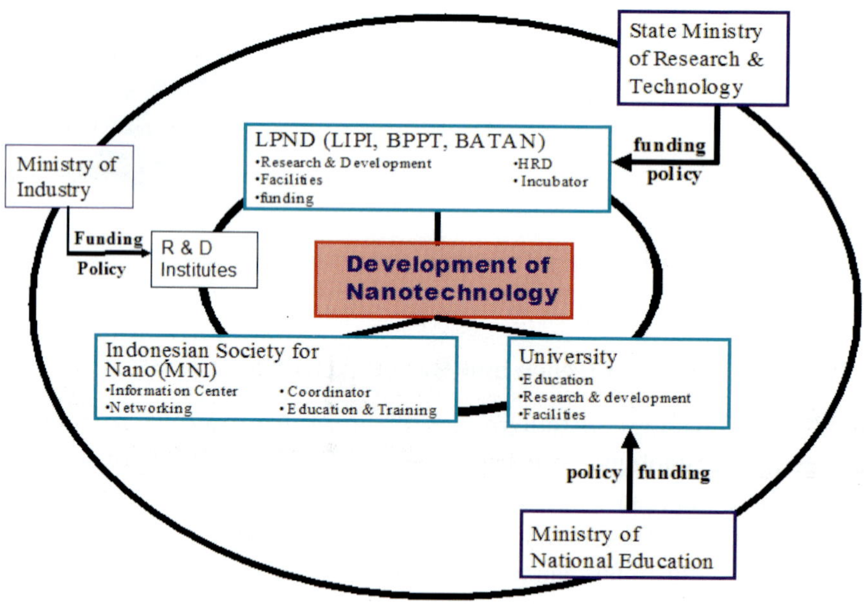

Fig. 11. Nanotechnology outreach strategy in Indonesia.

nanotechnology research. Meanwhile, universities act as information provider and educator need nanotechnology education resources. Role of Indonesian Society for Nano (MNI) is to coordinate and synergize all elements by developing programs that will accelerate goals achievement in educating public society.

Some MNI successful programs include:

- Providing information about nanotechnology development and activities in Indonesia (Newsletter, Indonesian Nanoletter, etc).
- Publishing Nano-Edu Books and Kit for nanotechnology education to students in-collaboration with Research Center for Physics, Indonesian Institute of Sciences as shown in Fig. 12.
- Supervising final project research supervision for last year college student.
- Organizing nanotechnology lecture and seminars.
- Publishing nanotechnology books and newsletters.

We are also working to gain support from government especially from State Ministry of Research and Technology and Ministry of National Education to accelerate nanotechnology development in Indonesia.

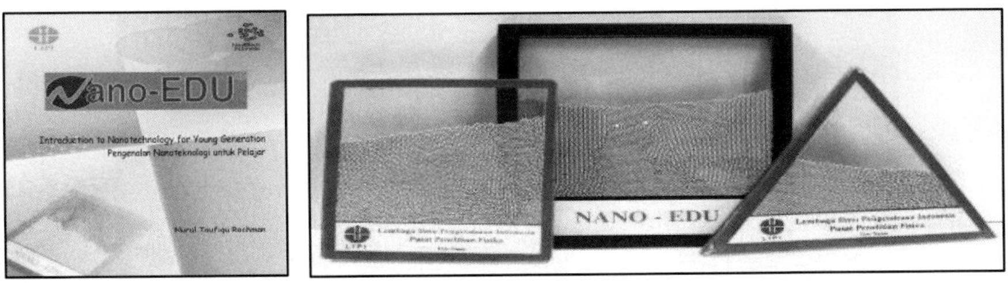

Fig. 12. Nano-Edu Kit developed by Research Center for Physics to introduce Nanotechnology since early age.

Various events and activities have been carried out to promote research collaboration among different groups and institutions in Indonesia as well as to promote international cooperation.

May 12th, 2006: Signing of Memorandum of Understanding In Nanotechnology between Indonesian Society for Nano (MNI) and Metallurgy and Material Department, Faculty of Engineering, University of Indonesia. This MoU was established to foster coordination between scientist and researcher in nanotechnology.

August 4th, 2006: Signing of Memorandum of Understanding between Sepuluh November Institute of Technology (ITS) and Indonesian Society for Nano (MNI). This signing was performed by Prof. Dr Mohammad Nuh, DEA (rector of ITS, now Indonesian Minister of Communication and Information) and Dr Nurul Taufiqu Rochman (chairman of MNI). This occasion has two objectives:

- To make closer relationship between educational institution (in this regard, ITS) and professional organization (MNI).
- To enhance the development of nanotechnology in Indonesia.

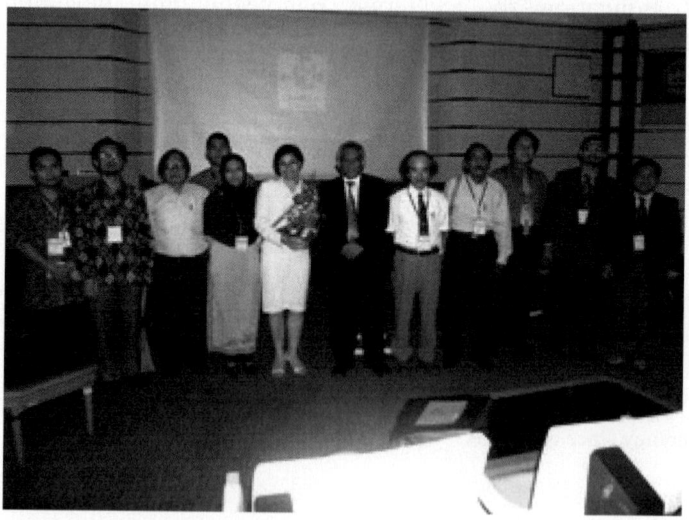

Fig. 13. 1st International Conference on Advanced Material and Practical Nanotechnology.

September 4th, 2006: 1st International Conference on Advanced Material and Practical Nanotechnology (as shown in Fig. 13). The aims of this event were to produce international track record, which shows Indonesia's awareness in nanotechnology, and to increase Indonesia's competitive advantages in nanotechnology. Some of the speakers were Prof. Kozo Obara (Kagoshima University - Japan), Dr Lerwen Liu (Zyvex Corporation - USA), Dr Ratno Nuryadi (Shizuoka University - Japan), Dr Brian Yuliarto (AIST - Japan), and Dr Haznan Abimanyu (KIST - South Korea).

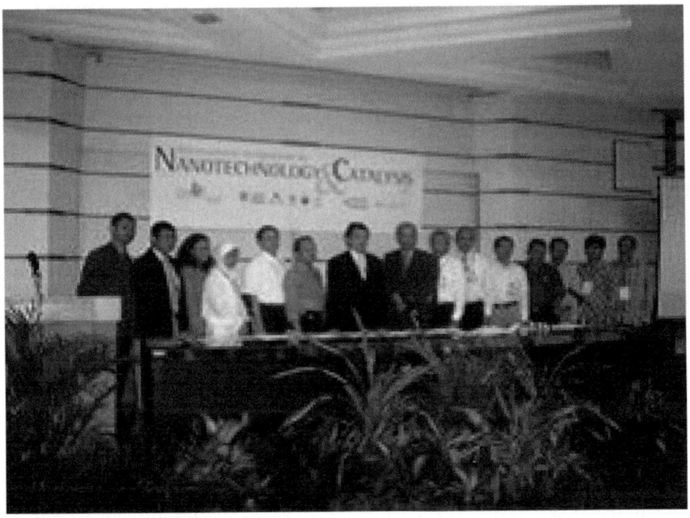

Fig. 14. International Symposium of Nanotechnology and Catalysis.

June 14th, 2007: Eleven Research Units from Indonesian Institute of Sciences, National Nuclear Agency, and Agency of Technology Assessment and Application that conduct Nanotechnology research have signed a Memorandum of Understanding (MoU) for collaboration in nanotechnology at International Symposium of Nanotechnology and Catalysis (as shown in Fig. 14). Recommendations given from this symposium are:

- To form nanotechnology research cluster between researchers across Institution inside PUSPIPTEK region in short term
- To expand current networking in nanotechnology and develop more collaboration not only inside PUSPIPTEK region but also outside of it.
- To propose and encourage government in support of developing regulation and allocating sufficient fund on enhancing nanotechnology research quality and quantity.

Also came as invited speakers were: Prof. Dr Kiyoshi Okada (Tokyo Institute of Technology), Prof. Dr Hermawan K. Dipojono (Bandung Institute of Technology), Prof. Dr Abdul Rahman Mohamed (Faculty of Natural Sciences - Universiti Sains Malaysia), and Dr C. D. Madhusoodana (Ceramic Technological Institute BHEL, Bangalore - India).

November 29th-30th, 2007: Indonesian Institute of Sciences (LIPI) as one of iron stock in nanotechnology research joined "The Advanced Material Research Group Seminar" in Bangi Selangor Malaysia, together with University of Indonesia and Universiti Kebangsaan Malaysia as the organizer. The program was conducted to share research information and to create research collaboration networking among them.

March 11th-12th, 2008: Our delegates represented by Dr Ir. Atih Surjati Herman, M.Sc (Senior Scientist, R&D Ministry of Industry, BPPI), Dr Nurul Taufiqu Rochman (Chairman of MNI) and Ir. Harjanto, M.Eng (Belgium Attache of Industry) participated in the 3rd Nanotech Dialog in Brussel, Belgium. The objective of this participation was to obtain wider and clearer ideas about recent nanotechnology development and its strategic issues especially in countries that are very concerned with nanotechnology.

March and August 2008: MNI organized Nanotechnology Workshop I and II respectively. Nanotechnology Workshop I was intended to introduce and educate public about nanotechnology, while Workshop II was to train public about the method in producing nanoparticles.

October 27th-8th, 2008: Indonesian Institute of Sciences (LIPI) organized 100 Nanotechnology Doctor Conference and Regional

Seminar Advanced Material Research Groups in PUSPITEK, Serpong. Some objectives of this event include:

- Creating awareness of nanotechnology importance for Indonesia development
- Presenting the recent R&D findings and breakthroughs in nanotechnology
- Facilitating the networking among scientists, industry players, and decision makers in nanotechnology in order to exchange ideas and information for beneficial and synergic cooperation

In this event, a book of 100 Nanotechnology Doctors Profiles was launched to acknowledge and appreciate their contribution to nanotechnology development in Indonesia.

5. Industry Status

A survey was conducted in August–September 2008 by the Ministry of Industry and MNI. The survey was intended to obtain some insights on current status of nanotechnology application in national industries. 30 industries were selected including textile, ceramics, chemistry, consumer goods, ICT and automotive industry. Findings of the survey include:

- 35% of the selected national industries has applied nanotechnology
- 89% of nanotechnology resources are imported
- Nanotechnology has been applied for raw materials as much as 41%, finishing process 40% and intermediate process 19%
- In future, most industries will implement nanotechnology in their production activities: 65% to use raw nanomaterials, 14% to implement in production process and 21% in final products.

In addition, based on the survey result, all industries acknowledged the importance of nanotechnology for overall quality improvement. However, they also pointed out several challenges for nanotechnology implementation in the national industry, as illustrated in Fig. 15. Therefore, the government is expected to take part more actively in

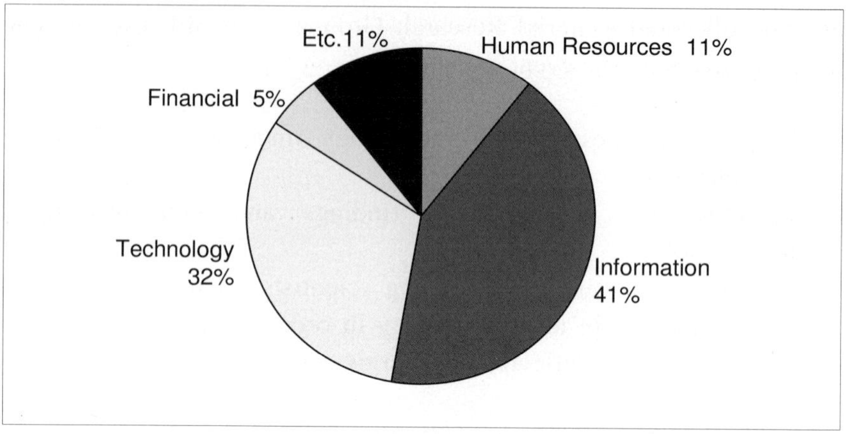

Fig. 15. Identified challenges in nanotechnology implementation in national industry (source: The Ministry of Industry, RI).

supporting the industry readiness in its capability improvement and enhancement.

At present, textile and ceramic industry in Indonesia are the most ready industries in implementing nanotechnology, considering the market criteria, technology, infrastructure, available fund, raw materials, human resources, and potential risk. Industries must be supported with capability in nanomaterials as it has been the main focus in nanotechnology implementation in industry activities. Other issues such as environmental issue, health and work risk, policy and strategic issue and nanotechnology outreach should also get attention to ensure that nanotechnology is integrally and comprehensively implemented in national industry level.

6. Conclusion and Recommendation

In summary, Indonesia has comparative advantages on its abundant natural resources and large population. Combined with nanotechnology, these can be the added values for Indonesia competitiveness. Interest in nanotechnology is increasing for the last few years, particularly from industry and government. Although nanotechnology R&D and policy are still in fragments, they will be integrated in the future.

There exist good synergy between academic, industry and government that will support the acceleration of nanotechnology development in Indonesia. Networking is still crucial especially with foreign institutions to further accelerate the progress of nanotechnology development.

Last but not least, more efforts are required to outreach and educate Indonesians, especially the industrial players about the importance of nanotechnology.

Author Biographies

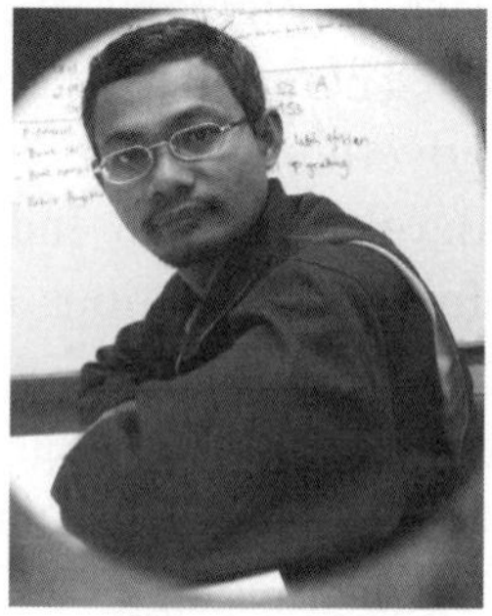

Dr Nurul Taufiqu Rochman was born in Malang, Indonesia, on 5 August 1970. He pursued his study in Advanced Material at Kagoshima University, Japan and finished his Doctoral degree in 2000 with *cumlaude* predicate. Following four years was spent on gaining experiences at Kagoshima Prefectural Institute of Industrial Technology as an invited researcher. He went back to Indonesia in 2004 after 14 years living in Japan. He has published more than 40 international papers, filed three patents, and numerous national publications. He was awarded the 'Best Young Researcher' by Indonesian Institute of Sciences 2004, the honored 'Adhidarma Profesi Award' from Indonesian Engineers Union and 'The Best Idea and Innovation Award' from Swa Magazine in 2005. He is also known as one of Nanotechnology pioneers in Indonesia, and established Indonesian Society for Nano (MNI) in 2005 where he has been the chairman until now. He's currently working at Research Center for Physics, Indonesian Institute of Sciences.

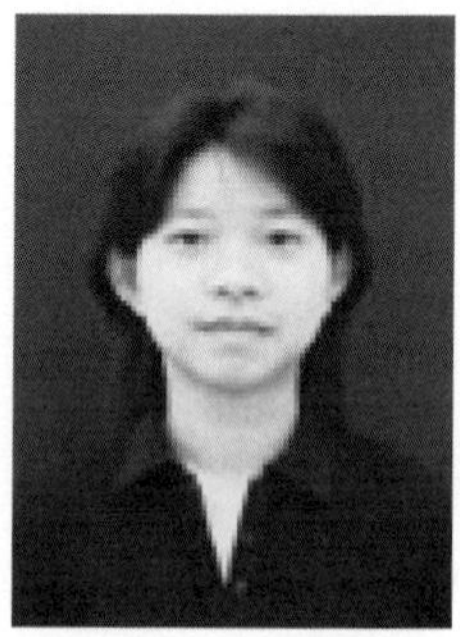

Yesie L. Brama was born in Semarang, Indonesia and was awarded Sembawang Scholarship to pursue her undergraduate study in Singapore in 2000. She graduated from School of Materials Science and Engineering, Nanyang Technological University in 2003 as First Class Honours. In 2004 she was awarded Prof Haresh Silver Award of NTU Business Plan Competition. She was an R&D Engineer in Nanofilm Technologies Intl. Pte Ltd, one of successful spin-off companies from NTU specialized in FCVA coating technology. She is currently a technology analyst of NanoGlobe

Pte Ltd and finishing her M.S degree in Industrial and Systems Engineering in National University of Singapore in June 2009. Her expertise is in nanomaterial coatings and characterization, as well as product quality and reliability.

CHAPTER 6

PART I JAPAN NANOTECHNOLOGY OVERVIEW: POLICY, INFRASTRUCTURE AND R&D

Mizuki Sekiya

Working Group on Strategic Area of Nanotechnology Advanced Industrial Science and Technology (AIST)
1-3-1, Kasumigaseki, Chiyoda-ku, Tokyo 100-8921, Japan
mizuki-sekiya@aist.go.jp

Masafumi Ata

Working Group on Strategic Area of Nanotechnology, Advanced Industrial Science and Technology (AIST), Japan
masafumi-ata@aist.go.jp

Kazunobu Tanaka

AIST and Japan Science and Technology Agency (JST), Japan
tanaka-kazunobu@aist.go.jp

Japan put great emphasis on nanotechnology research and investment is steadily growing since national initiative has been implemented in 2001. Japan is highly competitive in materials science and exploits its features successfully so far. Concerned issues, such as development of human resources, collaboration among various ministries and institutes, human and environmental impact of nanomaterials have carefully been addressed in recent years and these efforts help to further development of nanotechnology.

1. Introduction

Since the beginning of the 1990s, nanotechnologies have attracted high expectations as new technologies — not simply extensions of existing ones — that will have enormous impacts on society and the economy. Meanwhile, research and development activities have steadily gained momentum. A major revolution was anticipated in Japanese industry, which was already strong in precision and micro-scale processing technologies, but Japan's economic bubble happened to burst just then, resulting in difficult conditions for more than ten years. Private-sector research facilities suffered from a squeeze on investment in nanotechnologies, which require medium and long-term R&D programs. In the midst of these difficult economic conditions, public research institutes supported and drove nanotechnology R&D, quietly advancing certain core technologies — one example being the Atom Technology Project which ran for ten years from 1992. In 1995, Japan revealed its basic policy for the promotion of science and technology and enacted the Science and Technology Basic Law. On this basis, Japan implemented its First Science and Technology Basic Plan for the five years starting in 1996, working toward being a nation founded upon science and technology. A series of industrial technology strategies were also put into place to promote technology transfers to industry from universities and other public research institutes.

Nanotechnology R&D, which is still considered to be at an early stage, has changed dramatically after the start of the twenty-first century. The Science Council of Japan has for years represented scientists in this country, but a new entity joined the scene in 2001 when the government launched the Council for Science and Technology Policy (CSTP), an advisory body under the Cabinet Office, with the purpose of promoting science and technology policy. Japan's Second Science and Technology Basic Plan, compiled by the CSTP, entered into effect in April of that year, with the aim of reforming science and technology systems in Japan, based on the core points of promoting basic research, giving strategic priority to investments based on established priority areas, doubling competitive funding for research, and strengthening collaboration between industry, academia, and government. As part of this basic strategy, nanotechnologies/materials were selected as one of four priority areas, which were to receive a strategic allocation of resources. The ten-year economic slowdown eventually became known as the "lost decade", but finally began to improve exactly when private-sector investments in nanotechnology R&D began to perk up.

Meanwhile, comprehensive initiatives, which including the societal implication of nanotechnology, began in 2004. Even though these efforts began later than that in Europe and USA, initiatives in Japan to address the societal implications of nanotechnologies have made significant and steady progress. Exaggerated expectations for nanotechnologies have ebbed away, and Japanese nanotechnology has now entered a stage of sound efforts toward practical applications of fundamentally useful core technologies for nanotechnologies.

2. Japanese Resource Allocation Trends in Nanotechnology R&D

In Japanese science and technology policy, nanotechnology is classified as "materials science and technology area". The category "nanotechnology and materials" was first identified as one of four fields of strategic priorities set by the CSTP in its Second Science and Technology Basic Plan, which went into effect in April 2001. In the Third Science and Technology Basic Plan, which entered into effect in April 2006, this continues to be a priority area. The national R&D budget has increased each year since fiscal 2001, reflecting the government's stance on nanotechnologies.

National budget trends are shown in Table 1 and Fig. 1 with that of the United States and other countries. Since "nanotech and materials" were identified as a priority field in 2001, Japan' s budget in these areas started to grow significantly. It is precisely at that point that the commercialization of nanotechnologies began to appear frequently as an area of great interest for Japanese corporations.

Examining budget trends as a ratio of gross domestic product (GDP) as shown in Table 2 and Fig. 2 we see that before 2000, Japan budgeted for nanotechnology R&D at about the same level as that in the United States and European Union (EU), but five years later, the ratio

Table 1. International comparison of nanotechnology R&D investments (estimated)

Year/Area	US	EU	Japan	Other	Total
1997	116	126	120	70	432
1998	190	151	135	83	559
1999	255	179	157	96	687
2000	270	200	245	110	825
2001	465	225	465	380	1535
2002	697	420	720	550	2387
2003	862	650	800	800	3112
2004	989	950	900	900	3739
2005	1200	1050	950	1000	4200
2006	1351				
2007	1392				
2008	1445				

Source: Prepared based on the National Nanotechnology Initiative FY2008 Budget and Highlight. (unit: US$).

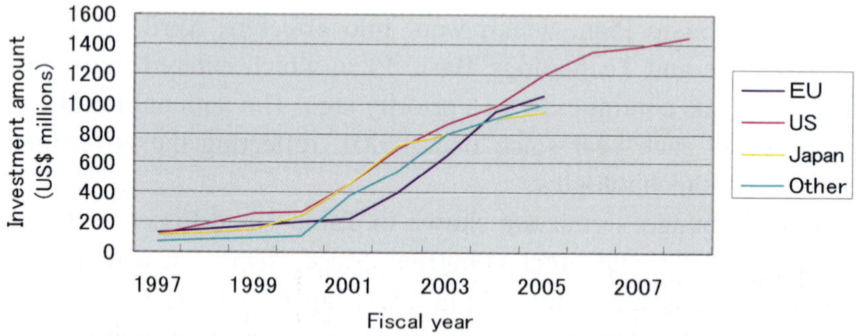

Fig. 1. International comparison of investment amount of nanotechnology R&D.

rose rapidly to about double that of the United States, and about triple that of the EU; these changes reflect the strengthening of the Japanese government policy of strategic and priority initiatives for R&D into the area of "nanotechnology and materials."

Table 2. International comparison of nanotechnology R&D investment using GDP (1997~2008, estimated)

Area	1997	1998	1999	2000	2001	2002	2003	2004	2005
US	1.4	2.2	2.8	2.8	4.6	6.7	7.9	8.4	9.6
EU	1.4	1.6	1.8	1.9	2.2	3.5	5.6	7.8	8.2
Japan	2.6	3	3.5	5.4	10.4	16.2	17.8	20	20.7

Source: Prepared based on the National Nanotechnology Initiative FY2008 Budget and Highlight etc. (GDP ratio 3-10%).

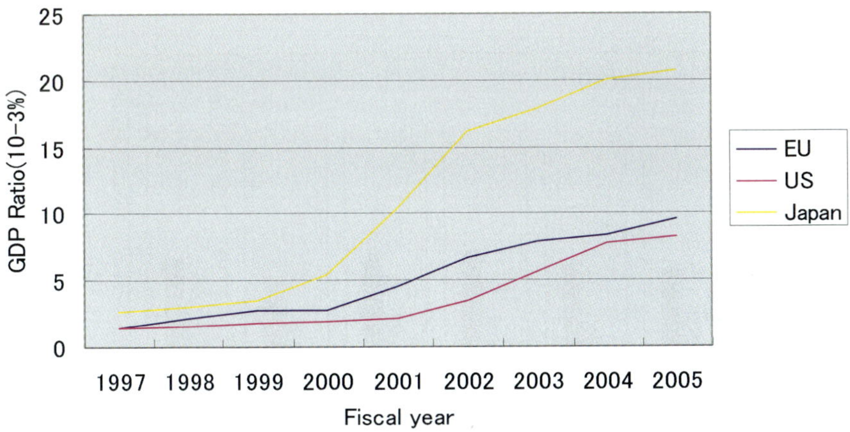

Fig. 2. International comparison of nanotechnology R&D investment using GDP.

Meanwhile, as indicators of the results of government R&D budgets, Table 3 and Fig. 3 show the numbers of "nanotechnology and materials" patents registered by Japan, the United States, and the EU. The total number of patents registered dropped slightly in fiscal 2005 for each of them, but the overall trend is rising. Japan is second after the United States, leading one to surmise that Japan joins the United States

Table 3. Trend in registration of nanotechnology-related patent

Year/Area	Japan	US	Europe	Total
2001	7459	12034	4049	23542
2002	6774	13243	5091	25108
2003	7239	14038	6375	27652
2004	7562	13837	6835	28234
2005	7135	11632	6836	25603
2006	8801	13471	7478	29750

Source: Prepared based on disclosed statistics of Japan Patent Office.

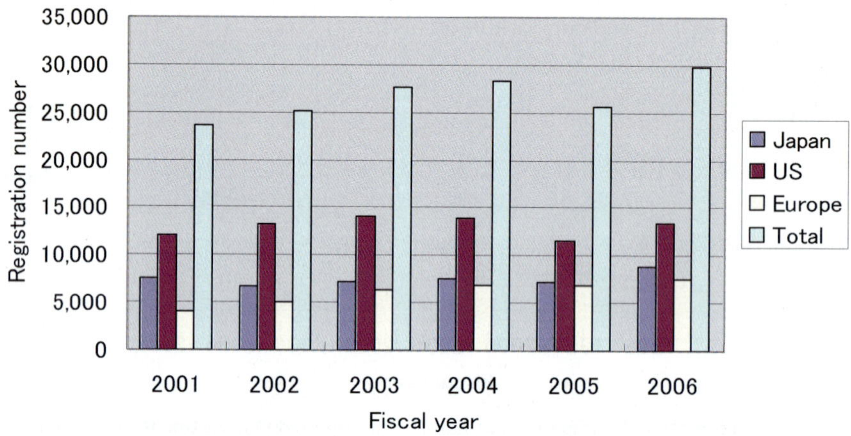

Fig. 3. Trend in registration of nanotechnology-related patent.

as a driving force in R&D for nanotechnologies and materials. Japan realizes that the number of EU patent registrations is increasing steadily and narrowing the gap with Japan.

3. Activities in Societal Implications of Nanotechnology

As shown in Table 4, many projects are being conducted on the societal implications of nanotechnologies in order to promote their public

Table 4. Research projects on societal implications of nanotechnologies

Project name	Implementing body	Period
Survey on Safety of Nanomaterials and Nanoparticles	Furukawa Research Inc.	2004
Open forum "Nanotechnology and Society"	AIST	2004
Research on Nanotechnology Risk Assessment, Risk Management Methodologies, and Standards	AIST	2005-2007
Research on Hazard Assessment of Nanomaterials, Particularly Neurological Impacts	AIST	2005-2006
Research on the Societal Implications of Nanotechnologies	AIST	2005-2007
Development of Standards for Measurement Methods of Airborne Nanoparticles, for Nanorisk Management	AIST	2005-2007
Development of Standards for Safety Assessment Methodologies of Nanoparticles (Standard Certification Project)	AIST	2005-2007
Costs for Japanese Organizations Deliberating on Development of International Standards (Tasks for International Involvement)	AIST	2005-2007
Research Project on Facilitation of Public Acceptance of Nanotechnology	AIST, NIMS, NIES, NIHS	2005
Multidisciplinary Expert Panel on Nanotechnology Impacts	NIMS, NIES	2006
Publishing draft Guidelines for Proper On-Site Handling Methods for Nanotechnologies in Research and Manufacturing	JEF Techno-Research Inc.	2007
Development of Risk Assessment Methodologies for Nanoparticles	AIST, University of Occupational and Environmental Health	2006-2010
Research for Development of Database Indicators to Serve as a Common Platform for Nanotechnology R&D	University of Tokyo, AIST, NIMS	2007-2009

acceptance in Japan. Even though Japan was relatively early to begin studying the biological impacts of nanomaterials, attempts to examine societal implications comprehensively were not particularly strong in 2001 when a new strategic focus was placed on nanotechnologies R&D. Some factors that may have delayed efforts to address societal implications are that, even though risk assessment and risk management are expected topics in the context of public acceptance, the word "risk" takes on a strongly negative connotation when translated into Japanese, people tend to fixate on the issue of "safety" rather than "risk," and the net effect is a resistance to conduct research for risk assessment and risk management. However, Japan was the 1st country that started the initiative of societal implication of nanotechnology among all the Asian economies.

In August 2004, the Technology Information Department of the National Institute of Advanced Industrial Science and Technology (AIST) initiated a comprehensive and coordinated forum titled "Nanotechnology and Society" to address societal implications of nanotechnology in Japan. This open forum was inspired by the First International Dialogue on Responsible Research and Development of Nanotechnology held in the United States in June 2004. The forum became a milestone in Japanese efforts to create a network for the exchange information relating to the societal implications of nanotechnologies and to systematically address related issues through collaborations among all stakeholders.

AIST held the first symposium on the societal implications of nanotechnology on Feb 1st, 2005. It was organized jointly with the National Institute for Materials Science (NIMS), the National Institute for Environmental Studies (NIES), and the National Institute of Health Sciences (NIHS), with support from the Ministry of Economy, Trade Industry (METI), the Ministry of Education, Culture, Sports, Science and Technology (MEXT), the Ministry of the Environment (MOE), and the Nikkei BP Nanotechnology, and the Nanotechnology Business Creation Initiative (NBCI). This cross-ministry cooperative relationship became an important factor in efforts to address the societal implications of nanotechnologies in Japan. The cooperative relationship has been a main driving force for the "Research Project on Facilitation of Public Acceptance of Nanotechnology" funded under the "Special Coordination

Funds for Promoting Science and Technology" started 1st quarter of 2005. In March 2006, our team at the Technology Information Department of AIST summarized the results of five workshops, international symposium, and research findings of the working groups into policy recommendations aimed at governments, public research institutes, and corporations.

The policy recommendations were incorporated into the Third Science and Technology Basic Plan started Japanese fiscal year of 2006, and the responsible R&D of nanotechnologies was identified as a priority research topic in the area of nanotechnology and materials. Thus, in December 2006, based on the Third Science and Technology Basic Plan, the CSTP announced that the "Public Acceptance of Nanotechnologies" would be a topic to address through inter-ministerial collaboration. In January 2007, Dr Junko Nakanishi, director of the AIST Research Center for Chemical Risk Management (AIST-CRM) was appointed as the coordinator of the Discussion Group, and a new set of science and technology collaborative efforts were launched relating to public acceptance of nanotechnologies.

4. Development of Standards for Nanotechnologies

The Japan Standards Association (JSA) is responsible for national standards in Japan and operates in cooperation with international standardization bodies such as the International Organization for Standardization (ISO) and the International Electrotechnical Commission (IEC). METI created the Nano-Standardization Panel (JSA-NSP) within JSA on November 11, 2004.

The ISO created new technical committee "TC-229" in May 2005. When ISO TC-229 was created, Japan made a proposal collaborating with the United States, recommending the formation of three subcommittees (SC): terminology, measurement/metrology, and health/safety.

On September 13, 2005, the first meeting of the Council on Nanotechnology Standards in Japan (JISC) was hosted by the Japanese Industrial Standardization Committee, marking the launch of domestic discussions relating to the setting of nanotechnology standards. AIST

Research Coordinator Dr Akira Ono was selected as the committee chair and AIST-CRM director, Dr Junko Nakanishi was selected as the vice-chair. An International Standardization Strategy Development Working Group created within JSA-NSP and chaired by Dr Shingo Ichimura continues its activities within this framework. In addition, a Terminology and Nomenclature Subcommittee, a Metrology and Monitoring Subcommittee, and an Environment and Safety Subcommittee were created.

The NBCI has played a leading role in the development of industrial standards for nanotechnologies as a forum to gather inputs from 300 private-sector nanotechnology business members.

AIST pointed out the importance of improving access for Asian countries to international discussions about standards, which are strongly affected by international politics. It was concluded that a working group on nanotechnology standards would be established within the Asia Nano Forum (ANF, asia-anf.org), and that this working group would function through external liaison with the ISO/TC229. Through such an arrangement, by participating in this working group, it will be possible to have access to international activities to develop standards even without being an ISO member country (See Fig. 4). Taiwan was delegated by ANF to be the liaison member representing ANF at ISO/TC229.

5. NEDO Project on R&D of Nanoparticle Characterization Methods

As a part of strategic efforts relating to the public acceptance of nanotechnologies, "R&D of Nanoparticle Characterization Methods" was launched as a project of Japan's New Energy and Industrial Technology Development Organization (NEDO) in fiscal 2006, and is being implemented as a five-year project through AIST-CRM and Japan's University of Occupational and Environmental Health. The R&D roadmap is shown in Fig. 5.

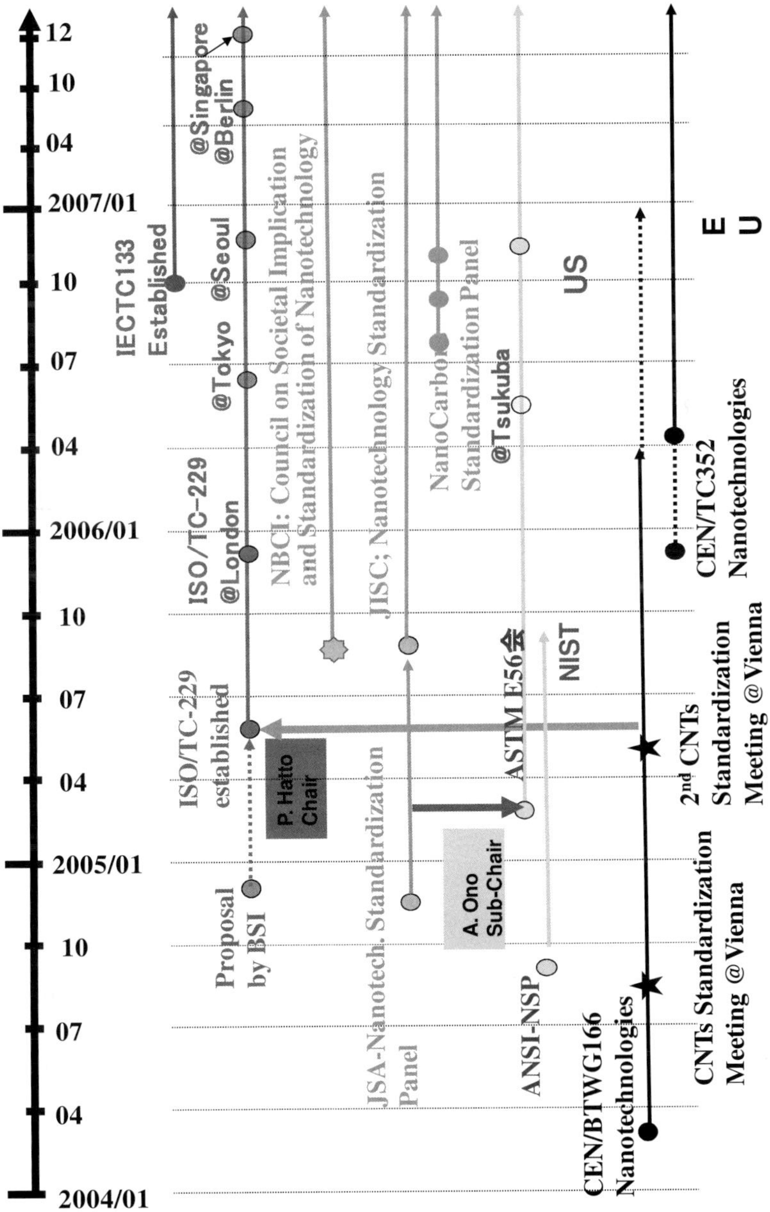

Fig. 4. Standardization of Nanotechnology.

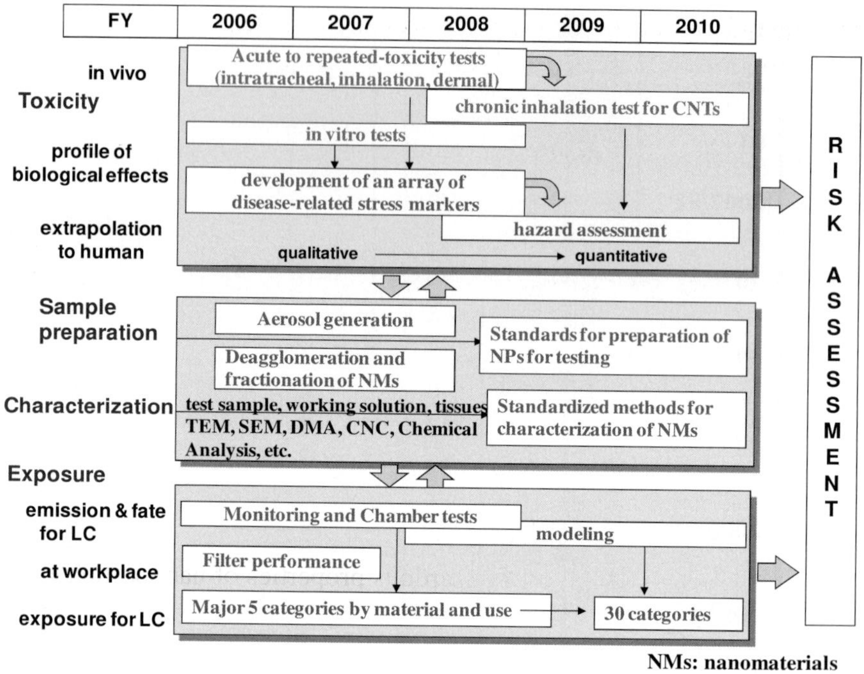

Fig. 5. The project R&D roadmap.

The aim of the project is to achieve progress in key areas: fundamental characterization methodologies such as for hazard assessments of industrial nanoparticles, exposure analysis, and risk assessments; exposure assessment methodologies including measurement of environmental concentrations, analysis of sources of releases into the environment, and fate and behavior in the environment; and basic hazard assessment methodologies that are matched with international standards, and to compile recommendations for proper management of industrial nanoparticles based on risk assessments.

The project includes plans to implement risk assessments of industrial nanoparticles such as carbon nanotubes, fullerenes, and titanium oxide, and to summarize the findings in a risk assessment report. In addition it is hoped that the project will establish approaches for risk assessment techniques and proper management that can be applied to other industrial nanoparticles.

6. Development of Guidelines for Voluntary Management of Nanoparticles

Until the above nanoparticle management policies are available, Guidelines for Voluntary Management have been developed as a provisional measure. These guidelines were prepared as a part of the "Research Project on Proper On-Site Handling Methods for Nanotechnologies in Research and Manufacturing," implemented in fiscal year 2006 by Dr Kenichi Otsuka of the JFE R&D Corporation. To begin with, the project collected and summarized data on the following topics:

(1) Research on assessment of hazardous properties of nanoparticles
(2) Status of research on assessment of exposure routes of nanoparticles
(3) Status of risk countermeasures for nanoparticles
(4) Trends in handling methods for nanotechnologies

In addition, "Guidelines for Proper On-Site Handling Methods for Nanotechnology in Research and Manufacturing" were prepared and approved by the project committee.

These guidelines are designed especially for industrial nanoparticles (including carbon nanotubes), and the following items were summarized as necessary measures to be implemented on-site, until the NEDO development project for risk assessment methodologies now being implemented in Japan has been completed and handling methods are compiled:

(1) Measures to prevent exposure during nanoparticle manufacturing and processing
(2) Measures to prevent exposure during handling in laboratories and workplaces
(3) Cleaning and waste treatment

(4) Preparation of handling manuals, as well as education and compliance
(5) Health precautions

These measures are further classified by the type of nanoparticle being handled, with reference to the control banding approach of the U.S. National Institute for Occupational Safety and Health.

7. Nanomaterial and Nanoparticle Management Policies in Japan

A variety of regulations and management methods are already in place for the control of chemical substances in Japan, reflecting the purpose of the controls, as shown in Fig. 6, but these regulations target substances for which a certain amount of hazard and exposure information is already available. The fact is that risk assessments have not been done yet for many substances defined as "existing chemical substances" under the Law Concerning the Evaluation of Chemical Substances and Regulation of their Manufacturer, etc.

The Subcommittee on Fundamental Issues of Chemical Substance Policy was created under the Chemicals and Bio-industry Committee of the Industrial Structure Council and the subcommittee has discussed appropriate responses to new topics such as nanoparticles, in the context of proper chemical management. (The Industrial Structure Council is an advisory body under METI, which has jurisdiction for the law concerning the evaluation of chemical substances and regulation of their manufacturing, etc. and the law concerning reporting, etc. of releases to the environment of specific chemical substances and promoting improvements in their anagement).

The subcommittee's interim report states as follows: "From the perspective of safety, Japan should try to take the necessary actions to deal with new issues such as nanoparticles that have scientific unknowns, in order to promote sound and responsible development of new technologies, by demonstrating leadership through strong support of further advances in scientific knowledge and assessments, and by making an effort to provide information to various stakeholders (businesses, the

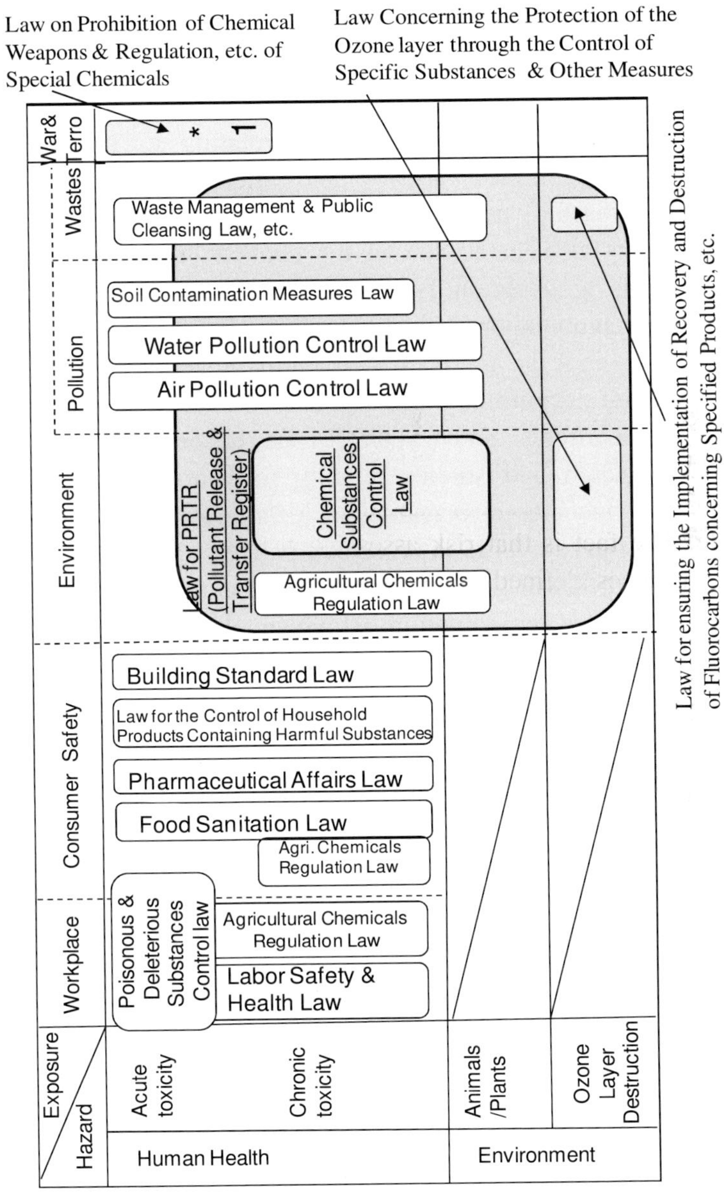

Fig. 6. Regulations relating to chemical management in Japan (Chemical Management Policy Division, Manufacturing Industries Bureau, METI).

public, NPOs and NGOs, etc.), while keeping a close link between CSTP discussions and international discussions with the Organization for Economic Cooperation and Development (OECD) and other bodies."

The first step is to promote R&D relating to risk assessment methodologies mentioned above, and then take action through regulatory approaches, as necessary, after obtaining the findings. The report summary also states that in the context of chemical substance management as a whole, it is important to seek an optimal mix of approaches in combination with regulations when promoting risk-based management, even if the approach is based mainly on voluntary action.

Japan has not yet put into place regulations relating specifically to the safety of nanoparticles. Some activities are happening, however, in a variety of research and studies relating to nanoparticle risk by the ministries and agencies responsible for a variety of legislation dealing with chemical substances. In the future, it is likely that the ministries and agencies concerned will give some consideration to which of these items should be controlled by regulation, which should be under a framework of voluntary controls, and what would be the optimal mix of the two approaches.

8. Japan Nanotechnology Network and Infrastructures

The Ministry of Education, Culture, Sports, Science and Technology (MEXT) launched in 2002 a five-year Nanotechnology Support Project that funds the Facility Use Support (universities and research institutions) and Nanotechnology Researchers Network Center of Japan (Nanonet) to provide coordinated infrastructure facilities for Japan nanoscience and technology R&D, information services for research community, and promote Japanese research activities internationally and facilitate research collaboration within Japan and overseas. In the case of Facility Use Support, it focuses on Shared Use of Equipments for Characterization and Fabrication including:

- High-voltage Electron Microscope
- Nano Foundry

- Synchrotron Radiation
- Molecular Synthesis and Analysis

involving 16 research institutions.

Nanonet provides information and networking services in both Japanese and English language for researchers in Japan and overseas. The services include:

- Web portal, Newsletter and Publication
- Symposium and Workshop
- Data-base
- Human Resource Development
- International Collaboration Facilitation

This project entered phase 2 in 2007 for another five years and updated its name to Nanotechnology Network Project. This project is managed by the National Institute of Materials Science (NIMS) and has expanded its network coverage to 13 center of excellence consisting of 26 research institutions across Japan. Figure 7 shows geographical distribution of the center and research institutions supported by the project. And Table 5 shows network centers and its facilities available. The philosophy behind phase 2 is shown in Fig. 8. Figure 9 shows the Nanotechnology Network Project organization including its facilities outline.

Compared with the Nanotechnology Support Project, the Nanotechnology Network Project continues to support Nanotech Users Facilities (see Fig. 7 and Table 5), disseminate information on nanotechnology development in Japan. There are new features in the Network project which include (a) service fees for facility usage, (b) the additional High Magnetic Field Station facility for Ultimate Analysis which includes the 930MHz Nuclear Magnetic Resonance (NMR), the strongest magnetic field for high resolution solid state NMR in the world, (c) enhanced support for young researchers (promoting visibility and mobility of young researchers especially through international exchange site visits).

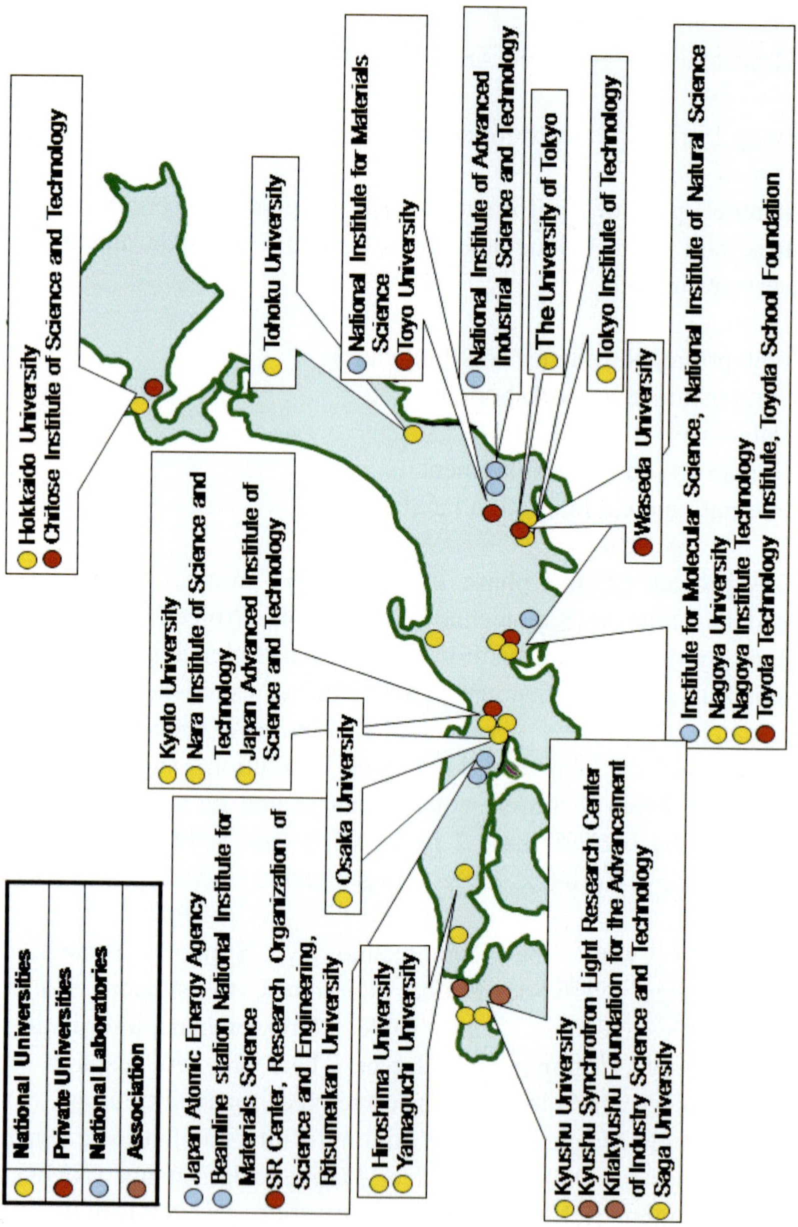

Fig. 7. Japan Nanotechnology Network Project started 2007 (Courtesy of NanotechJapan). Details of network centers are described in APPENDIX at the end of the Japan chapter-part I.

Table 5. 13 Network Centers and 26 Research Institutions in the Nanotechnology Network Project (Source from NanotechJapan)

Networks	Institutes	Nano Charatn	Nano Fab.	Molecular Synthesis and Analysis	Ultimate Analysis
1	Hokkaido University	•	•		
	Chitose Institute of Science and Technology	•	•	•	
2	Tohoku University	•	•		•
3	National Institute for Materials Science	•	•		•
	Toyo University		•		
4	National Institute of Advanced Industrial Science and Technology	•	•		
5	University of Tokyo	•	•		
6	Tokyo Institute of Technology	•	•		
7	Waseda University	•	•		
8	Institute for Molecular Science, National Institute of Natural Science	•		•	
	Nagoya University		•		
	Nagoya Institute Technology	•			
	Toyota Technology Institute, Toyota School Foundation	•	•		

Table 5 (continued)

Networks	Institutes	Nano Charatn	Nano Fab.	Molecular Synthesis and Analysis	Ultimate Analysis
9	Kyoto University	•	•	•	
	Japan Advanced Institute of Science and Technology	•			
	Nara Institute of Science and Technology	•		•	
10	Osaka University	•	•	•	
11	Japan Atomic Energy Agency	•			
	National Institute for Materials Science	•			
	SR Center, Research Organization of Science and Engineering, Ritsumeikan University	•			
12	Hiroshima University		•		
	Yamaguchi University		•		
13	Kyushu University	•		•	
	Kyushu Synchrotron Light Research Center	•			
	Kitakyushu Foundation for the Advancement of Industry Science and Technology		•		
	Saga Unviersity	•			

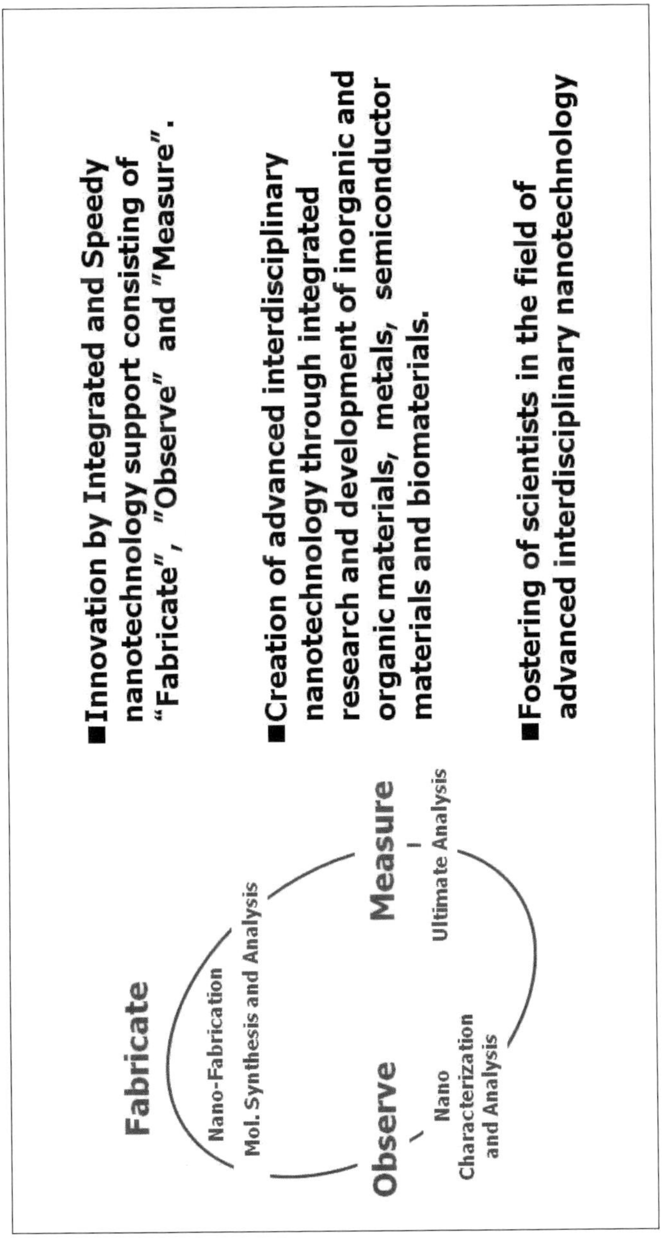

Fig. 8. Objectives of Nanotechnology Network Project (Courtesy of NanotechJapan).

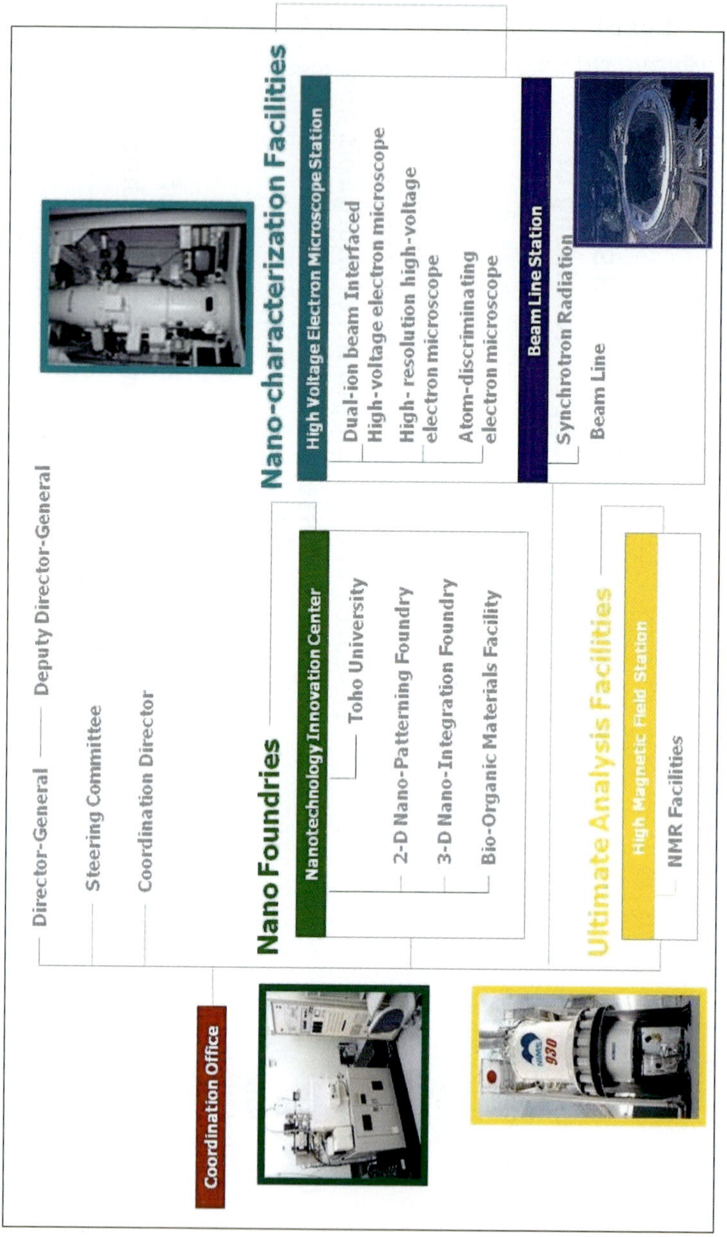

Fig. 9. Organization of Nanotechnology Network Project (Courtesy of NanotechJapan).

The information dissemination part of the project is called NanotechJapan and its web portal (nanonet.mext.go.jp) provides detailed information of project in both Japanese and English including (a) 13 network centers (b) Nanotech survey on popular/important research topics and trend reports in Japanese based on interviews of distinguished scientists (c) nanotech events in Japan and worldwide, (d) NanotechJapan Bulletin featuring articles on interview of young researchers, conference report, news on scientific breakthroughs and other noticeable development in Japan nanoscience and technology.

9. A Unique Industrial Endeavor — Nanotechnology Business Creation Initiative

The Nanotechnology Business Creation Initiative (NBCI) was established in Oct. 2003, as part of Japanese Ministry of Economy Trade and Institute (METI) nanotech policy, to provide the nanotechnology industry network for promoting nanotechnology commercialization and strengthening industry cooperation. NBCI was founded by a number of Japanese industry leaders including NEC, Mitsubishi, Hitachi, Olympus, Sumitomo, Cluster Technology, Chubu Electric Power and others, and has over 60 corporative executive members. It is supported by 13 distinguished advisory board members from Japanese universities and national research institutes. Currently NBCI has about 300 corporate and research institute members. Figure 10 shows the distribution of NBCI membership. NBCI formed the following committees to carry out its various activities:

- Business
- Technology
- Societal Acceptance and Standardization
- Networking
- Planning and Promotion
- National and International Collaboration Task Force
- Policy Strategy Advisory and Survey Task Force

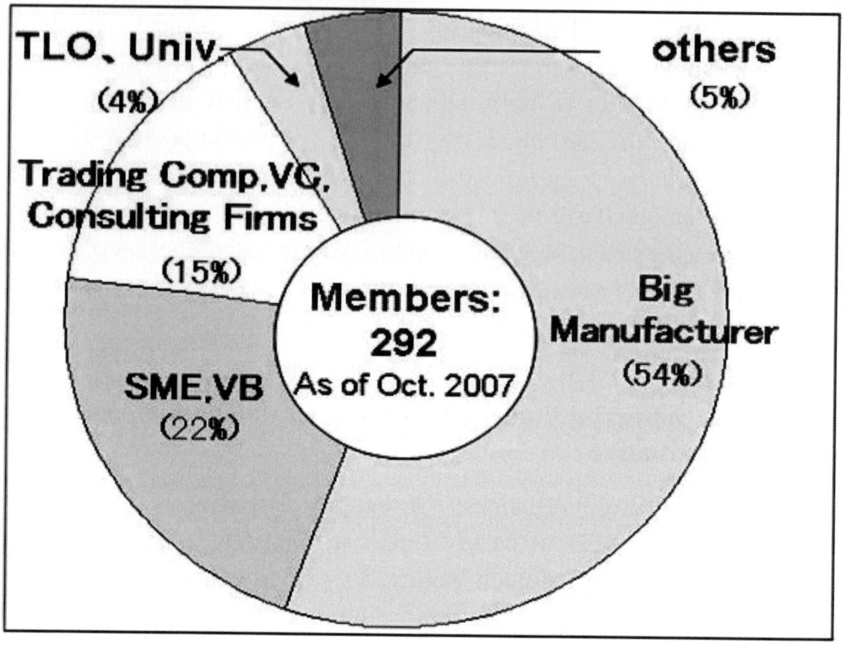

Fig. 10. NBCI membership break-down (Source from NBCI).

NBCI has been playing an important coordinated role for the following milestones.

(1) Nanotech Business Matching - There have been 32 nanobusiness matching forums organized by NBCI on different nanotech application topics presented by start up companies, small and medium size enterprise and large corporations in Japan and overseas.

(2) Nanotechnology Business Strategic Road-map - NBCI coordinated its industry and research members and compiled the Nanotechnology Business Strategic Road Map which was launched Oct 2007. This roadmap covers 8 areas which include Electronics, Bio, Fuel Cell/Energy, Environment, Ultra-precision Manufacturing/Processing, Catalysis/Coating/

Materials, Aerospace/Transportation Machinery, and Measurement/ Characterization Instruments.

(3) Standardization and Societal Implications - NBCI established a working group on social implication and standardization of nano materials in August 2005 (Fig. 11). This working group aims to provide a forum for information sharing among nano materials manufacturers, users and other stakeholders and to gather opinions on definition, measurement, and safety test methods from industry for making recommendation to the ISO.

(4) Globalizing Japan Nanotech Business - NBCI has so far established strategic alliances with Australia, Canada, South Korea, Taiwan and UK; It organize BIZMATCH at nanotech 2007 and 2008, the world's largest nanotech show to facilitate international business cooperation between Japanese and overseas industries. See details of BIZMATCH at 2007 at Part II of this chapter. Details about NBCI activities can be found at their website: www.nbci.jp.

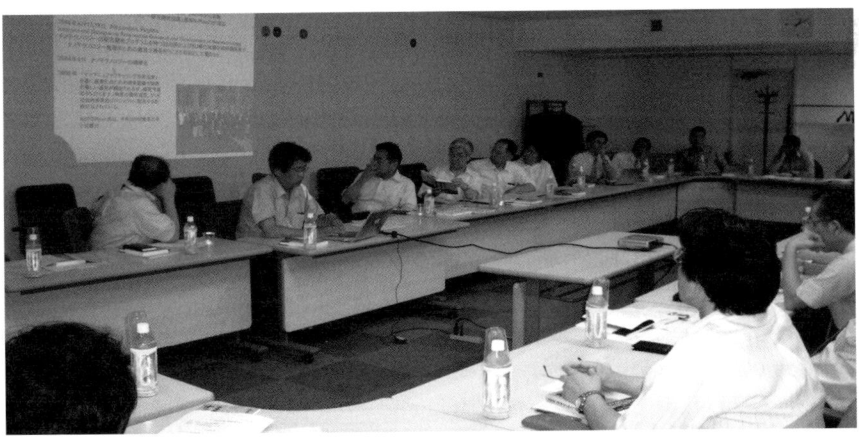

Fig. 11. Kick off meeting of NBCI Nanotechnology Societal Implication and Standardization Workshop on Aug. 4th, 2005.

10. Academic Laboratories

Session 8 has introduced the Japan Nanotechnology Network Project which supports 13 network centers total of 26 research institution in Japan. We selected the following five interdisciplinary centers of excellence and introduce briefly their research activities. For more details on research highlights, we recommend readers to visit http://nanotechjapan.jp/.

10.1. *Nanoscience and Nanotechnology Center at Osaka University*

The Institute of Scientific and Industrial Research (ISIR) was established in 1939 as part of the Osaka University. It adapts to the changing social needs such as industrial structure and develops into current size. ISIR's high research potential and historical significance contributes to creation of new research center for nanotechnology. In 2002, ISIR has started Nanoscience and Nanotechnology Center which focuses its research on nano materials and devices, beam science for nanotechnology and industrial nanotechnology. Nanoscience and Nanotechnology Center is made up of four departments which including various research fields: Nanomaterial and devices research department: Nano-quantum beam research department: Nanotechnology industrial applications department: Nano-structure and evaluation research department.

Nanoscience and Nanotechnology Center: http://www.sanken.osaka-u.ac.jp/labs/nano/english/index.html

10.2. *Nanoelectronics Collaborative Research Center, Tokyo University*

Nanoelectronics Collaborative Research Center (NCRC) was established within Research Center for Advanced Science and Technology for realizing core technologies towards the development of ubiquitous information devices based on nanotechnologies. NCRC is jointly operated by Institute of Industrial Science and is aiming at becoming one of the Center of Excellence (COE) for advanced nano-photonics and electronics. NCRC conducts research closely cooperated with industry, domestic and international universities. Currently, two collaborative researches are underway: NanoQUINE and Japan-Italy Global Cooperative Research Center.

NCRC: http://www.ncrc.iis.u-tokyo.ac.jp/e/index.html

10.3. *The University of Tokyo Center for NanoBio Integration*

Center for NanoBio Integration (CNBI) is founded for understanding the function and structure of human body on nano scale, developing bioinspired nanomachines and creating methodologies to integrate biomolecules and cells into nano devices. CNBI contributes to establish intellectual foundation and interdisciplinary science and technology system and to create innovative nano medical system and new nano-bio industry in the end of five-year implementation period.

Following three research themes are set for increase efficiency:

- creation of bioinspired nanomachines.
- creation of an accurate spatiotemporal control-type biosensing system.
- creation of nanotechnology and materials technology for nanoscale cell therapy.

CNBI: http://park.itc.u-tokyo.ac.jp/CNBI/e/index.html

10.4. *Esashi, Ono, and Tanaka Laboratory at Tohoku University*

Esashi, Ono, and Tanaka Laboratory at Tohoku University (Esashi Laboratory) is within the Department of Nanomechanics. Esashi Laboratory focuses its researches on socially beneficial technologies by using Micro Electro Mechanical Systems (MEMS). Esashi Laboratory has developed active catheters for noninvasive procedure and small size electrical power generator is under development as a micro energy source which will be essential for mobile information equipments. Esashi Laboratory's research themes are: Packaging, Wireless, Sensors;

Multiprobe data storage, Micro probes, Nanomachining; Energy Sources, Harsh environment technology; Medical, Optics.

Esashi Lab: http://www.mems.mech.tohoku.ac.jp/index_e.html

10.5. *Shinohara Research Laboratory at Nagoya University*

Shinohara Research Laboratory (Shinohara Laboratory) is within the Department of Chemistry at Nagoya University and its research interests are:

metallofullerenes
- fabrication and structural analysis of novel metallofullerenes
- magnetic, electronic, and optical properties of metallofullerenes
- applications of metallofullerenes for MRI contrast

carbon nanotubes
- synthesis of carbon nanotubes for high quality, high selectivity
- characterization with spectroscopy and microscopy
- device application focuses on double wall carbon nanotubes

peapods
- encapsulation of various atoms and molecules into carbon nanotubes
- characterization of peapods.

Shinohara Lab: http://nano.chem.nagoya-u.ac.jp/index.html

11. Conclusion

We provided an overview of policy, R&D infrastructure and network of nanotechnology in Japan. Nanotechnologies have the potential to offer solutions to a variety of problems facing the world we live in today. In order to contribute to sustainable development of Japan and the world, Japan can be expected to continue nanotechnology R&D in close collaboration with international organizations such as OECD and ISO and regional network such as the Asia Nano Forum, while continuing to build upon its own strengths in R&D.

References

1. Japan Patent Office, 2007. Trend Report on patent application technologies in eight selected areas: nanotechnology and materials.
2. AIST, 2004-2005. Open Forum "Nanotechnology and Society," Tokyo, Japan,
3. http://www.aist.go.jp/aist_j/research/honkaku/symposium/nanotech_society/index.html
4. AIST, NIMS, 2006. The Second International Dialogue on Responsible Research and Development of Nanotechnology, June 26-28, Tokyo, Japan.
5. AIST, NIMS, NIES, NIHS, 2006. Research Project on Facilitation of Public Acceptance of Nanotechnology, Summary and Policy Recommendations.
6. AIST, 2007. "Symposium on the Future of Nanotechnology and its Diffusion in Society", Symposium Report, February 5, Tokyo, Japan.
7. Government of Japan, 2001. The 2nd Science and Technology Basic Plan (FY2001-2005), Tokyo, March 30.
8. Government of Japan, 2006. The 3rd Science and Technology Basic Plan (FY2006-2010), Tokyo, March 28.
9. NEDO, 2006. Basic Plan for Research and Development of Nanoparticle Characterization Methods.
10. METI, 2007. Results of FY 2006 survey of proper on-site handling methods for nanotechnologies in research and manufacturing
11. METI, 2006. Fundamental Issues Subcommittee on Chemical Policy, Chemicals and Bio-industry Committee, Industrial Structure Council, Tokyo.
12. Ishizu S., M. Sekiya, K. Ishibashi, Y. Negami and M. Ata, 2007. Toward the responsible innovation with nanotechnology in Japan: our scope, Journal of Nanoparticle Research, Volume 10, Number 2, pp. 229-254.

Appendix

Hokkaido Innovation through NanoTechnology Support (HINTS)

Hokkaido University (National University Corporation)
Chitose Institute of Science and Technology (Incorporated Educational Corporation)

Website: http://hints.es.hokudai.ac.jp/
Main research domains: "Nano-Characterization and Analysis", "Nanofabrication"

Main shared-use facilities:

- Ultra-high precision electron beam writer
- Ultra-thin film evaluation device
- Focused ion beam processing observation device
- Multi-beam UHV electron microscope system
- Scanning near-field microscope
- Coherent anti-Stokes Raman microspectroscopy system

HINTS is operated in collaboration between Hokkaido University and Chitose Institute of Science and Technology. Linking "cutting-edge equipment and technology" based on the key phrases of "nanofabrication" and "nano-characterization and analysis", we conduct research on the creation of new nanodevices that control light, electrons and spin, and the fabrication of new functional nanomaterials. We also support the development of technologies and products. At Hokkaido University, as well as nanofabrication devices such as the ultra-high precision electron beam writer, we have an array of ultra-thin film evaluation devices and other nano-characterization and analysis devices. With these, we support R&D on cutting-edge nanodevice materials like photonic crystal devices and plasmonic devices. Moreover, we not only support this kind of top-down research and development, but also bottom-up technologies in the development of self-assembled monolayer fabrication methods and the design and synthesis of catalysts with new functions. Chitose Institute of Science and Technology, drawing on its unique research in light nanotechnology, supports diverse forms of nano-characterization and analysis with its various scanning probe microscopes, time-resolution spectrometry devices and others. At HINTS, as well as sharing the use of these cutting-edge research facilities, we also provide concrete support through the concerted efforts of our staff, who have a wealth of experience and achievements in research and strategies for developing technologies and products.

Center for Integrated Nanotechnology Support (CINTS)

Tohoku University (National University Corporation)

Website: http://www.rpip.tohoku.ac.jp/cints/index.htmMain research domains: "Nano-Characterization and Analysis", "Nanofabrication", "Molecular Synthesis and Analysis", "High Magnetic Fields"

Main shared-used equipment:

- UHV ultra-high resolution transmission electron microscope (JEMARM-1250)
- Double-sided aligner for bonding (EVG620)
- 4-axis X-ray structural analysis device (Mercury CCD)
- Tera-Herz ESR device (TESRA-IMP-P)
- RIE device (original)

CINTS has set out to build a system whereby researchers with high-level specialist knowledge in four fields (structural analysis, micro-processing, molecular synthesis and high magnetic fields) laterally support the needs of industry. Specific support work includes (1) Evaluating textures and composition in real space, and effective crystalline structures of metallic, semiconductor, organic and other material groups at nuclear level, mainly using a transmission electron microscope; (2) Allowing broad access from inside and outside the university to related facilities and devices in the possession of Tohoku University, with added technology and knowhow, with a view to developing technology and making prototypes based on MEMS technology, a basic technology for next-generation "product manufacturing", and supporting upgrades of MEMS; (3) Linking technology for the creation of new functional molecules with a complex structure to micro-molecular structure analysis using NMR and X-ray diffraction, and synthesizing new functional molecules in small quantities and with high precision; and (4) Supporting analysis of nanospin systems, particularly nano-magnets using a Tera-Hertz high magnetic field ESR device, and supporting nanomaterial processes using high magnetic environments, such as magnetic levitation and magnetic field orientation.

As well as making best use of this University's nanoscience potential in various fields, we also aim to take advantage of its characteristics as a general university, and to create new innovations in combined domains that could not be obtained in individual fields of technology.

NIMS Nanotechnology Support Network (NIMS)

National Institute for Materials Science (Independent Administrative Institution)
Toyo University (Incorporated Educational Corporation)

Website: http://www.nims.go.jp/nsnet/
Main research domains: "Nano-Characterization and Analysis", "Nanofabrication", "High Magnetic Fields"

Main shared-used equipment:

- High-resolution electron beam writer
- 5 NMR devices including 930MHz for individual high resolution
- Ion in situ observation UHV electron microscope
- Clean rooms with a total area of 1,000m^2, including bio facilities

The NIMS Nanotechnology Support Network was formed through collaboration between the National Institute for Materials Science and Toyo University. The nanofabrication domain is handled by the 2-D

Nano-Patterning Foundry, the 3-D Nano-Integration Foundry and the Bio-Organic Materials Facility (the three divisions of the Nanotechnology Innovation Center) and the Bio-Nano Electronics Research Center of Toyo University. The 2-D Nano-Patterning Foundry supports lift-off processes for fabrication of micro-electrodes and fluid device structures suited to organochemical and biomechanical materials, semiconductors, etc. The 3-D Nano-Integration Foundry supports three-dimensional processing of compound semiconductors, optical materials, oxides, superconductors, metal materials, and others ranging from nano to millimeter scale. The Bio-Organic Materials Facility supports the fabrication and evaluation of soft materials such as organic, high-polymer and biomechanical materials, as well as their application to the field of biotechnology. Finally, Toyo University supports nano-fabrication of silicon semiconductors and others to a minimum dimension of 10nm, as well as bio-nano related technology, etc. The High Voltage Electron Microscopy Station is in charge of nano-characterization and analysis. Among others, it supports moving image observation using an in-situ observation UHV electron microscope, EDS/EELS analysis, nanostructure fabrication via in-situ ion injection, ultra-high resolution (0.1nm) observation using an ultra-high resolution UHV electron microscope with maximum acceleration voltage of 1300kV, and ultra-low temperature observation (15K (liquid helium) or 90K (liquid nitrogen)). Finally, the High Magnetic Field laboratory is in charge of the extreme conditions domain. Among others, it supports various measurements of physical properties using 16 devices, including 930MHz NMR for individual high-resolution. The latter can even measure quadripolar nuclei.

NanoProcessing Partnership Platform (NPPP)

National Institute of Advanced Industrial Science and Technology (Independent Administrative Institution)

Website: http://www.nanoworld.jp/nppp/

Main research domains: "Nano-Characterization and Analysis", "Nanofabrication"

Main shared-use facilities:

- line stepper
- Electron beam writer
- Focused ion beam (FIB) process observation device
- Multipurpose reactive ion etching machine
- Semiconductor parameter analyzer
- Sputtering system/8-series electron beam evaporation and other film-making device
- Organic element/mass analyzer

The National Institute of Advanced Industrial Science and Technology (AIST) is one of the largest public research institutes in Japan. As a site for promoting fusion across sectors such as electronics, molecular nano-electromechanical systems, biotechnology, environment and energy, we offer a Nanoprocessing Facility (NPF) equipped with a series of cutting-edge devices that facilitate diverse material analysis, nanofabrication, configuration, measurement and evaluation, and device prototyping. The Analysis Section of our Technical Service Center (Research Environment Improvement Division) is also available for use by industrial, academic and governmental researchers. The main services that characterize this support work are (1) technical counseling and support in the creation of results, drawing on the wide spectrum of research resources owned by AIST, and (2) generous human resource development and practical training services for researchers who wish to use the devices or acquire skills.

The NPF not only provides technical advice from experts, use of devices, technical support, and results creation support services, but also offers training for young researchers and advanced technicians, while building networks with related facilities inside Japan and overseas. In doing so, we draw on our track record of active support as an administrative institution in the previous Nanotechnology Support Project, and our experience of running advanced technician training programs. In the Analysis Section of the Technical Service Center, we carry out analytical work such as organic element quantification analysis, mass analysis, high-resolution NMR analysis, ICP emission spectro-analysis, X-ray crystal structure analysis and FE-SEM observation, mainly targeting organic molecule and complex molecules that have various functionality and structures. We also give technical advice on analysis, as well as supporting research through decisions on molecular structure, component analysis and purity analysis.

Center for Nano-Lithography and Nano-Characterization

The University of Tokyo (National University Corporation)

Website: http://nanotechnet.t.u-tokyo.ac.jp
Main research domains: "Nano-Characterization and Analysis", "Nanofabrication"

Main shared-use facilities

- Variable shaping electron beam writer (modified Advantest F5112)
- Silicon deep etching device (Alcatel AMS100)
- UHV electron microscope (JEM-ARM-1250)
- Analytical transmission electron microscope (JEM-2010F)
- High-resolution FE-SEM (JSM 7000F)

In this Center, the Nano Laboratory of the Institute of Engineering Innovation (School of Engineering, The University of Tokyo) and VLSI Design and Education Center (VDEC) of Tokyo University collaborate closely to provide support.

The Nano Laboratory possesses a UHV electron microscope that has the highest resolution in the world, making it possible to analyze nanomaterials to a resolution of 0.1nm or less and directly observe lightweight elements such as oxygen, nitrogen. It is extremely strong in observation and analysis technology, and many of its results have been announced in academic journals of high international prestige. VDEC, meanwhile, offers the most up-to-date large-scale high-speed electron beam writer, based on a track record of shared use going back more than 10 years. This device can handle various sizes and shapes of substrate up to a maximum 8-inch wafer, achieving a high throughput with an average writing time of less than one hour using the variable shaping beam method. Though depending partly on the writing conditions, this makes it possible to write micropatterns to the scale of tens of nm. This writer is installed in a Class 1 clean room, where nanostructures can be formed by combining lithography with etching and other micro-processing technology. The silicon deep etching device is particularly adept at forming complex structures such as micromachines.

Support for Nanostructure Formation and Observation Using Electron Beams

Tokyo Institute of Technology (National University Corporation)

Website: http://www.pe.titech.ac.jp/rcqee/nano_support/index-j.html
Main research domains: "Nano-Characterization and Analysis", "Nanofabrication"

Main shared-use facilities

- Electron beam exposure device (JBX-5FE)
- Electron beam exposure device (JBX-6000FS)
- Scanning electron microscope (S-5200)
- Reactive ion etching device (RIE-1ONR)
- Transmission electron microscope (H1250)

Support from Tokyo Institute of Technology mainly consists of three-dimensional nanostructure formation based on two electron beam exposure devices, and support through a combination of sample fabrication and observation for electron microscopes.

In the nanofabrication domain, we provide process technology for nanostructure formation based on two electron beam exposure devices at the world's highest level, thanks to continuous replacement of device components. On a foundation of micropatterning technology in the 20nm class, we provide comprehensive technology for building three-dimensional nanostructures, in accordance with user needs. This includes transcription onto thin film, which is important for device structures such as semiconductors, metals and insulators. Also, as support aimed at building a system for increased use of the electron beam exposure devices owned by major universities, research institutes, etc., we hold technical guidance, schooling and other sessions inside the institution in question.

In the nano-characterization and analysis domain, we provide support for observation using an electron microscope (1,250V) that can keep samples in conditions of ultra-high vacuum and ultra-low temperature. This is the only UHV transmission electron microscope that can do this in the world. For support applicants who have insufficient experience in using transmission electron microscopes, we provide support in nanostructure observation including preparation of observation samples, an important factor in obtaining better results.

Waseda University Custom Nano-Formation and Device Evaluation Support

Waseda University (Incorporated Educational Corporation)

Website: http://www.all-nano.waseda.ac.jp/custom_nano/shien.html
Main research domains: "Nano-Characterization and Analysis", "Nanofabrication"

Main shared-use facilities

- Electron beam writer (ELS-7700W)
- Reactive ion etching machine (MUC-21 SR)
- Hot embosser (520HE)
- Precision plating device (original)
- Scanning electron microscope (S-4800)

To meet a wide variety of needs, we support the formation of nanostructures at cutting-edge level in three domains, making full use of the technologies, facilities and equipment accumulated here. (1) Precision plating technology support: Technical support for the formation of thin film with various electrical, magnetic, mechanical and optical functions, using the highest level of precision plate formation technology based on our own knowhow. In conjunction with nano/micro-processing technology, we also conduct selective deposition formation and three-dimensional microstructure formation for nano domains combining lithography processes, and analyze surfaces. (2) Nano/micro-processing technology support: We form microstructures with a high aspect ratio on a scale ranging from nm to μm, targeting silicon, glass, plastics and a variety of other materials. We form and process cutting-edge nanoscale structures, combining pattern formation by EB writers and other high-precision pattern forming devices with isotropic and anisotropic etching technology using various dry etching devices. (3) Technical support for device elementary structure formation: We offer technical support and property evaluation support for the formation of elementary structures (for example, the formation of microterminals for probers) that are indispensable for precise evaluation of device properties. This mainly involves support for the elementary structure formation needed in high-frequency property measurement and output property measurement.

Chubu Region Nanotech Support: Nanomaterial Creation and Processing, Advanced Equipment Analysis

Institute for Molecular Science
National Institutes of Natural Sciences, Inter-University Research Institute Nagoya University (National University Corporation)
Nagoya Institute of Technology (National University Corporation)
Toyota Technological Institute (Incorporated Educational Corporation)

Website: http://nanoims.ims.ac.jp/
Main research domains: "Nano-Characterization and Analysis", "Nanofabrication", "Molecular Synthesis and Analysis"

Main shared-use facilities

- 920MHz NMR
- Ultra-high speed near-field micro-spectroscope (original)
- Direct-writing electron beam writer
- 2-cycle excitation plasma etching device
- Plasma gas condensation cluster deposition apparatus
- Special-type scanning probe microscope
- Carbon nanotube generator
- Spray film-forming device

Four institutions in Aichi Prefecture collaborate to provide support for nano-characterization and analysis using revolutionary or creative equipment, nanofabrication based on the world's highest level of plasma technology, and synthesis of functional molecular substances. In the field of nano-biotechnology, for example, cross-sector lateral research on the expression and functional analysis of a single molecule inside a single cell of disease-related protein is made possible by using nanofabrication, plasma processing and unicellular molecule manipulation from Nagoya University, near-field imaging from the Institute for Molecular Science, and cell structure analysis from Nagoya Institute of Technology.

The Institute for Molecular Science meets a wide variety of needs, including the design and synthesis of functional molecules, large-scale quantum chemical calculation and point-contact current imaging. It also uses ultra-strong magnetic field NMR, revolutionary electron microscopes, and the world's first UV magnetic circular dichroism photo-emission electron microscope that does not depend on radiation. Nagoya University provides technical support for semiconductor processes, nanomagnetic devices, MEMS, nano-bio elements, and others based on plasma technology, as well as nanodevice structure analysis and advanced element evaluation led by electron microscopes. Support by Nagoya Institute of Technology takes the form of development and evaluation of new nanomaterials, cell structure evaluation at molecular level, and the development and functional evaluation of elements and sensors using carbon nanotubes, among others. Finally, Toyota Technological Institute carries out processes and evaluation with emphasis on hybridization with compound semiconductors, carbon, metal and various other nanostructures, on a foundation of silicon process technology.

Kyoto University Nanotechnology Support Network

Kyoto University (National University Corporation)
Japan Advanced Institute of Science and Technology (National University Corporation)
Nara Institute of Science and Technology (National University Corporation)

Website: http://eels.kuicr.kyoto-u.ac.jp/nano/aim.html
Main research domains: "Nano-Characterization and Analysis", "Nanofabrication", "Molecular Synthesis and Analysis"

Main shared-use facilities

- Mass spectrometry device (FT-ICR-MS)
- Electron probe microanalyzer (EPMA)
- Transmission electron microscope (TEM)
- Rutherford backscattering spectrometry, high-energy ion injecter
- Micro-Raman device

In the collaboration between three universities – Kyoto University, Japan Advanced Institute of Science and Technology (JAIST), and Nara Institute of Science and Technology (NAIST) – we provide support for general R&D in the fields of nano-characterization and analysis, nano-fabrication, and molecular synthesis and analysis. As a center for innovation creation closely linked to the surrounding region, we also support regional promotion and cultivate venture business. In its Advanced Research Institute of Nanoscale Science and Engineering, Venture Business Laboratory, and Institute for Chemical Research, Kyoto University promotes various related projects as a research center for nanotechnology and nanoscience. The University has a considerable track record of support in the previous Nanotechnology Support Project over the last five years. JAIST is Japan's first independent national university for graduate studies. In its School of Materials Science and Center for Nano Materials and Technology, it undertakes research and education on materials science and nanotechnology fusing the fields of physics, chemistry and biology, with a basic foundation in atomic and molecular level science. NAIST has a track record of supporting nanotechnology research for a large number of companies in its Research and Education Center for Materials Science. This is home to cutting-edge nanotechnology research equipment and nine full-time technical employees. NAIST is ranked No. 1 in Japan (FY2005) for expenditure on industrial-academic collaboration per member of teaching staff.

Osaka University Compound Function Nano Foundry

Osaka University (National University Corporation)

Website: http://www.uhvem.osaka-u.ac.jp/nanonet-index.html
Main research domains: "Nano-Characterization and Analysis", "Nanofabrication", "Molecular Synthesis and Analysis"

Main shared-use facilities

- UHV electron microscope (H-3000)
- Electron microscope for tomography (H-9500)
- Laser MBE device (PLD-020R)
- Electron beam exposure device (Beam Draw + JSM-6500F)
- Focused ion beam device (SMI-2050)

Osaka University Compound Function Nano Foundry was formed as a collaborative amalgamation between the Institute of Science and Industrial Research and the Research Center for Ultra-High Voltage Electron Microscopy, both of Osaka University. To support nanotechnology research, it combines and links the three domains and functions of (1) nano-characterization and analysis, (2) nanofabrication and (3) molecular and thin film synthesis. It also provides opportunities for technical guidance and human resource development while enhancing its consultancy functions and intensifying organic links with companies and research institutes.

(1) In nano-characterization and analysis, we provide support for structure and texture analysis using various types of electron microscope, the principal example being a UHV electron microscope with the world's highest accelerated voltage of 3 million volts. We support UHV-EM observation, sample preparation, analysis of EM images and electron diffraction patterns, and others in which the inside of a sample on a thickness scale of micrometers is analyzed three-dimensionally with nanoscale resolution. (2) In nano-fabrication, we support nano-fabrication and devicing of thin film samples, measurement of thin film properties, evaluation of device characteristics, and so on. (3) In molecular and thin film synthesis, we support the design and synthesis of materials that have optimal atomic and molecular sequences for expressing the targeted functions, synthesis of functional artificial lattice film using an excimer laser, synthesis of organic thin film using K cells, and so on.

Measurement and Analysis of Nano Structures and Functions Using Radiation

Japan Atomic Energy Agency (Independent Administrative Institution)
National Institute for Materials Science (Independent Administrative Institution)
SR Center, Research Organization of Science and Engineering, Ritsumeikan University (Incorporated Educational Corporation)
Cooperating institution: High Luminance Optical Science Research Center (Incorporated Foundation)

Website: http://www.harima.jaea.go.jp/
Main research domain: "Nano-Characterization and Analysis"

Main shared-use facilities

- 5 SPring-8 dedicated beam lines (BL11XU, BL14B1, BL15XU, BL22XU, BL23SU)
- 8 SR Center beam lines

Radiation is a light source that excels in the attributes of luminance, directionality and polarization across the broad wavelength domain from infrared to gamma rays. It facilitates the measurement and analysis of nano structures and states (such as the localized structure of selected atoms in material non-destruction) that are impossible with other methods. At SPring-8, the world's largest radiation facility, revolutionary technologies for the use of radiation are being developed, while wide-ranging research is conducted on substances and materials based on these. Ritsumeikan University, meanwhile, has developed radiation facilities ahead of other universities, and is making use of independently developed radiation technology to promote analysis of electron states and crystal structures (particularly light element materials), as well as the development of nanotechnology-fabrication technology. This Center is an amalgamated entity that has the highest potential in the world as a group of institutes boasting a strong record of performance in the previous Nanotechnology Support Project. SPring-8 (a soft to hard X-ray source) and the compact radiation of Ritsumeikan University (an ultrasoft to soft X-ray source) complement each other in supporting nano-characterization and analysis of crystalline structures, electronic or magnetic structures, elements, composition, and others. They do so by measuring diffraction, absorption, spectration and emission targeting a wide range of elements from light to heavy. The priority fields of this Center are next-generation magnetic recording materials, energy conversion and storage materials, nanotechnology-electronics materials, and, as advanced new domains, new nano-particle functional materials, nano thin film functional materials, nano region measurement technology and nano fusion regions. On this basis, we support nano measurement and analysis using radiation.

Support for Silicon Nano Processing and High-Quality Vacuum Utilization Technology (Research Center for Nanodevices and Systems)

Hiroshima University (National University Corporation)
Yamaguchi University (National University Corporation)

Website: http://home.hiroshima-u.ac.jp/nanotech/
Main research domain: “Nanofabrication”

Main shared-use facilities

- Electron beam writer (50nm)
- Class 10 clean room including chemical filter
- High-speed heat treatment device
- ECR sputter device
- 3-dimensional magnetron sputter device

In this Center, we offer support that won the “Manufacturing Grand Prize – Encouragement Prize” in the previous Nanotechnology Support Project, support mainly for silicon materials in Hiroshima University, which boasts a strong background including academic awards, and support on non-silicon materials in Yamaguchi University, which has a track record in high-quality vacuum utilization technology.

At Hiroshima University, as well as support that draws on design and manufacturing technology for nano transistors with a gate length of tens of nm using an electron beam exposure device, we also offer technical consultation on nanostructure formation processes and the nanodevices that use these. The support consists of (1) nanostructure process design, (2) thin film formation, introduction of impurities, (3) nanostructure pattern design, and (4) silicon nanodot related support. At Hiroshima University, the support mainly targets high-quality vacuum utilization technology, using a series of micro-processing devices including an electron beam writer, a mask aligner, and an ion beam etching device installed in clean rooms, etc. The main targets are non-silicon materials such as superconducting materials, inductor materials and magnetic materials.

Kyushu Regional Nanotechnology Center Network

Kyushu University (National University Corporation)
Saga Prefectural Regional Industry Support Center, Kyushu Synchrotron

Light Research Center (Incorporated Foundation)
Kitakyushu Foundation for the Advancement of Industry Science and Technology (Incorporated Foundation)
Saga University Synchrotron Light Application Center (National University Corporation)

Website: http://nanoscience.cstm.kyushu-u.ac.jp/
Main research domains: "Nano-Characterization and Analysis", "Nanofabrication", "Molecular Synthesis and Analysis"

Main shared-use facilities

- Dynamic secondary ion mass spectrometry device
- SQUID magnetic susceptibility measuring device
- Electron spectroscopic multi-dimensional analysis UHV electron microscope
- Lorenz electron microscope
- Synchrotron radiation utilizing equipment (4 beam lines)
- Laser beam writer
- Ion injector

At the Kyushu Regional Nanotechnology Center Network, four institutions – Kyushu University, Kyushu Synchrotron Light Research Center, Saga University and Kitakyushu Foundation for the Advancement of Industry Science and Technology – collaborate to support (1) molecular synthesis and analysis, (2) ultra-microscopic analysis, (3) synchrotron radiation measurement and analysis, and (4) MEMS measurement and analysis. Kyushu University oversees two of these domains. In "molecular synthesis and analysis support", it provides general technical support for synthesis, as well as structural and functional analysis and evaluation aimed at creating nanomolecular materials and devices of organic, inorganic, and organic/inorganic hybrid types. In "ultra-microscopic analysis support", it allows access to its UHV electron microscope, special function electron microscope, sample preparation devices and others. It also gives general support in four priority fields, namely "sample preparation", "nano domain analysis", "electron microscope in situ analysis" and "difficult-to-observe material analysis". Kyushu Synchrotron Light Research Center and Saga University Synchrotron Light Application Center cooperate with each other, the former offering general support foj solving problems in research on advanced nanomaterials, using equipment and analytical techniques such as XAFS, X-ray diffraction, X-ray small-angle scattering and reflection rate measurement. The latter uses new spectrometry methods combining photo-electronic spectroscopy and radiation with laser photo-activation and spectroscopy, to support analysis of electronic states and material surface states of photo-functional materials, among others. Finally, Kitakyushu Foundation for the Advancement of Industry Science and Technology promotes technical support concerning semiconductor device manufacturing technology and MEMS/NEMS, and general support for cell electronics development using advanced integration.

Author Biographies

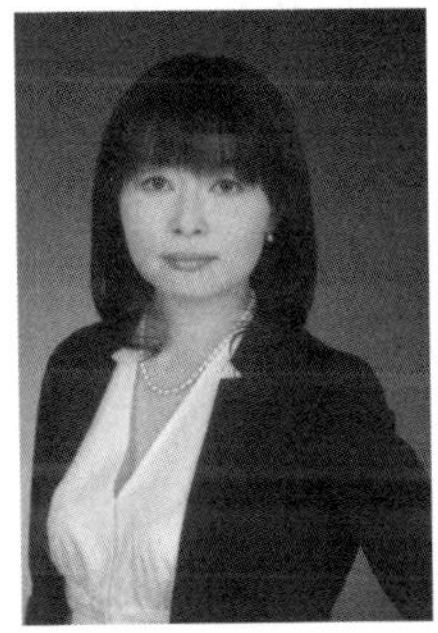

Ms Sekiya Mizuki is a member of the Working Group on Strategic Area of Nanotechnology and has joined the AIST in 2004. Ms Sekiya was also member of the fiscal 2006 project "Research on Facilitation of Public Acceptance of Nanotechnology". She actively carries on research on domestic and international situation of social implication of nanotechnology.

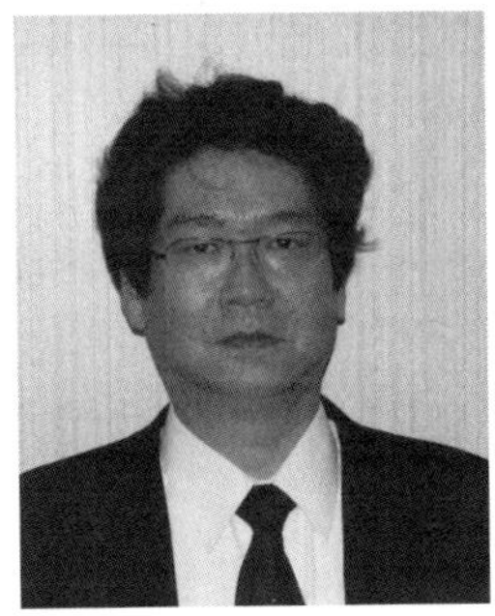

Dr Masafumi Ata has been an Assistant Director of Technology Information Department of AIST since 2004. He leads a Working Group on Strategic Area of Nanotechnology in AIST currently. He led a project "Research on Facilitation of Public Acceptance of Nanotechnology" in fiscal 2006, which was the first attempt to address societal implication of nanotechnology in Japan. He also has been a Program Director of JST for the new research priority area since 2007. He engages on various issues for smooth and successful commercialization of nanotechnology with knowledge form many years of experience in private company.

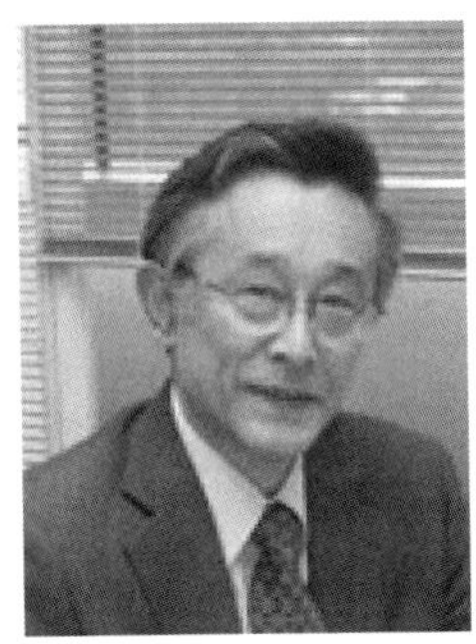

Dr Kazunobu Tanaka Graduated from the Department of Electric Engineering at the Faculty of Engineering, the University of Tokyo in 1963. He joined Tokyo Laboratory of Matsushita Electric Industrial Co. in 1963, joined the Physical Science Division of the Electrotechnical Laboratory in 1971, became Director of Amorphous Materials Laboratory in 1980, became Director of Materials Science Division in 1988, became Supervising Researcher at the National Institute for Advanced Interdisciplinary Research in 1993, became Board Trustee of the Angstrom Technology Partnership in 2000, and became Vice President at the National Institute of Advanced Industrial Science and Technology in 2001. He has been in his current post since 2005. Before that he had been Professor of Department of Chemical Engineering at Faculty of Engineering, the University of Tokyo (1984-88), UNIDO Project Advisor (1988-1992), Professor of Cooperative Graduate School at the Institute of Material Science, the University of Tsukuba (1992-2000), Project Leader of the Joint Research Center for Atom Technology (JRCAT) (1997-2002), expert member on Nano-technology Committee at the Council for Science and Technology Policy. He has been Senior Fellow at the Center for Research and Development Strategy, the Japan Science and Technology Agency since 2005.

Dr Kazunobu Tanaka discovered light induced structural change of chalcogenide-glass and firmly verified its reversibility. This has been accepted as a phenomenon specific and common to overall amorphous materials. He led the research into amorphous silicon as a material for solar batteries and showed typical examples of material design including process diagnosis. This has been widely applied not only to solar batteries, but also to the production of image device TFT arrays for image device scanning. He played a leading role in two industry-government-academia strategic projects; "Amorphous Thin-layer Solar Battery Project (Sunshine Project)" (Distributed Research Theme, 1978-88) and "Atom Technology Project (NEDO)" (Concentrated Research Theme, 1992-2002), also providing academic human resources for them. The NEDO Project became one of the incentives for the USA to start their National Nanotechnology Initiative (NNI).

CHAPTER 6

PART II JAPAN NANOTECHNOLOGY OVERVIEW: COMMERCIALIZATION HIGHLIGHTS

Gimpei Sato

Free-lance Technological Writer, Japan
gimpei@s05.itscom.net

Nanotechnology, a key technology of the 21st century, is the engine of the electronics, information and communications, bioscience, and environmental/energy technologies. It does not only contribute to innovative manufacturing, but also becomes the leading technology on which one can build competitive advantage. This paper provides the overview of Japanese nanotechnology industry and picks up some of the latest topics.

1. Nanotechnology-Related Market: A Growth Engine

Several forecasts of the nanotechnology-related market have been published by research groups, such as think-tank organizations. The forecasts include leading industries in today's market, including electronics, bioscience/healthcare/cosmetics, and fuel cell/energy industries. The forecasts assume that in all industry segments nanotechnology-based products will replace existing products and/or will open a new market.

Nanotechnology is, indeed, expected to be utilized in all industries, expand the existing market, replace existing products, and establish a new market segment. However, measuring accurately the current size of nanotechnology-related market is difficult because of the fluidity of the market and the infancy status of nanotechnologies, most of which is still

Table 6. Market Size Trend of Nanotechnology-Related Market

	2000	2005	2010	2020	2030
Electronics Area	1,640,740	1,934,670	2,429,100	8,049,000	18,013,500
Bioscience/ Healthcare/ Cosmetics Area	62,966	115,000	195,500	658,000	1,140,000
Fuel Cell/ Energy Area	302,150	482,060	866,960	3,167,000	4,430,000
Environment Area	3,000	5,000	8,000	45,000	80,000
Ultra Precision Manufacturing	18,900	94,200	209,300	440,600	679,200
Catalyst/Paint/ Material Area	29,650	86,950	353,100	822,400	1,526,200
Measurement Area	29,250	70,650	91,850	154,900	245,600
Aerospace/ Transportation Area	10,000	20,000	60,000	100,000	150,000

in the transitional stage from R&D to applications. Nonetheless, it is certain that it becomes the foundation that supports industries in the 21st century.

It is, in fact, undeniable that the nanotechnology-related market will grow and become a giant. What is the current market size estimate? The report from the Ministry of Economy, Trade and Industry and Fuji-Keizai — "Market Size Trend of Nanotechnology-Related Market" (March, 2006) — summarizes as follows: 'Electronics Area,' 'Bioscience/Healthcare/Cosmetics Area,' 'Fuel Cell/Energy Area,' 'Environment Area,' 'Ultra Precision Manufacturing and Fabrication Area,' 'Catalyst/Paint/Material Area,' 'Measurement Area,' and

'Aerospace/Transportation Area.' Each market size is analyzed for the period until 2005 and forecasted for 2010, 2020, and 2030. (Table 6).

The largest market is Electronics with the expected size of ¥2,429,100 million in 2010, ¥8,049,000 million in 2020, and 18,013,500 million in 2030. The next is the Fuel Cell/Energy sector with ¥866,900 million in 2010, ¥3,167,000 million in 2020, and ¥4,430,000 million in 2030. The fastest growth is expected to be in the Catalyst/Paint/Material sector, which will grow four-fold from ¥86,900 million in 2005 to ¥353,100 million in 2010, more than nine times to ¥822,400 million in 2020, and more than 17 times to ¥1,526,200 million in 2030.

2. Rapidly Growing Applications of Nanotechnology

The Electronics market continues to be large because the flat panel display applications, such as LCD TV and PDP, have expanded rapidly, and new displays, such as organic EL and FED, are expected to become commercially viable. Unsurprisingly, other electronics, such as a cell phone, digital camera, digital camcorder, and DVD recorder, are also growing rapidly. The parts of these devices, such as various processors, DRAMs, and flash memories, also show sustaining growth. Furthermore, telecom-related nano-devices, such as photonic crystal and quantum-dot, are expected to significantly contribute to the growth.

In the area of device manufacturing, the development of ultra-fine pattern fabrication technology and equipment is active. New technologies are used for mass manufacturing. For example, the transfer etching of a circuit on wafers by lithography is transitioning from conventional exposure technology to liquid immersion lithography technique using liquid with high refractive index. The finer the circuit line width of semiconductor becomes, the more precise the measurement devices become.

Next, the dominant focus in the area of Fuel Cell/Energy is the application of fuel cell in an automobile. Its commercialization will be still down the road. In contrast, the direct methanol fuel cell (DMFC) technology, which is used in notebook PC and cell phone applications, is near complete, expected to be commercialized in 2008. The solid oxide fuel cell (SOFC) technology, which utilizes natural gas, has been rapidly developed and is commercialized soon. This technology not only reduces the CO_2 emission, on-site power generation via SOFC also co-generates electricity and heat, resulting a highly efficient energy alternative.

Thin film type solar cells are increasing, moving away from the polycrystalline silicon type, which has been volatile depending on the supply and demand of semiconductors. The next generation solar batteries – compound semiconductor CIGS (Cu-In-Ga-Se) and spherical silicon – are in commercial mass production. Film type dye-sensitized solar cells are almost entering the commercial production stage.

The areas of preventive medicine, diagnosis, and therapy in the Bioscience/Healthcare/Cosmetics segment are growing, which echoes today's strong health consciousness. New products, such as a DNA chip that analyzes genes, a polymer micelle that packages anticancer drug, artificial red blood cell using liposome, are in the clinical test stage. The technologies that utilize nanotechnology, such as cell sheets, scaffolding drug for cell growth, and synthetic bone formation using inkjet, are multiplying.

Cosmetics have been employing nanotechnology aggressively, such as microparticles of titanium oxide and zinc oxide. The examples include the surface-modifying technology to form a ultra thin film of less than 1nm on a fine particle surface, which enables water-resistant foundation, the development of functional cosmetics that soften and give nutrient to the skin, skin-lightening agent in a nano-size capsule, and, Drug Delivery System (DDS) technology that regulates constant flow of medicinal properties to avoid pigmented spots and wrinkles, the manipulation of nano-level barium sulfate crystals to diffuse reflections, which disguises sagging and fine wrinkles. Thus, nano cosmetics, which reach the cell level, are born one after another. It is not an exaggeration to say that nano cosmetics will soon become the standard in the market.

So far, nanotechnology-related market has been looked at in the order of segment size. At this point I would like to mention one of the everyday products that utilize nanotechnologies as well. It is clothing that uses nanotechnologies. Some of them are sold in local supermarkets. We owe to nanotechnologies without noticing. In fact, almost all domestic textile manufacturers use nanotechnology material and fabrication methods.

Superior water/stain repellent and wrinkle- and shrinkage-free pants and skirts for women, men's pants that quickly absorb sweat and humidity and get dry by diffusing them, special fabric with a thin film of nano particles on the surface, which makes it difficult for pollens to stick to it as well as deactivates pollen allergen. The application was originally developed for linens; however, it is also used in spring coats and dress shirts as anti-pollen products.

Morphologically stable shirts became possible by making a resin cross-linking agent react to cotton fiber at the level of sub-nanometer. The fiber with thermal-storage insulation uses nano particles fixed in the polyester clothing fabric that absorbs infra-red light from the sun and transfer it to the heat. The temperature inside of the garment could be almost three degrees higher than the outside of the fabric. Silver particles of several nm in size that are penetrated and attached in the fabric have antibacterial effect. In this fashion, nanotechnology applications in clothing are numerous.

Although it might be an overused example in Japan, it is worth mentioning a popular nanotechnology fiber product, called Morphotex produced by Teiijn Fibers. It is a biomimetics technology that mimics the phenomenon that the wing of Morpho Butterfly in the South/Central America shows bright coloration as a result of its air layers and protein layers (lamellar interface) that interfere light. Actual fiber has 70nm-100nm in thickness, constructed with 61 layers of polyester and nylon with different refraction index. Multiple refractive spectra produce multiple colors. The colors are more vivid and translucent than the colors produced by the usual dyeing technique. The colors change depending on the strength of light and viewing angle. Because the manufacturing process does not use dye, this nanotechnology product contributes to the reduction of wastewater and energy consumption.

3. Steady Advance of Nanotechnologies in the Semiconductor

Among Electronics, for sure, the micro fabrication of semiconductors and magnetic discs most significantly receive the benefit of nanotechnology. Both are data storage applications.

For instance, until recently the micro-processor was the leading semiconductor product. However, the current leader is the NAND flash memory, which is increasingly used in the cell phone, digital audio player such as iPod, digital camera, digital camcorder, and other applications.

Accordingly, Toshiba's Yokkaichi factory has begun mass manufacturing using the 56nm process since April 2007. This micro fabrication technology is most advanced in the world without a peer. Toshiba is going to begin the mass manufacturing of the next generation 43nm during the 1st quarter of 2009. It has succeeded the development of a MLC (Multi-Level Cell) 4 bits/cell prototype chip, which can process 16 separate data per cell. Toshiba is not only the leader in the MLC chip technology but also in the miniaturization technology.

The Yokkaichi factory has realized this microscopic miniaturization by utilizing Nikon's liquid immersion lithography equipment, which employs the immersion exposure technology. The immersion lithography equipment transcribes micro patterns by utilizing the ArF light source of 193nm in wave length and the liquid, whose refractive index is higher than the air, positioned between the projector lens and wafer. When the double-exposure technology is combined with the immersion exposure technology, the next generation 32nm can be achieved. The use of this technology in commercial production is expected to begin in 2009.

Then, what kind of technology awaits us beyond it? There are three major paths: EUV (extreme ultraviolet light with its wave length of 13.5nm) exposure technology, nanoimprint, and near-field light lithography that uses near-field light. All of these technologies are promising to make the fabrication of smaller-than 32nm possible.

The EUV exposure technology has achieved 26nm line width in the R&D stage; however, there remain some difficulties toward commercial applications. Nanoimprint transcribes circuit patterns by pressing a silica glass template onto resin. Toshiba and Dai Nippon Printing have achieved the formation of 18nm patterns. However, several issues remain for the commercial application, such as the consistency of circuit pattern dimensions and the precision of superimposition. The commercial use will have to wait for a few more years.

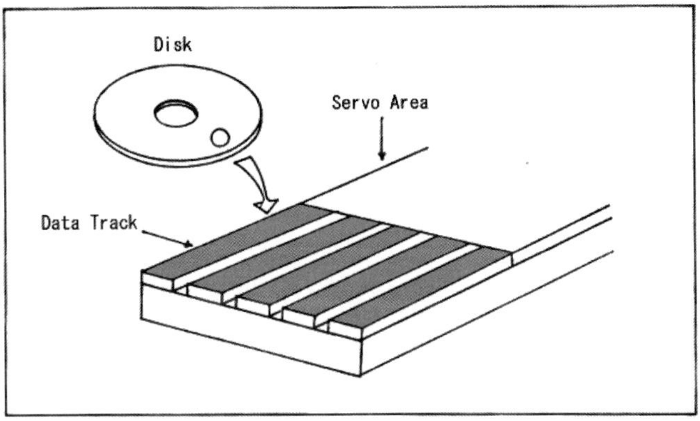

Fig. 12. Discrete Track Media.

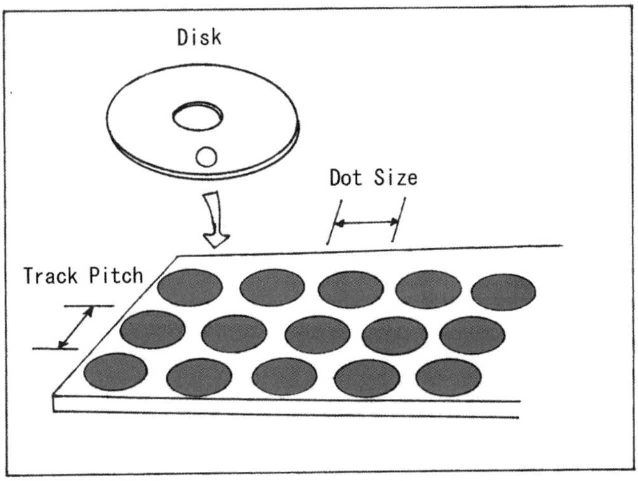

Fig. 13. Nanopatterned Media.

The third alternative, near-field lithography, uses near-field light. Different from normal diffused light, near-field light emerges from the surface of micro-objects and is not affected by the diffraction limit of wavelength. Professor Motoichi Ohtsu of Tokyo University and Canon achieved the formation of smaller-than 50nm patterns in August 2007 by using a near-field light lithography machine, which they jointly developed. This technology is hopeful for further miniaturization of semiconductors because it supports conventional light sensitive films and engineering material for optical wavelength and because it allows high resolution and accurate measurement and position control.

4. Nanotechnology's Great Strides in the Storage Technology

The development of the super-high-capacity feature in the next generation HDD (hard disk drive), which competes with flash memory, is currently taking place. Its market introduction is coming up soon in 2008-2009. There are three types: next-generation perpendicular recording, discrete track media (Fig. 12), and nanopatterned media (Fig. 13).

Fuji Electric Device Technology developed the next-generation perpendicular recording technology, which allowed high density recording of over 800 GB per square inch by layering magnetic and nonmagnetic substance on the disc.

Both discrete track media and nanopatterned media are structured in such a way that adjacent magnetic substance does not affect each other. These technologies theoretically achieve the density of 1TB per square inch. Discrete track perpendicular recording media can significantly increase recording density by etching grooves between tracks to prevent inter-track interference and narrowing the track width. Toshiba, Showa Denko, and TDK are working on the commercial applications of this technology.

Nanopatterned media records 1 bit per magnetic particle. Particles are arranged along the track. The specification of the disc that is prototyped by Fujitsu has the diameter of magnetic substance of 10nm, the interval of 3nm, and recording density of 4TB per square inch. It is about 20 times larger than the capacity of today's disc. The commercial introduction of nanopatterned media is expected to come in several years. Japanese focus of DNA analysis has been changing from exhaustive gene analysis to specific gene analysis. The gene analysis that uses DNA chips has been dominated by American and European researchers. Currently, many of their researches are shifting from R&D to clinical trials.

5. Robust Application Development of Nano-Biology

Japanese enterprises in their effort to prevent American and European companies from sweeping the worldwide market established "Japan MicroArray Consortium" in October 2007. The promotion of industrialization of biochips, particularly DNA chips, is the objective of the consortium. A total of 67 companies from electronics, mechanical, and chemical industries joined. Companies, otherwise competitors, joined under the consortium to commit themselves to the problem-solving necessary to commercialize biochips. They will together establish DNA chip analytical methods for specimen and reliability measures. That is, they will promote the standardization of DNA chip analysis.

One of the most advanced in the DNA chip operation is found at Toshiba, which submitted an application for monitoring human papillomavirus (HPV) infection via a DNA chip in June 2007 jointly with its partners, Daiichi Pure Chemicals and Toshiba Hokuto Electronics. HPV, which causes cervical cancer, can be cured completely if detected in the early stage. The mass manufacturing line for DNA chips is already built and ready for operation as soon as the application is granted. This product will be included in the standard healthcare technology that medical insurance covers as early as in 2008.

Toray with Matsushita Environmental and Air-conditioning Engineering and Matsushita Ecology Systems developed a highly sensitive DNA chip that detected the microorganism that broke up contaminants and introduced it in May 2007. This one chip can simultaneously detect 22 types of VOC (volatile organic compounds) degrading microorganism. It is a new and valuable application to identify VOC degrading bacterium which is beneficial for soil and underground water bioremediation treatment with microorganism, and to quantitatively detect for each type of microorganism.

Yokogawa Electric developed its integrated cartridge in June 2007. For the first time in the world it succeeded in conducting a fully-automated gene analysis from the extraction of DNA from acid-fast bacteria, culture, formation, to detection. The gene analysis system allows safer and easier gene analysis operations in the clinical setting.

6. Japanese-Born Nanotechnologies that Contribute to the Environmental Improvement

Photocatalyst technology is Japanese origin and is an example of nanotechnology promising commercial success in the environmental arena.

When Mr Akira Fujishima (present Chairman and Director at Kanagawa Academy of Science and Technology) was a graduate student under Assistant Professor Kenichi Honda (present Professor Emeritus at Tokyo University) at Tokyo University in 1967, he discovered that ultraviolet light helped produce oxygen from the titanium-oxide electrode and hydrogen from the platinum electrode when it was applied in the water tank during the electrolysis of water. It was later called "Honda-Fujishima effect."

It was hoped-for method to produce clean hydrogen energy from the sunlight and water. However, the actual energy conversion efficiency was too low at 0.3% for commercialization. Later, the strong oxidizing power of photocatalyst was utilized in removing the odor and disinfecting tiles in the toilet by way of organic decomposition. The discovery was made in 1994. Then, the superhydrophilic nature of photocatalyst was discovered a year later in 1995. When titanium oxide is coated on exterior building material such as tiles, the sunlight decomposes light greasy dirt into carbon dioxide (CO_2), and rain water washed off heavy greasy dirt, which is lifted by the superhydrophilic coating. This is the photocatalystic self-cleaning effect, and is currently used by construction companies, such as PanaHome, Sekisui House, and Daiwa House in Japan.

In 2001 Toyota Central R&D Labs developed a new photcatalist that sensed the visible light. By doping a small quantity of nitrogen in titanium oxide, they made the photocatalytic reaction to the visible light possible although the visible light contains small amount of energy. At present, the research and development of photocatalyst is taking place to make it operable at the wavelength of 450nm-500nm. This type of photocatalyst can decompose formaldehyde, which causes the sick house syndrome, even in the indoor area where ultraviolet rays do not usually reach.

Also, building's exterior surfaces that have photocatalytic coating maintain wetness for long period of time. This propensity to form a water film on the surface of building and a roof can be used to lower the air temperature surrounding the building by releasing the evaporative latent heat using deposited rainwater. The application could be used to ease the heat-island effect in high-rises during mid-summer. This technology will provide a high energy-saving effect.

The talk of the town during this year was the new Tokaido and Sanyo Shinkansen fleet N700, which was put in operation on July 1st, 2007. All seats of the N700 fleet are nonsmoking, and there are four smoking areas where units of photocatalytic deodorizing equipment were installed. This device, which was equipped with a photocatalytic filter — titanium-oxide coated ceramic surface — and ultraviolet light source, decomposed the content of cigarette smoke, such as acetaldehyde, into water and carbon dioxide.

The windows of the green car section of the fleet use titanium-oxide nanosheet, which have self-cleaning effect. The thickness of the nanosheet is about 1nm, the size of several atoms. Its surface is smooth, and the nanosheet adheres firmly and unlikely to peel off. Thanks to the coating with the nanosheet, green car windows are always clean and maintain clear view for the passenger.

Another type of nanotechnology other than photocatalyst that clean the soil was recently developed. The developer is Toda Kogyo, and its technology utilizes iron nanoparticles of 70nm in diameter and $30m^2/g$ in specific surface area. The iron nanoparticles have the core-shell structure. When the aqueous solution of them is injected in the soil, the film of ferric oxide (Fe_3O_4) on the surface of nanoparticles cleans up the soil by absorbing contaminants, such as lead and cadmium. This purification technology has been employed at more than 40 domestic sites. In 2000, NASA in the U.S. also proved the positive outcome of the technology. Toda Kogyo announced that in May 2007 it would start a research project in Europe to determine the effectiveness of the soil purification technology. In Germany the soil contamination of the former factory sites is a serious problem. Various methods have been tried to solve the contamination problem. Therefore, Toda Kogyo conducted an experimental study from August to December. Based on the result, it plans to install a full-fledged business in Europe.

Because environmental issues tend to be low-profile and take time to apply and commercialize, the development of environmental technologies have been assigned mainly to governmental projects. However, in recent years emerged some businesses that saw potential prospects in nanotechnology applications and took up related projects.

One of the examples of the environmental nanotechnology projects would be the "Development of Small-Scale Environmental Measurement Devices Using Nanomaterial," which has been carried out by the National Institute of Environmental Studies. It is a project that intends to develop products and systems that monitor environmental pollution and improvement using nanomaterial and ultra fine fabrication technology. The micro electron beam source measures the concentration of contaminants in aerosol, and the x-ray source analyzes the content. Notable participants in this project from its inception in 2003 are Stanley Electric and Horiba.

The number of projects run by the National Institute of Environmental Studies was four in the beginning. It has grown to nine and had nine participating companies as of April 2007. Let's list the participants and the names of their development projects.

- "The Development of Cell-Phone Size Environmental Monitoring Device" by NTT.
- "The Development of Environmentally Responsive Artificial Tissues and Nano Structures for Response Detection" by Nippi.
- "The Development of Reusable Nanostructure Molecular Recognition Film that Easily Complements and Measures the Distance of Substance" by Toyo Ink MFG and Shimadzu.
- "The Development of Environmental Restoration Technologies Using Rapid Microbe Extraction Technique" by KRI.
- "The Development of a Toxic Chemical Compound Analysis System without the Use of Organic Solvent" by CellSeed.
- "Boron/Fluorine Removal Technology in Discharged Water Using Reverse Osmosis Membrane" by Toray.

Needless to say, there is no boundary in environmental pollution. The case in point is the global warming. The contribution to the global environmental improvement by one nation alone is not effective. Therefore, it is necessary to cooperate on the worldwide basis, including activities such as information sharing and joint research.

7. The Latest Nanotechnology Messages to the World in Nano Week

In 2002 the first Japan International Nanotechnology Exhibition, nano tech 2002 was held. The number of participating enterprises was 91 and that of visitors was 10,000. In the latest "nano tech 2007 International Nanotechnology Exhibition and Conference," 317 domestic and 167 foreign companies and organizations from 22 countries/regions participated, which suggested the phenomenon of rapidly growing worldwide nanotechnology R&D and commercialization. A total of 48,565 people visited the three-day event, which made the conference the world's largest (Fig. 14).

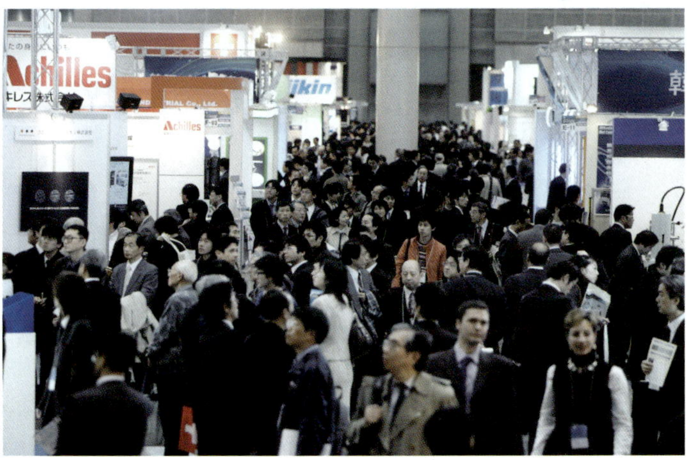

Fig. 14. Nanotech 2007 Exhibition site.

In fact, organizations from the following countries participated in nano tech 2007. A total of 99 companies from Europe came from the UK, Germany, France, Finland, Switzerland, Belgium, Russia, Denmark,

Sweden, the Netherlands, Italy, Spain, and Israel. Asian companies counted 25 from South Korea, Taiwan, China, Hong Kong, and Singapore. As many as 23 companies joined from Oceania, which included Australia and New Zealand, and 27 North American companies from Canada and the U.S. also participated.

Also, the week in which the International Nanotechnology Exhibition and Conference is held is called nano week. In 2007, for example, 20 major Japanese conferences to discuss nanotechnology-related issues were hosted simultaneously during that week. They included "JAPAN NANO 2007," "BIZMATCH @nano tech 2007," (Fig. 15) "Nano Bio International Symposium," "Workshop on International Nanotechnologies Standardization," "Nanotechnologies in France," "Nanotechnology Cooperation with UK: Successful Cases in Japan and South Korea," and others.

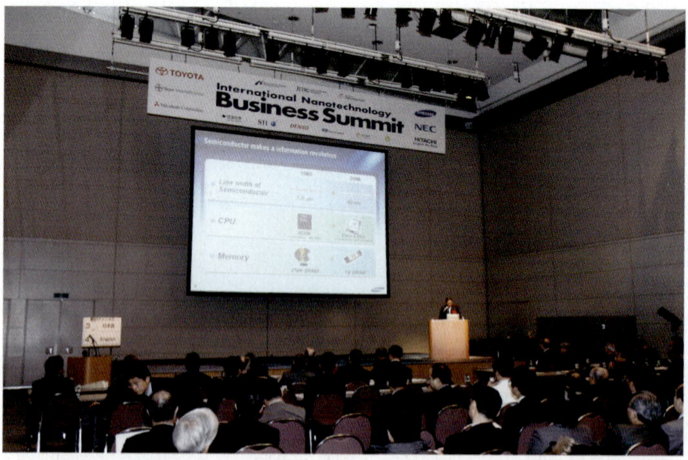

Fig. 15. BIZMATCH @nano tech 2007.

The conference has been valued as a place for information sharing, information launching, and networking through promoting interdisciplinary research exchange that goes beyond the conventional industry-government-academia framework. It is reported that there is a steady increase of participants whose objective is to attend some of the conferences. For them it is attractive and convenient that simultaneous conferences to discuss the comprehensive theme of the latest nanotechnologies from home and abroad are organized in one location.

The discussion of basic research and applied development were central in nano tech 2002, and the connection to business applications was rather weak. However, in recent years an increasing number of exhibitions begin to show concrete products and commercialization technologies and suggest potential for business applications. The participants are not only able to obtain the information of leading nanotechnologies of the world, but also find new business partners and discover many business opportunities. The nano tech 200X information can be found at website (www.nanotechexpo.jp/en/) and all previous event statistics and other details can be found at this site.

The evidence is seen from every corner, even from prize-winning examples of "nanotech grand prix". For example, Toshiba was awarded for the grand prix of nanotech 2006. Apparently, the key criterion of the prize winning was the appraisal of the comprehensive application of nanotechnologies. In particular, with the nanopatterned media that uses self organization, Toshiba overcame the noise, which had been the persistent issue of that time, and achieved the memory capacity of approximately 25 times that of perpendicular magnetic recording. In addition, Toshiba was able to demonstrate its comprehensive approach by exhibiting products, such as the refrigerator that was equipped with the deodorization and anti-bacterial functions by titanium-oxide photocatalyst and high-capacity NAND flash memory, which could become the enabler for the mass production of the current 56nm technology.

The winner in the biotechnology category was NanoCarrier with its drug delivery system (DDS) for anti-cancer medicine using polymer micelle. It would substantially reduce adverse side effects and increase beneficial medicinal effects, which were not feasible with the conventional drug delivery systems. The project is presently in the clinical trial.

The winners of the grand prix of nanotech 2007 were Fujitsu and the Fujitsu Laboratories. With the collaborative research with Tokyo University, they developed a quantum dot laser by self organization, which could lead to further applications such as the development of a quantum dot light amplifier and high-speed optical communication devices. The grand prix was awarded to them because of their effort to link the basic discovery to practical applications. The fact that a new company was established to promote the commercialization of the quantum dot devices, while the basic technology research continued, was also a big plus to win the grand prix prize.

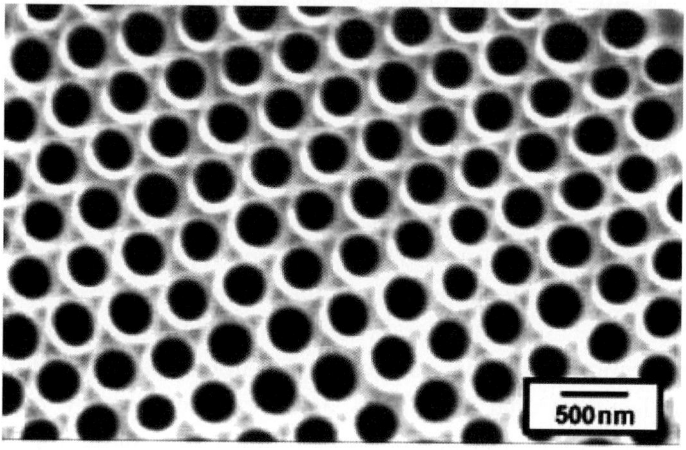

Fig. 16. Honeycomb-patterned Film.

Likewise, in the biotechnology category NOASTEC was the winner for its keen viewpoint of commercialization. The reason for the award was said to be that NOASTEC developed its technology to the point of possibility commercial production. A team under the leadership of Professor Masatsugu Shimomura (presently at Tohoku University) of Hokkaido University developed the honeycomb-patterned film (Fig. 16), and it became ready for commercial production by the research partner Fujifilm.

This honeycomb-patterned film regulated its surface pore diameter between 3μm and 10μm (technically feasible up to 50nm-80μm) by self organization. The film is expected to improve the safety of regenerative medicine in such applications as adhesion prevention film used in surgical procedures and scaffold material for cell growth. It is no doubt that the International Nanotechnology Exhibition and Conference in Japan will play an increasingly important role as a valuable opportunity to interface with leading-edge nanotechnologies.

Author Biography

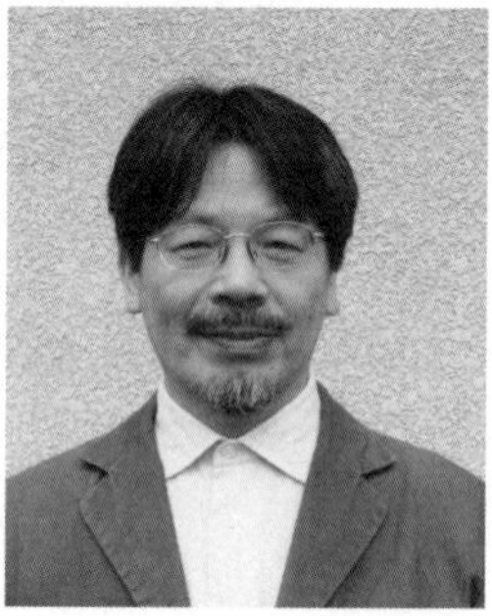

Mr Gimpei Sato was born in Miyagi Prefecture in 1951. He graduated from the Department of Humanities and Social Sciences, Faculty of Letters, Chuo University and received the Bachelor of Philosophy. Mr. Gimpei Sato is a technological writer of Nikkei Nanotechnology and Nikkei NanoBusiness Magazine in Japan. Mr Sato has been a free lance writer since 1985 working as an editor/writer of advertising and science magazine. He published four books in Japanese titled *'Illustration nanotechnology: Zukai Nanotechnology', 'Those who create it originally lives: Dokusosha Retsuden', 'The works series of chemistry Understand household electrical appliance'* and *'Kagaku-no-Hataraki series Kaden-seihin ga wakaru'*.

Major works in nanotechnology-related field

- Handai Frontier Research Organization "Web introductory course in nano-engineering learn" (http://www.fre.jp/)
- "Illustrated Guide of Nanotechnology" (Gijutsu-Hyohron Co., Ltd 2005)
- Writing articles as technical writer for "the Nikkei Nanotechnology" (from August 2004 to September 2005)
- Writing articles as technical writer for "the Nikkei nano-business" (from September 2005 to December 2006)
- Writing articles as technical writer for "the Nikkei Micro Devices", (from December 2006 to present)
- Writing articles as technical writer for "the Electronics seminars" (from December 2006 to present)
- Interviews and writing for "the NIMS NOW magazine" of the National Institute for Materials Science (from December 2006 to present)

Other Works

- "The future of Japan which e-community changes" (2003 cowrite NTT publishing Co., Ltd)

- "Dokusousha Retsuden-Biographies of creative persons" (2005 NTT publishing Co., Ltd)
- "Function of Chemical series — Understanding Household electrical appliance I" (2008 Tokyo Shoseki Co., Ltd)
- "Function of Chemical series — Understanding Household electrical appliance II" (2008 Tokyo Shoseki Co., Ltd)

CHAPTER 7

KOREA NANOTECHNOLOGY: POLICY, INFRASTRUCTURE, R&D AND COMMERCIALIZATION

Sang-Hee Suh

Center for Nanostructured Materials Technology, P. O. Box 130
Cheongryangri Seoul, Korea
shsuh@kist.re.kr

Kyung-Ho Kim

Korea Institute of Science and Technology Information 206-9
Cheongryangri-dong, Dongdaemun-gu Seoul, Korea
kimkho@kisti.re.kr

Korea is one of the several leading countries in the world actively developing and commercializing nanotechnologies. Korean government has increased its investment on nanotechnologies ever since its Nanotechnology Initiative was launched in 2001, with astounding accomplishments on every aspect of technologies. Now, Korea is in the 2nd phase of the Initiative, aiming to join the world top three nations on nanotechnology. The Initiative's vision and road map and the country's current status of nanotechnology development, commercialization and infrastructures will be given in the paper.

1. Introduction

Korea is aiming to lead the world in nanotechnology R&D, and related products and industries. At the heart of the Korean nanotechnology drive is the Korea Nanotechnology Initiative (KNNI), which was launched in 2001 with its first phase master plan, and the KNNI revised its master plan at the end of 2005. Figure 1 shows the milestones in Korea

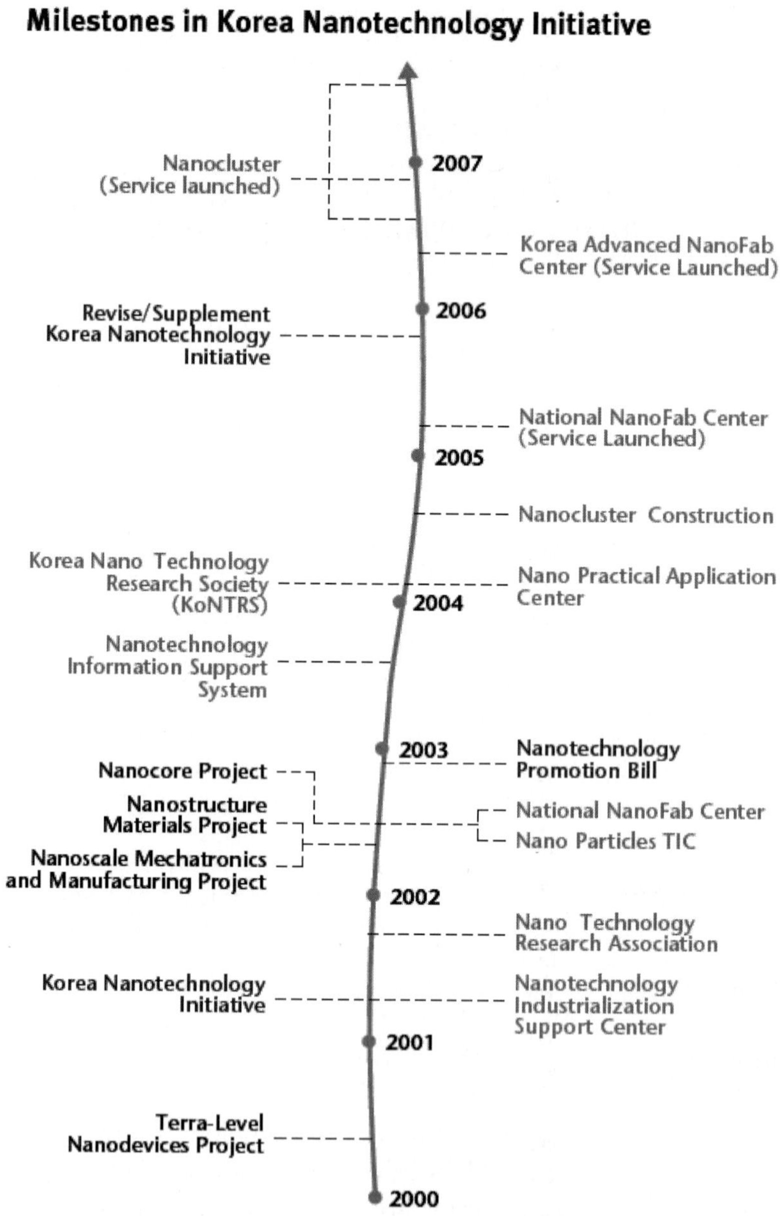

Fig. 1. Milestones of Korea Nanotechnology Initiative.

Nanotechnology Initiative[1]. According to the KNNI, Korea aims to join the world top three nations in global nanotechnological competitiveness by 2015. In order to coordinate and harmonize nanotechnology-related policies among the key Ministries, the Nanotechnology Coordination Committee is being operated by the Ministry of Education, Science and Technology.

Since 2001, Korean government has been increasing its investment on nanotechnology every year to 287 Million US Dollars in 2006. Number of papers published and patents filed has increased to more than three times during this time of period. Thanks to this aggressive investment by the government and focused efforts by researchers, Korea has become ranked 4th in nanotechnology competitiveness since 2005, as noted by Lux Research[2].

In Korea many nanotechnology-related research projects are being performed by all sectors concerned. At universities, various basic researches on nanodevices (CNT-FET, single-electron transistor, nano-bio device, etc.), nanoanalysis (scanning probe microscopy, atomic force microscopy, etc.), nanomaterials (nanoparticles, nanowires, nanotubes, and materials for energy and environment), and nanoprocessing technology (including top-down and bottom-up processing technologies) are being performed with support of government project funding. In the industrial sector, the research centers of Samsung Electronics and LG Electronics are exploring various nanotechnologies related to organic/inorganic nanomaterials, nanodevices, and nanoprocessing for next generation semiconductors and data storages, and energy storage. Nanoelectronic devices such as carbon nanotube-based transistors are being investigated as terabit memory devices, and research on next-generation information storage systems based on scanning probe microscopy and perpendicular magnetic recording is being conducted, aiming at a terabit storage density.

Various nanotechnologies have been commercialized. Samsung Electronics recently announced development of 30 nano 64 giga nand flash memory and has commercialized various nanoscale memory devices including 60 nano 8 giga bit nand flash memory. LG Life Science developed carbon nano balls which are being applied for many uses such as deodorants in refrigerators. ILJIN nanotech is producing carbon nanotube powders of various grades, and several application technologies such as metal matrix CNT composites and CNT transparent conductive films were commercialized by several companies. Many small/medium-sized companies have developed variety of consumer products such as cloths, soaps, and toothpastes, using nanosilver powders.

Since the first phase of the KNNI, five nanofabrication centers have been built across the nation to help universities, research institutions and companies develop nanotechnologies and nanoscale devices in various regions of the country.

2. Vision and Technology Road Map for the Second Phase KNNI[1]

2.1. *Vision*

On November 2005, Lux Research noted that Korea became the world's fourth-ranked Nanotechnology (NT) nation. Now Korea aims to become one of the world's top three nations in the nanotechnology field, securing sustainable growth potential with development of related new technologies and products. Korea is also pursuing a dominant market share of new technology in the world by facilitating synergy effect through fusion with Information Technology (IT), Biotechnology (BT), Environmental Technology (ET), etc. With this vision, Korea is aiming to realize a safe and affluent, environment-friendly society.

2.2. *Nanotechnology Development Roadmap*

Korea's roadmap for realizing its vision of becoming one of the world's top three nations in the nanotechnology field focuses on three principal areas: R&D, research infrastructure and manpower cultivation. The main strategy is to create synergies by facilitating innovations in existing fields of excellence such as IT field.

2.2.1. *R&D Effort*

During the first phase of the KNNI (2001~2005), Korean government launched three 21C Frontier Programs for nanotechnology development: National Program for Tera-Level Nanodevices, Center for Nanostructured Materials Project and Center for Nanoscale Mechatronics and Manufacturing. These three big national programs are playing as pivots for developing keen nanotechnologies in Korea.

The major R&D efforts in the 2nd phase (2006~2015) are divided into two stages. The first stage (2006~2010) pursues basic prerequisite activities that lay the groundwork for commercialization and advanced technologies in the second stage. During the first stage comparatively advantageous fields are being identified and expanded. Application technologies are being developed, and their industrial utilization is being pursed.

During the second stage (2011~2015) system level technologies by nanotechnologies will be identified and promoted. Commercialization of the first stage R&D results will be accelerated. As a result, strong technology base is expected to be built, leading to international cooperation.

2.2.2. *Research Infrastructure Development*

Research infrastructure development focuses on establishing efficient facilities to utilize R&D results as well as enhancing the ability of nurturing venture-type firms and related fusion technologies.

Infrastructure facilities will play as hubs of nanotechnology development, assisting early commercialization of nanotechnologies held by venture firms and small/medium size companies, providing IPs (Intellectual Properties) secured by infrastructure facilities as well as specialized technologies to promising enterprises or supporting business start-ups. Small scale function-oriented fabrication facilities are proposed to complement existing large scale infrastructure facilities to cope with NT-IT-BT-ET fusion technologies and disperse from integrated, comprehensive facilities to specialized fields.

2.2.3. *Manpower Cultivation*

Manpower cultivation for nanotechnology development covers the entire spectrum from elementary school through professionals already in the workforce.

Short-term plan for manpower cultivation is to operate intensive education programs for undergraduate/graduate school students and businessmen, conduct short-term re-education programs for elementary/ middle/high school teachers, develop education programs for special activities at elementary/middle/high schools and for general public, and activate nanotechnology re-education programs for industrial R&D manpower and designate/nurture nanotechnology education footholds at technical high schools and colleges for cultivation of engineers and a skilled workforce.

Mid- and long-term plan is to designate/operate nanotechnology education centers by foothold, assist nanotechnology departments at undergraduate schools and cooperative learning courses under a multi- and interdisciplinary system at graduate schools, support nanotechnology researchers (new/high-end manpower), and attract excellent foreign and Korean scientists from abroad to Korea and provide support for their R&D activities.

3. Statistical Data on Korea Nanotechnology Development[3]

3.1. *Government Investment*

Investment by Korean Government on nanotechnology development has increased steadily since the KNNI was launched in 2001. In 2006, it amounted to 287 Million USD, more than three times compared to that of Year 2001. Figure 2 shows Korea nanotechnology government investment during 2001-2006. In terms of investment per capita Korea is 3rd ranked just after Taiwan.

3.2. *Publications*

In terms of SCI publications, Korea ranked 8th in the world in 2001 with 408 papers, but jumped to 5th in 2004, 2005 and 2006. In 2006, Korea has 1,845 SCI publications, increase of more than three times since 2001 (see Fig. 3 for details).

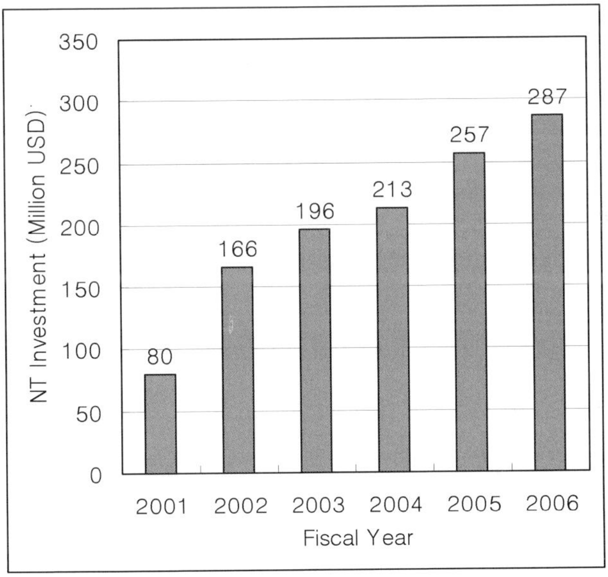

Fig. 2. Korean government investment on nanotechnologies.

3.3. *Patents*

Korea filed 80 nanotechnology related patent applications in countries around the world, ranking fifth in the world when the EU as a whole is included. In 2005, the number increased to 354, about three times increase compared to that of 2001. Figure 4 shows the number of patents filed during 2001-2005.

3.4. *Research Manpower*

In 2000, Korea's nanotechnology research manpower totaled around 1,000. By 2004, however, the research workforce had increased about four-fold to approximately 3,900. Nanotechnology-related departments at colleges also jumped about 13-fold to 38 in 2004 from just three in 2001. According to a nanotechnology R&D manpower demand projection, Korea will need about 20,000 specialists in 2015, implying an expected shortage of about 12,000.

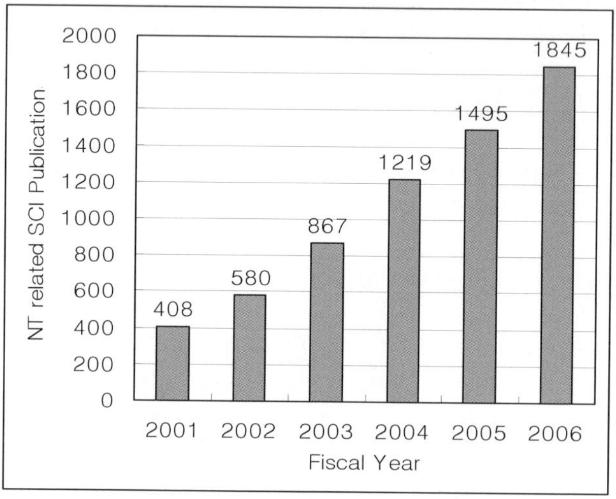

Fig. 3. Number of SCI publications on nanotechnologies by Korean researchers.

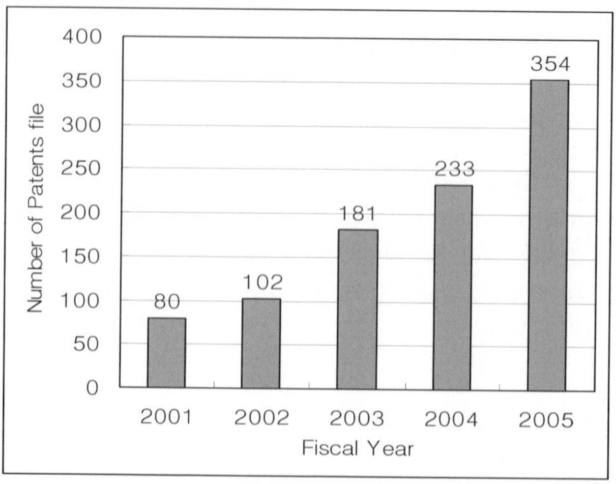

Fig. 4. Number of patents filed on nanotechnologies.

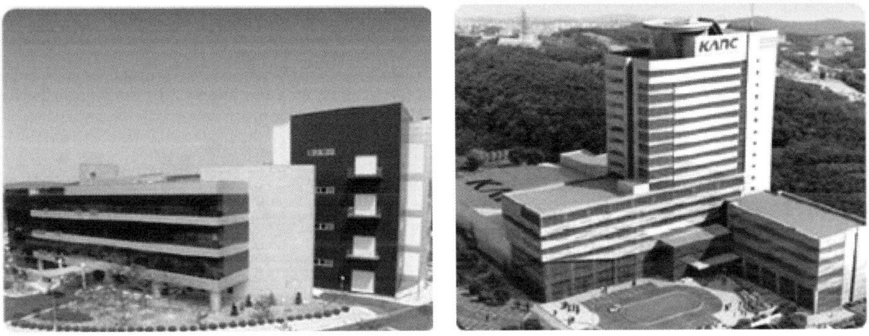

Fig. 5. Two integrated nanofab centers: National NanoFab Center (left) and Korea Advanced NanoFab Center (right).

3.5. *Infrastructure Establishment*

Before the establishment of the first-phase Korea Nanotechnology Initiative, there was no integrated nanotechnology-specialized facility at all in the country. Since the first-phase initiative, five nano support facilities including two integrated nanofab centers (Fig. 5), three nanoclusters) and related support facilities (Nano Practical Application Center, Nanotechnology Industrialization Support Center, etc.) have been established by the Ministry of Education, Science and Technology (MEST) and Ministry of Knowledge Economy (MKE). Table 1 summarizes the five major nanotech facility centers in Korea, while Fig. 6 shows the location of the facility centers.

Although aforementioned three nanoclusters have been built, taking into consideration of regional characteristics and functions, facilities with specialized functions (nano bio) has not been sufficiently supported yet.

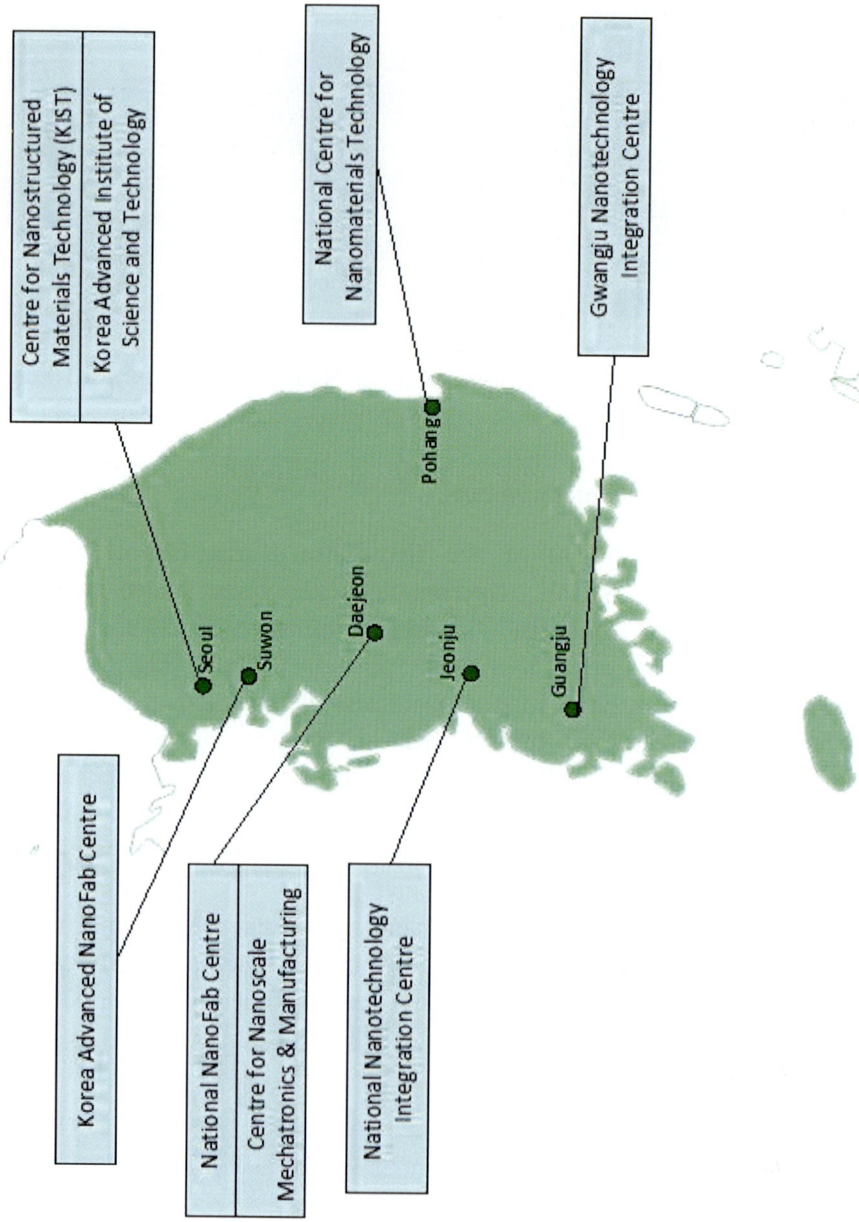

Fig. 6. Location of the main nanotechnology research institutes in South Korea.

Table 1. Infrastructures for nanotechnology in Korea

Organization	National NanoFab Center	Korea Advanced Nano Fab Center	National Center for Nano-materials Technology	National Nanotechnology Integration Center	Gwangju Nanotechnology Integration Center
NT Field	Silicon-based	Compound electronics/ photonic devices	Semiconductor/ display nano materials	Nano process and equipment	Nano process and equipment
Clean Room	$2,370m^2$	$3,368m^2$	$2,370m^2$	$2,346m^2$	$1,195m^2$
Facilities	8 inch CMOS package line	4~6 inch electronic device line	8 inch-based equipment	Establish module process-oriented equipment	Establish module process-oriented equipment
Service Start	Mar. 2005	May 2006	July 2006 (scheduled)	Dec. 2007 (scheduled)	Nov. 2006 (scheduled)
Location	Daejeon	Suwon	Pohang	Jeonju	Gwangju
Support Ministry	MEST	MEST	MOCIE	MKE	MKE
Webpage	http://www.nnfc.com	http://www.kanc.re.kr	http://www.nano.or.kr	http://www.nnic.re.kr	http://gnic.kitech.re.kr

4. R&D Activities by the Three 21C Frontier Programs for Nanotechnology Development

4.1. *National Program for Tera-Level Nanodevices*

The National Program for Tera-level Nano devices (http://www.nanotech.re.kr) was launched in 2000, aiming for innovative performance improvements in nanoelectronics through the development of new devices at the nano-level. They have been developing ultra fast, ultra highly integrated, and ultra low-power nano devices to overcome the technological and manufacturing limitations of semiconductor devices facing within 5 to 10 years. Among many astounding research outcomes are Charge Trap Flash (CTF) core technology for 32G Nand Flash Memory and application of neutral beam to Atomic Layer Etching (see Fig. 7).

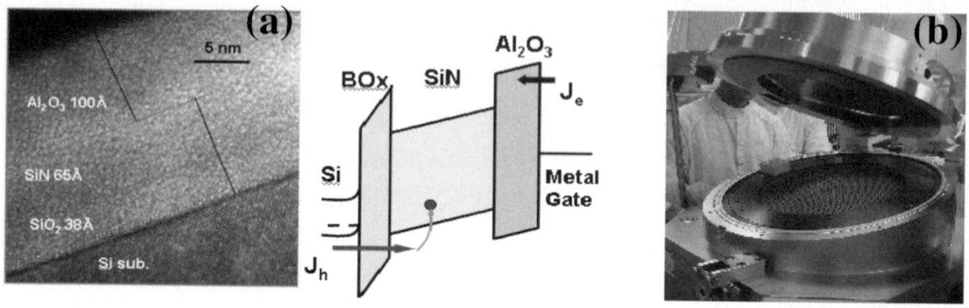

Fig. 7. (a) Cross section of Charge Trap Flash (CTF) Memory device and structure and (b) Atomic Layer Etching by neutral beam.

4.2. *Center for Nanoscale Mechatronics and Manufacturing*

Center for Nanoscale Mechatronics and Manufacturing (http://www.nanomecca.re.kr) is very committed and dedicated to developing nanomanufacturing processes and related equipments in order to produce commercial nanoproducts smarter, cheaper, faster, and environmentally friendlier. Its current achievements are as follows.

- Sub 50nm Nanoimprinting Lithography (NIL) and its application (Fig. 8(a))
  * Resistance RAMs on Si wafer and on flexible substrate
- Sub 50nm hybrid NIL equipment
  * 6 inch wafer, ultraviolet NIL, thermal NIL
- Transparent CNT film and its application (Fig. 8(b))
  * Touch screen, heater, LED
- Nanoinjection molding technology and its application (Fig. 8(c))
  * Next generation data storage, blu-ray disk
- Metrology and its standardization
  * Micro/Nano Dynamic Testing Machine and ellipsometry

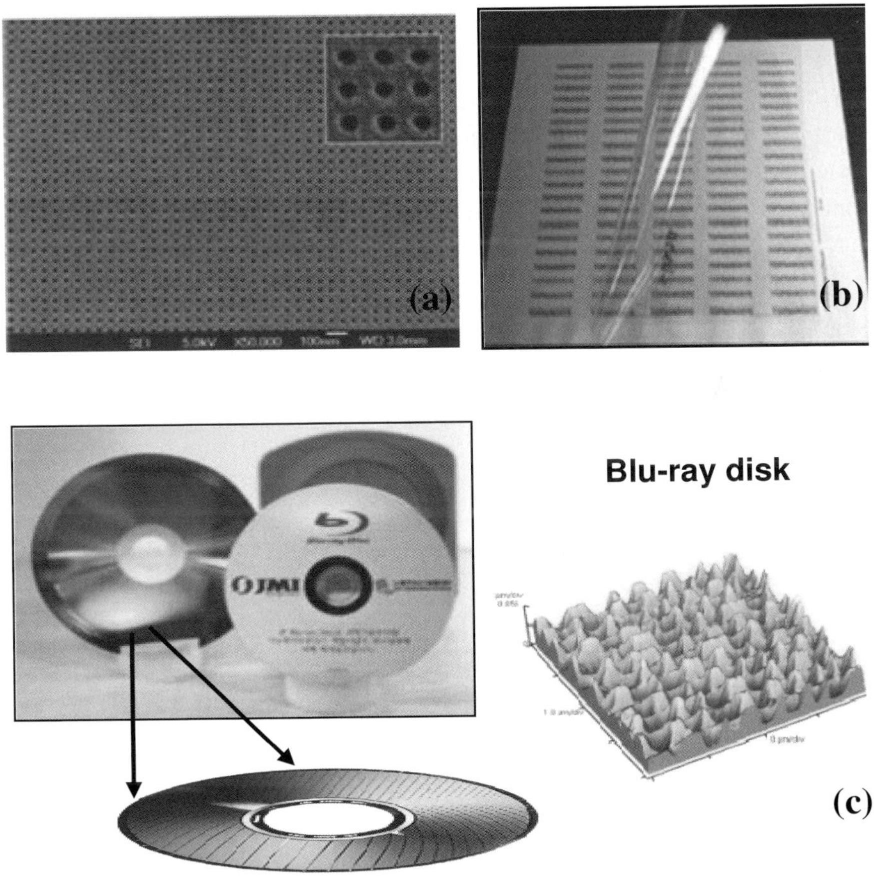

Fig. 8. (a) Pattern made by Nanoimprinting Lithography, (b) Conductive CNT thin film, and (c) Blue-ray disk with its surface morphology.

4.3. *Center for Nanostructured Materials Technology*

Center for Nanostructured Materials Technology (http://cnmt.kist.re.kr) has been managing one of the biggest programs for developing nanotechnologies in Korea. This program aims to develop various nanostructured materials with superior properties by creating new materials or by applying nanotechnology to the already existing materials. The R&D areas covered by the Program include nanostructured materials for structural applications, environment and energy applications, and information technology applications. The important accomplishments by the center include super hard cutting tool with nano-layered coating (Fig. 9(a)), carbon nanotube/metal matrix nanocomposite, amorphous structured thermal spray ceramic coating (Fig. 9(b)), UV-responsive coating (Fig. 9(c)), nanomembranes with dual functionalities of catalytic reaction and separation (Fig. 9(d)), nano-cathode materials for Li rechargeable batteries, single crystalline nanorods, and nanomaterials for artificial bones.

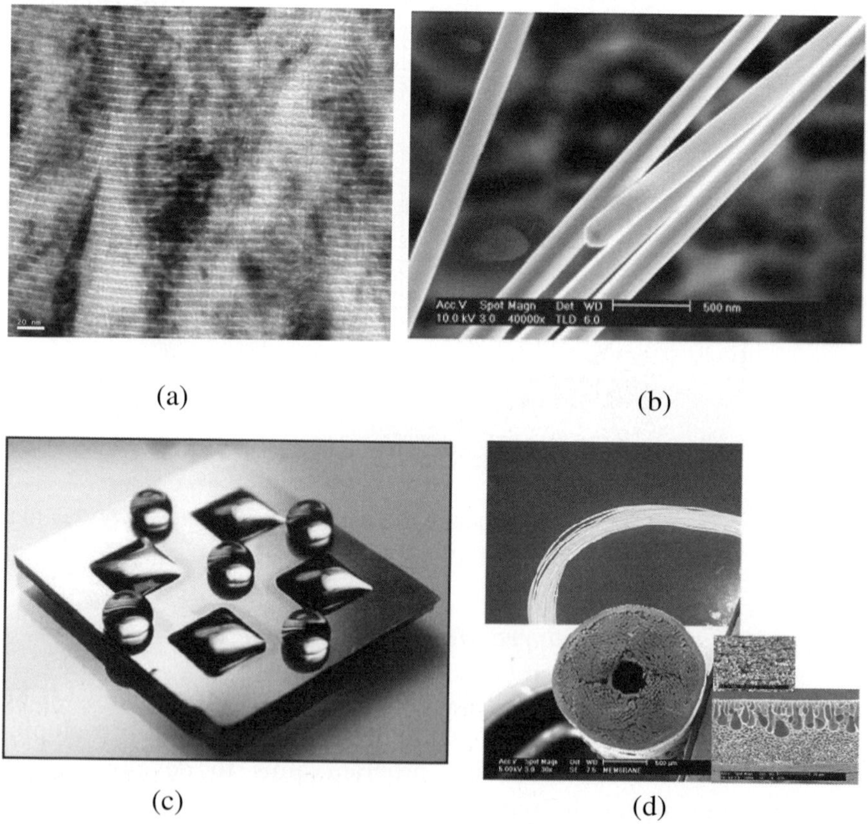

Fig. 9. (a) Cross-section TEM of Nano-layered coating on a cutting tool, (b) Ag nanorods, (c) UV-responsive coating, and (d) nanomembrane with catalytic reaction coating.

5. Highlights of Commercialization of Nanotechnology[3]

5.1. *Nano DRAM, NanoFlash Memory*

Samsung Electronics and Hynix have developed a series of nanometer scale memory devices. 80 nano 512 Mega bit DDR2 DRAM and 60 nano 8 giga bit nand flash memory are being mass produced by Samsung Electronics. 40 nano nand flash memory with CTF (Charge Trap Flash), 60 nano 2 giga bit oneNAND, and 40 nano SONOS (Si/Oxide/Nitride/Oxide/Si) devices have been developed by Samsung Electronics. Recently (October, 2007), Samung Electronics announced world first development of 30 nano 64 giga nand flash memory (see Fig. 10). Hynix also developed 60 nano 1 giga bit DDR2 DRAM with 800 MHz speed.

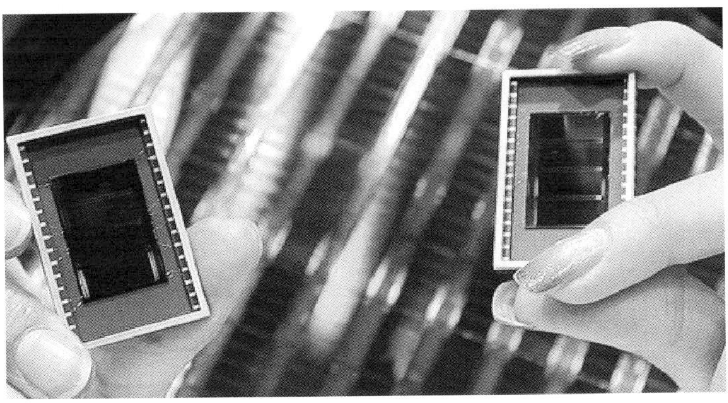

Fig. 10. 30 nano 64 giga nand flash memory by Samsung Electronics.

5.2. *Nanosilver Ink for RFID*

NPK (www.npk.co.kr) has developed conductive nanosilver ink which can be cured at temperatures lower than 100°C. Nanosilver ink is expected to be applied to PDP electrode, RFID antenna, EMI shielding, solar cell, and flexible display. InkTec (www.inktec.com) also developed nanosilver ink and applied it for RFID tag antennas (Fig. 11).

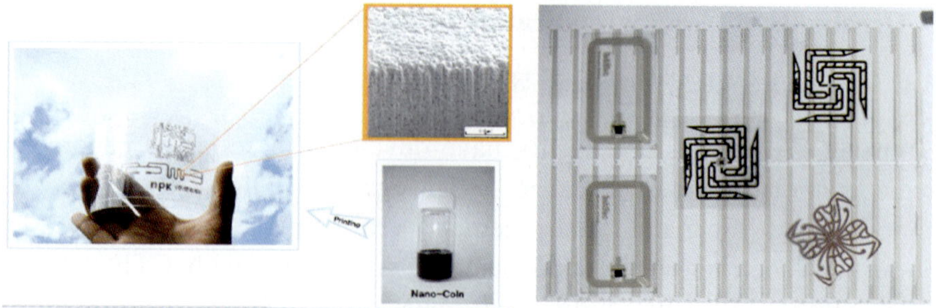

Fig. 11. Conductive nanosilver ink by NPK (left) and RFID using nanosilver ink by InkTec (right).

5.3. *Antiglare Coating*

Doosan Electro-Materials (www.dse.co.kr) developed an anti-glare film (Fig. 12) which prevents reflection when used in conjuction with a polarizing film. By controlling surface roughness on the nanoscale, reflectivity was reduced to half of the conventional films.

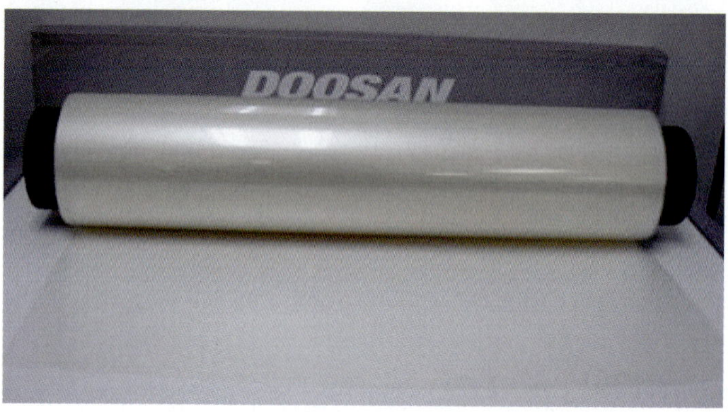

Fig. 12. Anti-glare film by Doosan Electro-Materials.

5.4. *Color Film with Nanolayered Structure*

SKC (www.skc.co.kr) developed a color film with several hundred layers of nano meter thick resin (Fig. 13). The color film shows different colors depending on the viewing direction. This color film would be used for packaging high end cosmetic or liquor bottles.

5.5. *CNT Transparent Film*

Topnanosys (www.topnanosys.com) developed a wide area transparent film using CNT touch pad screen panels (Fig. 13). This transparent electrode film is manufactured by coating several tens of nanometer thick CNT on plastic or glass substrates. Conductive and transparent CNT films are expected to replace high cost ITO (indium tin oxide) films for touch pad screen panels.

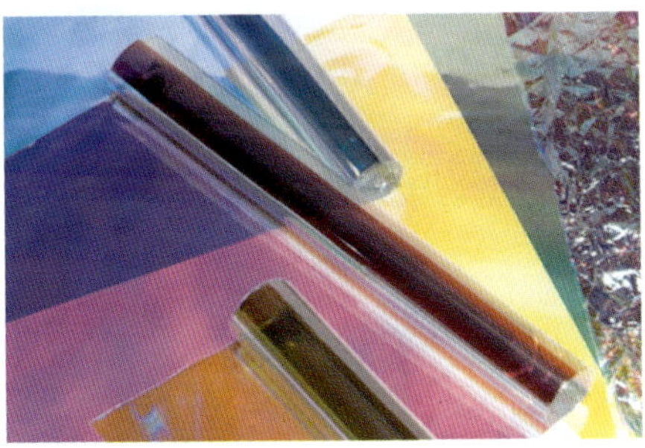
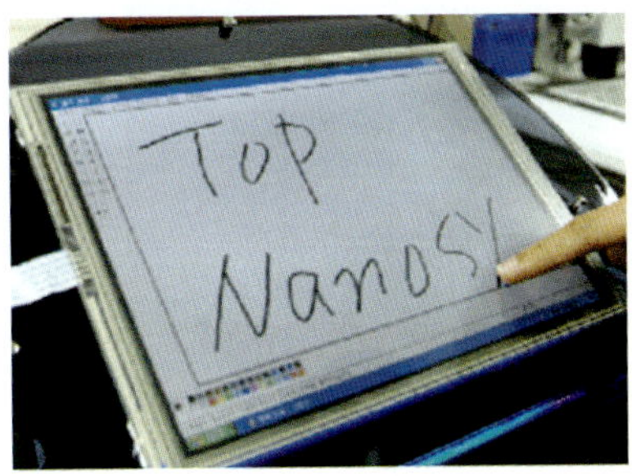

Fig. 13. Color film by SKC (left) and CNT transparent film by Topnanosys (right).

5.6. *Pharmaceutical Products for Atopic Dermatitis*

Nanohybrid (www.nanohybrid.com) succeeded in stabilizing Gamma-linolenic acid which is very effective in treating atopic dermatitis by using nanocapsule technology. Using this technology, the company introduced atopy cosmetics in the market (Fig. 14). A joint team of HanAll Pharmacutical (www.hanall.co.kr) and Korea Research Institute of Chemical Technology used nanoliposome technology to increase effectiveness of adenosylcobalamin which is know to be very effective in treating atopic dermatitis.

5.7. *Antibacterial Nanosilver Powders and Applications*

NPK (www.npk.co.kr) has developed antibacterial nanosilvers in aqueous and powder forms with 10 nm particle size. Being in aqueous and powder forms, nanosilvers could be used both for plastics and aqueous products.

NPTech and Saehan (www.saehan.co.kr) developed textile products coated with nanosilver particles. These textiles were proven to have very effective antibacterial property (Fig. 14).

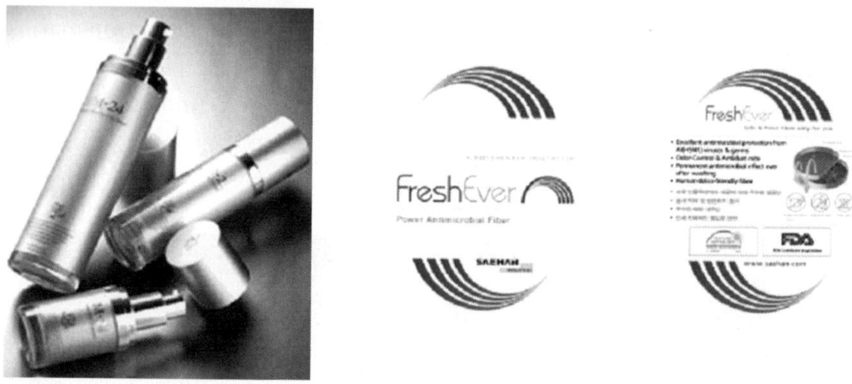

Fig. 14. Atopy cosmetics by Nanohybrid (left) and a textile coated with nanosiliver powders by Saehan.

6. Nanotechnology Societies in Korea

There are two major nanotechnology societies in Korea. Korea Nano Technology Research Society (http://www.kontrs.or.kr) supported by the MEST is the one mainly for networking among nanotechnology researchers not only within the Korea nanotechnology society but also with foreign countries. Other functions of the Korea Nano Technology Research Society are developing educating tools and text books; providing education for students and general public both by hosting intensive workshops, seminars and on-line classes; and setting up and updating plans for KNNI including technological roadmaps. The Nanotechnology Research Association (http://good.nanokorea.net) is the one mainly for supporting industries involved in nanotechnology development. It helps MKE set up a nanotechnology development plan, and it is now working as a PI of a nanotechnology program sponsored by MKE. Two societies have jointly hosted the Nano Korea Exhibition and Symposium which has been held every year since 2003. In 2007, 153 institutions and companies not only from Korea but also from foreign countries of USA, Japan, Netherlands, UK, and Germany attended the exhibition attracting 6,438 visitors (see Fig. 15).

Korea Institute of Science and Technology Information, or KISTI (http://www.nanonet.info) has been playing a signficant role of distributing information on nanotechnology including symposiums, public forums and domestic/international news. It has been publishing a weekly webmagazine, *Nano Weekly*, since 2003. It also publishes Korea Nanotechnology Annuals.

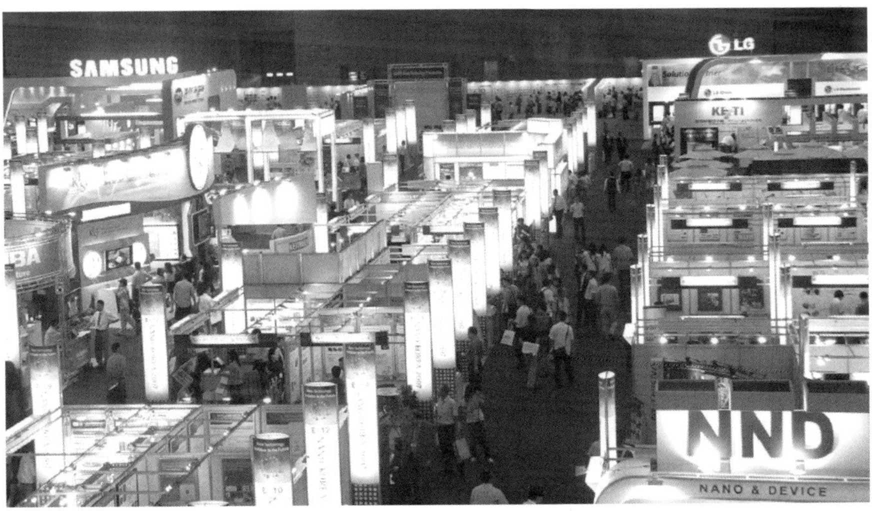

Fig. 15. Bird eye view of Nano Korea 2007.

Acknowledgements

Many works of KNNI activities including setting its master plan and publishing *Nano Weekly* have been supported by the Korea Ministry of Education, Science and Technology. Nano Korea Exhibitions and most of R&Ds on nanotechnologies have been supported by Korea Ministry of Education, Science and Technology and Ministry of Knowledge Economy.

References

1. Nanotechnology Korea published by Korea Ministry of Science and Technology, 2006.
2. http://www.luxresearchinc.com/press/RELEASE_NationsRanking2007.pdf.
3. Korea Nanotechnology Annual published by Korea Institute of Science and Technology Information, 2007.

Author Biographies

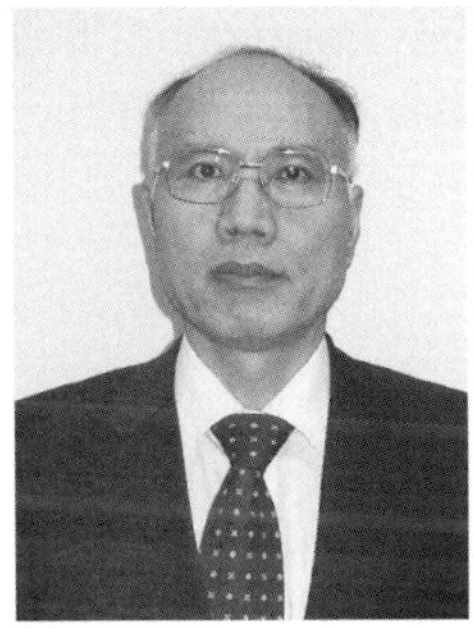

Dr Sang Hee Suh is currently the Director in the Center for Nanostructured Materials Technology, Seoul, Korea. He graduated from the department of Materials Science and Engineering in Northwestern University in 1982. His Research field and interests focuses on the Nanostructured Materials Technology, Compound Semiconducting Materials and Devices as well as Infrared Detectors and Materials. In recent years, Dr Suh received awards and honors including 1st Songgok Science and Technology award in Materials Engineering, Korea in 1998, KIST's best researcher of the year 2003 and 2006 Medal of Merit for Science and Technology by Korean Government. From 2006 to present, he has been the vice president in Korea Nano Technology Research Society.

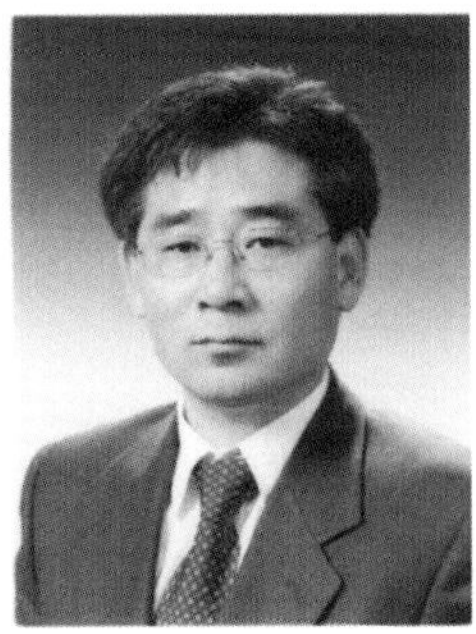

Kyung Ho Kim is a principal researcher and Director of the Nanotechnology Information Division at KISTI (Korea Institute of Science and Technology Information). He graduated from the Department of Chemical Engineering at Seoul National University in 1979 and received a master degree in Chemical Engineering from KAIST (Korea Advanced Institute of Science and Technology) in 1981. In mid-1990s, he partially completed a doctoral course in environmental engineering at University of Seoul. He spent three years as a researcher at KRICT (Korea Research Institute of Chemical Technology) in 1981-1984. He has been working at KISTI since 1984. He presently services as a program leader of the national nanotechnology information support system program. He has published lots of technical books, information analysis reports and review papers in various areas of technology including nanotechnology.

In recent years he is pursuing strong interest in worldwide trends of policies, R&D, and commercialization of nanotechnology as well as environmental, health and safety issues surrounding nanotechnology. He is a member of the Korea Nano Technology Research Society.

CHAPTER 8

NANOTECHNOLOGY IN MALAYSIA: A SMALL STEP TODAY TOWARDS A GIANT LEAP TOMORROW

Halimaton Hamdan

Academy of Sciences Malaysia, 902-4, Jalan Tun Ismail
50480 Kuala Lumpur, Malaysia
hali@kimia.fs.utm.my

Nanotechnology not only enables further technical and scientific advancement of existing technologies, rather creates new technology and imposes drastic impact on industries, economy and society in the coming decades. This means that investment in nanotechnologies is imperative for a developing nation like Malaysia in order to remain competitive and ensure a future growth industry in its own right. The National Nanotechnology Initiatives of Malaysia (NNIM) was officially launched by the Deputy Prime Minister in September 2006 with specific functions, strategies and future plans. The established research areas, research path and business opportunities in various areas of nanotechnology in Malaysia are proposed.

1. Introduction

Nanotechnology is defined as "the science of materials and systems with structures and components which display improved novel physical, chemical and biological properties; phenomena that exist in the nano size scale (1-100 nm)". A nanometer (nm) is one billionth of a meter. For comparison, a single human hair is about 80,000 nm wide.

People are interested at the nanoscale because it is at the scale that the properties of materials can be very different from those at a larger scale.

The difference in properties of materials at the nanoscale is due to two main reasons. First, nanomaterials have a relatively larger surface area when compared to the same mass of materials produced in a larger form. This can make materials more chemically active and affect their strength or electrical properties. Second, quantum effects can begin to dominate the behaviour of matter at the nanoscale particular at the lower end, affecting optical, electrical and magnetic behaviour of materials. The promise is of systems which are smaller, lighter, less expensive and requiring less power to run.

The beginning of the 21st century is remembered as the era in which nanotechnology flowered. New technologies, ideas and things emerge in profusion, all of which are destined to bring about big changes in our everyday lives. Nanotechnology is basic scientific knowledge recognized having big potential to bring benefits to new emerging technology and attracting rapidly increasing investments from governments and from industries in many parts of the world. The worldwide annual industrial production in the nanotech sectors is estimated to exceed USD1 trillion in 10~15 years from now, which would require about 2 million nanotechnology workers.

Nanotechnology not only enables further technical and scientific advancement of existing technologies, rather creates new technology and imposes drastic impact on industries, economy and society in the coming decades. Nanotechnology provides us the ability to manipulate matter at atomic and molecular scale leading to a wealth of innovative new technologies across a vast array of fields including agriculture, healthcare, information technology, energy production and utilization, homeland security and national defense, biotechnology, food and agriculture, aerospace, materials manufacturing and environmental improvement. Nanotechnology will transform and displace many of the products with better property, better functionality, energy saving, environmentally friendly and manufacturing processes that are the basis of our existing industries. This means that investment in nanotechnologies is imperative for a developing nation like Malaysia in order to remain competitive and ensure a future growth industry in its own right.

2. National Nanotechnology Initiatives of Malaysia (NNIM)

Malaysia's involvement in nanotechnology began in early 2000 with the establishment of a number of research centers at public universities. Malaysia officially joined the Asia Nano Forum (ANF) in 2004 after having participated in a number of Nanotechnology Meetings and Summit held in Japan and Australia. Following that in 2005, Ibnu Sina Institute UTM organized the Malaysian Nanotechnology Forum and at the same time hosted the ANF Meeting in conjunction with the official opening of the Institute. During that important occasion, the Deputy Prime Minister proposed the formation of the National Nanotechnology Initiatives.

A year later, the National Nanotechnology Initiatives of Malaysia (NNIM) was officially launched by the Deputy Prime Minister YAB Datuk Sri Najib Tun Razak on 19 September 2006 (see Fig. 1) with the mission: Nanotechnology for sustainable national development of science, technology, industry and economy. The functions of NNIM are to integrate all existing local nanotechnology activities, coordinate and plan the R&D activities, prepare a platform for commercialisation and transfer of new technology to generate economic return for the general public, develop educational resources, skilled labour, expertise and nanotechnology infrastructure and provide facilities and nanotechnology research support services.

The strategies of NNIM focus on:

(a) Improve Malaysian economic competitiveness to face global challenges,
(b) Accelerate scientific breakthrough on selective beneficial nanotechnologies,
(c) Enhance societal and environmental contribution.

Although Malaysian nanotechnology is still at infancy stage, the approval of NNIM by the Government would drive Malaysia towards forming the National Nanotechnology Centre (NNC). This would stimulate and accelerate the progress of the development of home grown nanotechnology into beneficial technologies.

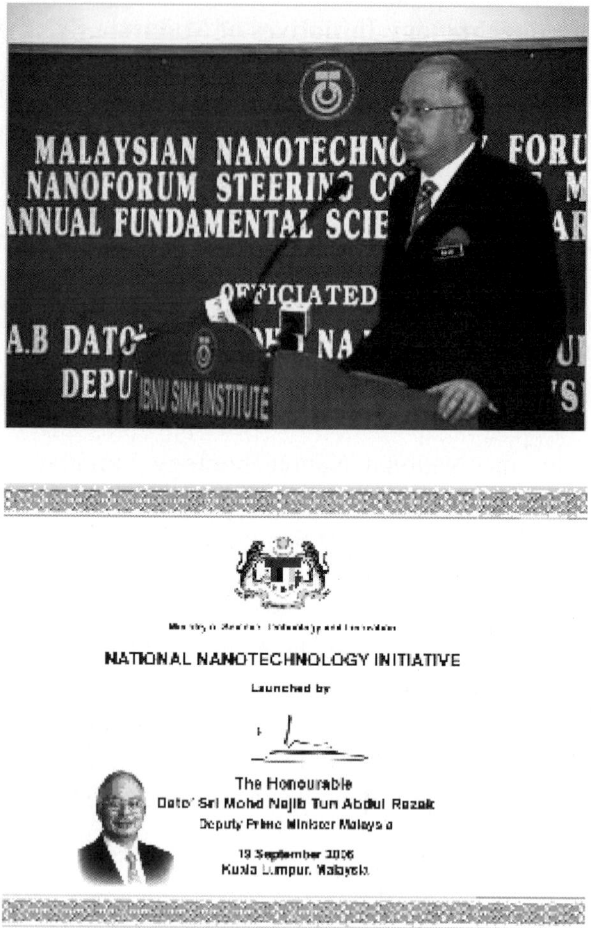

Fig. 1. Deputy Prime Minister at the MNF and ANF Forum at UTM (2005).

3. Current Development of NNIM

The Ministry of Science Technology and Innovation (MOSTI) is now entrusted to spearhead the planning and development of the NNIM. A number of actions implemented include the incorporation of nanotechnology as a national priority in the Ninth Malaysia Plan by the Cabinet, proposed establishment of National Nanotechnology Centre

(NNC), the Brain Gain Program for Nanotechnology and conducting a study on Business Opportunity and R&D in Nanotechnology.

The short term strategy of Malaysia is geared towards identifying researchers in various areas of nanotechnology with specific expertise; upgrading and equipping nanotechnology laboratories with state-of-the-art facilities; and to prepare a comprehensive human resource development programme for producing nanotechnologists. Malaysia has for decades trained scientists capable of contributing to the national development in science and technology (S&T), where some pioneering work in nanotechnology were initiated since the Seventh Malaysia Plan (7MP) in the late 1990's. At the end of Eighth Malaysia Plan, Ministry of Science, Technology and Innovation (MOSTI) has awarded more than RM1 billion to S&T research projects.

National Budget (2006-2010)

- Enhancing the effectiveness of government financial management, efficiency and competitiveness
- Developing human capital as a catalyst for growth
- Ensuring the well-being of the population by improving their quality of life
- Accelerating the shift to a higher value-added economy

In his 2006 Budget Speech, the Prime Minister of Malaysia, YAB Datuk Sri Abdullah Ahmad Badawi among other words stated that: "To further strengthen and diversify the sources of economic growth, the Government will intensify its efforts to encourage the private sector to venture into new areas with high growth potential and competitive edge. These include modern methods for agriculture, biotechnology, nanotechnology, high-technology manufacturing as well as services, especially ICT, education and tourism". His speech reiterated the need to intensify the pursuance of new areas of technology with a high growth potential such as nanotechnology. Hence, the inclusion of nanotechnology as a priority area under Intensified Research Prioritized Areas (IRPA) for 8&9 MP is timely, and is poised to position the country in the long term to nurture a nanoscience research culture among researchers and develop world class nanotechnology laboratories in Malaysia. The federal investment for nanotechnology research is expected to increase significantly, following an increase in allocation for R&D in the Ninth Malaysia Plan (2006-2010).

Under the current National Science and Technology Policy (NSTP), nanotechnology is included in the strategy of building competence for specialization in key emerging technologies, and has been identified as a key technology area to support the local industry. In order to move forward, the industry participation is strengthened by encouraging the private sector to be directly involved in the development of nanotechnology. The government is encouraging local industries, including government linked companies (GLCs) to use local R&D products or processes. In addition, the gross expenditure on R&D (GERD) and number of scientists to population will be gradually increased from the current 0.5% to 1.5% and 18 to 60 per 10,000 respectively by year 2010. The underlying objective is to enhance commercialization of local products and increase the critical mass in the focus areas.

Some recent developments of Malaysia in nanotechnology are summarized below:

- Establishment of well-equipped nanoscience/nanotechnology research centres, for example: Zeolite and Nanostructured Materials Research at Ibnu Sina Institute for Fundamental Science Studies (IIS), Universiti Teknologi Malaysia; Nanoelectronic Devices Research at Institute of Microengineering and Nanotechnology (IMEN), Universiti Kebangsaan Malaysia; Nanomanufacturing at Advanced Materials Research Centre (AMREC) of SIRIM Bhd; and Nanomaterials and catalysts at the Nanotechnology Research Centre, University Malaya.

- Increased number of postgraduates in nanoscience/advanced materials. The government has introduced the National Science Fellowship (NSF) scheme, which is open to postgraduate studies in nanoscience and technology. There are more than 300 graduate students in the country actively pursuing research in nanotechnology.
- Commendable number of journal publications, patents and awards.
- Organization of national seminars on nanoscience and nanotechnology, for example, Symposium on Nanotechnology by MOSTI (2003), Nanotechnology R&D Infrastructure Workshop (2007), Palm Oil International Congress (PIPOC), Electron Microscopy Conferences, Advanced Technology Congress, Green Chemistry and Malaysia Nanotechnology 2007.
- Categorization of nanotechnology as a priority area under IRPA of 8 Malaysia Plan (MP), where RM1 billion is available to IRPA under 8MP; and 9MP where a total of 2.5 billion is allocated.
- Establishment of the Malaysian Technical Nanotechnology Committee ISO229 under SIRIM.
- Active involvement with Asia Nano Forum (ANF) and collaborations with international nanotechnology research organizations by participating in Nanotechnology Exhibition and Conferences in Japan (2003-2007), Australia (2004-2005), Thailand (2004), Italy (2005), Hong Kong (2006), Taiwan (2007) and China (2005).

4. Nanotechnology R&D Activities

The potential benefits of nanoscience and technology are pervasive, as illustrated in the burst of interest and effort worldwide in several fields outlined below: materials and manufacturing; nanoelectronics and computer technology; medicine and health; aeronautics and space exploration; environment and energy; biotechnology and agriculture; national security; and science and education.

There are growing number of groups actively involved in nanotechnology R&D in Malaysia, majority of which are from the Institute of Higher Learning and Government Research Institutes. The committed support from the government, seen in the increased R&D funding to MOSTI for nanotechnology R&D, enabled the growth of a significant number of research centers pursuing nanoscience and technology.

4.1. *Material and Manufacturing*

One of the major areas pursued by Malaysian scientists, in tandem with governmental funding through research and educational sponsorship by the MOSTI is the generation of nanomaterials. Table 1 summarizes some of the major effort undertaken by various research groups in Malaysia and Fig. 2 shows several achievements in nanomaterials R&D.

4.2. *Nanoelectronic and Computer Technology*

This area of nanotechnology is the main driver of the field. It is embedded into the current production strategy of the industry and is a major single source of innovation. It is thus important for the country for Malaysia to pursue some research activities with international visibility. Table 2 lists several major projects in electronics and communications, while Fig. 3 illustrates the R&D achievements in nanoelectronics.

4.3. *Life Sciences/Medicine and Health*

Living systems are governed by molecular static and dynamical properties at nanometer scales, where the disciplines of chemistry, physics, biology, and computer simulation all now converge. There is one area of development in life-sciences with Malaysia's active participation: the design and synthesis of biologically functional nanostructures by genetically modified living systems. Table 3 tabulates the major projects in medicine and health.

Table 1. Examples of projects in nanostructured materials and sensors

Areas	Applications	Head/Institute
Metal Industry	Metal Composites	AMREC UTM UiTM
Chemical industries Pharmaceuticals Nanomaterials Medical Renewable Energy	Catalysts, Adsorbent Aerogel, Mesoporous materials, Zeolites, CNT, DDS, Membrane	UM, UTM,UPM USM
Nanoelectronics	Chemical sensor SET Nanodevices	UTP, UKM UNIMAP, UTM MMU

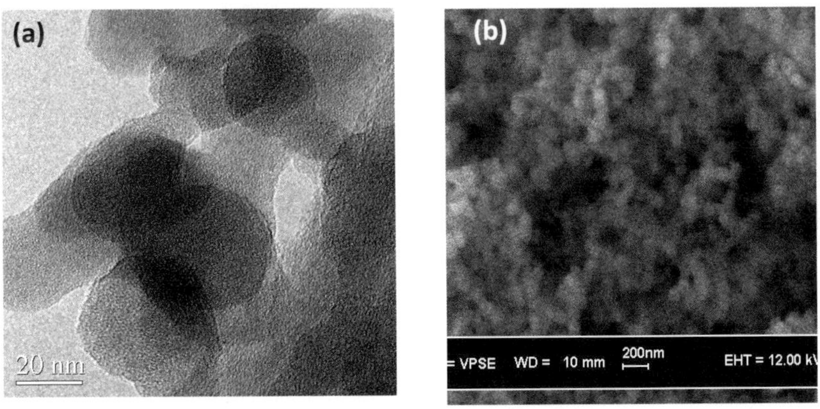

Fig. 2. (a) FESEM image of nanozeolite by UTM and (b) TEM image of Maerogel: Silica aerogel from rice husk by UTM.

Table 2. Examples of projects in electronics and communications

Areas	Applications	Head/Institute
MEMS and Microsensors/ Organic Electronics	Automotive	IMEN UKM, UPM, VLSI
Electronics	GMR	AMREC
Electronics	Blue Light Emitting Devices	USM
Electronics	Advanced optical crystal for elctro-optic application	UTM, UM, UPM

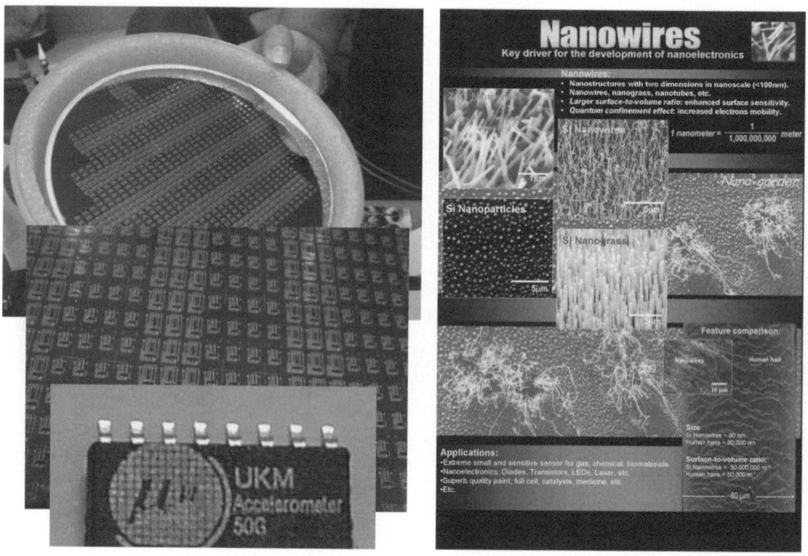

Fig. 2 (continued)

Table 3. Examples of projects in medicine and health

Areas	Applications	Head/Institute
Nanomedicine	Biopharmaceutical proteins for human therapeutics drugs and vaccines	UPM
Nanomaterial	Bone graft substitutes	MINT, SIRIM, USM, UKM, UIA
Nanodevice Molecular Nanotechnology	Diagnostic kits for infectious diseases	USM

5. Establishment of National Nanotechnology Centre (NNC)

With the establishment of NNC, a management team will be appointed to administer and plan the National Nanotechnology Initiative. NNC shall be entrusted to drive the NNI, coordinate national R&D in nanotechnology, strengthen present nanotechnology research centres to become National Nanotechnology Research Centre (NNRC) with state of the art research facility and liaise with industries to address business and economic agenda and develop international networking. Detailed functions and coordination activities of NNC Working Committees as illustrated in Fig. 1 are as follows:

- Form the organizational structure of NNC
- Draft a National Nanotechnology Policy
- Formulate financial implication strategies for R&D, education and national nanotechnology management grants
- Establish national nanotechnology niche areas and conduct nanotechnology foresight exercises
- Manage nanotechnology research and development activities
- Provide, enhance and monitor the national nanotechnology infrastructure and research facilities
- Training and human resource development

- Update the national nanotechnology database
- Create and strategize national nanotechnology education programme
- Assist the national nanotechnology commercialization and investment activities
- Extend the national and international collaboration and networking to develop local human capital and expertise in nanotechnology
- Monitor the potential health, environmental and societal impacts of nanotechnology
- Formulate nanotechnology standards and specifications
- Manage nanotechnology Intellectual Properties and legal matters
- Undertake commercialization and industrial collaboration activities

5.1. *National Nanotechnology Research Centres (NNRC)*

In Malaysia, a series of internationally-renowned research centres in nanotechnology exist in the present key areas of technological relevance: nanoelectronics; nanobiotechnology; and nanomaterial science (see Fig. 3). The NNRCs are Zeolite and Nanostructured Material Research at Ibnu Sina Institute for Fundamental Science Studies (IIS), Universiti Teknologi Malaysia; Nanoelectronics and Devices Research at Institute of Nanoelectronics and Microengineering (INEM), Universiti Kebangsaan Malaysia; Nanomanufacturing Research at Advanced Materials Research Centre (AMREC) of SIRIM Bhd; and Catalyst and Nanomaterials at Nanotechnology Research Center, Universiti Malaya.

The NNRCs will be given full support to develop nanotechnology research, facilities, laboratory space and networking at the national and international level.

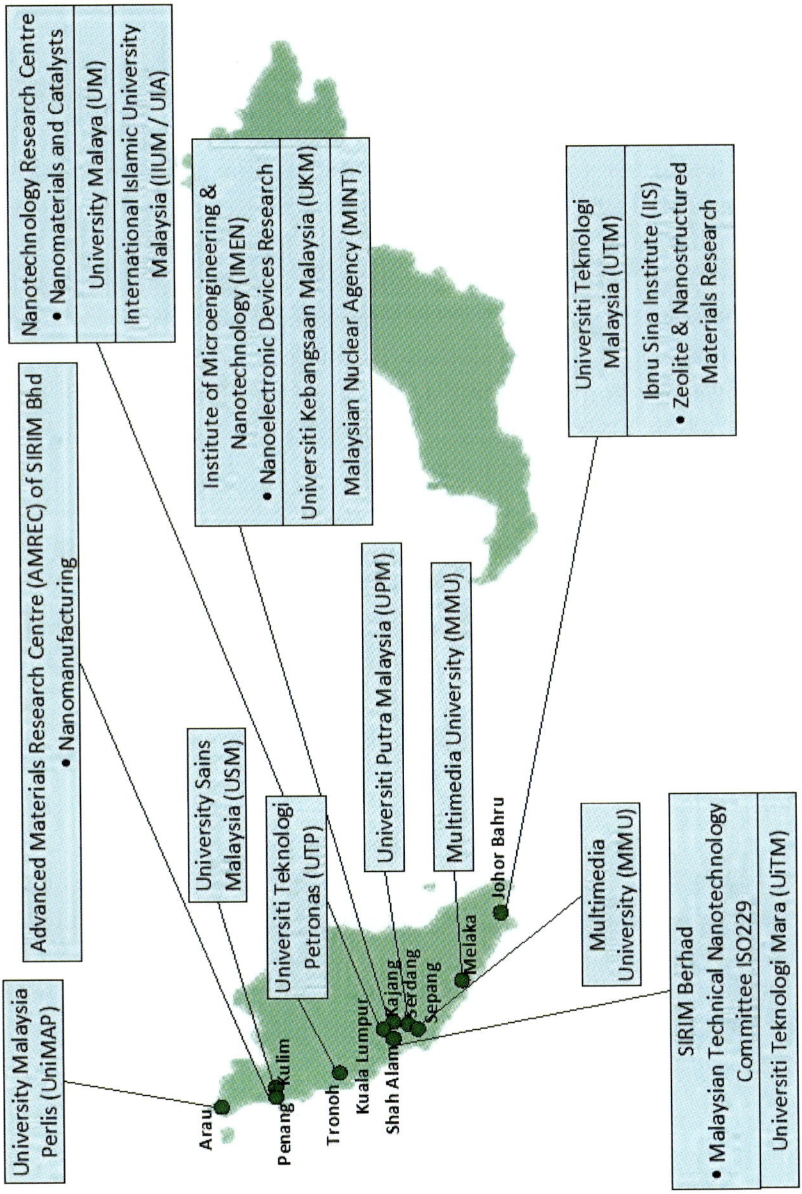

Fig. 3. Location of the main nanotechnology research institutes in Malaysia.

5.2. *Action Plans of NNC*

With the establishment of NNC, a management team will be appointed to administer and plan the National Nanotechnology Initiative. NNC shall be entrusted to drive the NNIM, coordinate national R&D in nanotechnology, strengthen present nanotechnology research centres to become National Nanotechnology Research Centre (NNRC) with state of the art research facilities and liaise with industries to address business and economic agenda and develop international networking. A long list of activities will be implemented in phases in order to ensure that adoption, adaptation and innovation of nanotechnology is a gradual process such that transformation and dissemination of the technology is well infused to the Malaysian way of life.

The key actions are implemented in phases to ensure that actions taken are reviewed and evaluated. The phases are Short Term (5 years), Medium Term (5 to 15 years) and Long Term (above 15 years) action plans. The detailed activities and action plans proposed for each term is shown in Fig. 4.

5.2.1. *Short Term Action Plans*

Action I: Identify and Strengthen Existing Nanotechnology Research Centres to become National Nanotechnology Research Centres (NNRC)

In Malaysia, a series of internationally-renowned research centres in nanotechnology exist in the present key areas of technological relevance: nanoelectronics; nanobiotechnology; and nanomaterial science. These research centres will be identified by the NNC and given full support to develop nanotechnology research and activities at the national level.

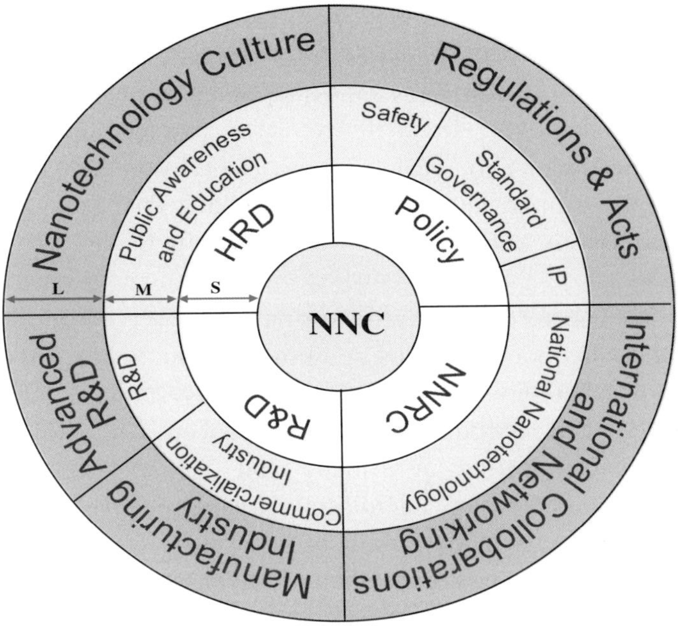

Fig. 4. Detailed activities and action plans proposed for short-term, medium-term and long-term for NNC.

Action II: Identify and Embark on Strategic Research
The NNRCs should actively pursue 3 categories of research namely:

(i) New fundamental research that will lead to breakthrough in nanotechnology,
(ii) Commercialization of research with emphasis of Malaysian primary industry and;
(iii) Advancement in existing manufacturing into nanomanufacturing and industrial applications

The outcome of the nanotechnology foresight and niche areas identification studies will prioritize the nanotechnology areas for the country.

Action III: Integration of Technical and Human Resource Infrastructure

Most crucial recipe to the success of the R&D is the human resource factor. In summary NNC requires internationally-competitive level of technical and human resources infrastructure to warrant success in the R&D.

Recommendations to create a knowledge based society that will generate interest in advance scientific research especially in the field of nanotechnology are as follows:

(i) Promote Greater Public Awareness on the Impact of Nanotechnology in Future Economy, Societal and Environment. Successful campaign in nanotechnology will generate similar interest by the public.
(ii) Develop and introduce nanoscale concepts into mathematics, sciences engineering in Primary and Tertiary Education. It is imperative to develop interdisciplinary perspective to students to enhance their disciplinary skill.
(iii) Educate and train new generation of scientists and supporting teams in nanotechnology which include societal implications.
(iv) Inculcate the concept of mobility within nanotechnology research community. The infrastructure, quality and efficiency of NNRC must be enhanced to attract researchers from diverse expertise to perform prioritized nanotechnology research.
(v) Integration and collaboration among institutes of higher learning and between institutes of higher learning and the industry to combine resource and expertise.
(vi) International strategic partnerships must be intensified by pursuing collaborative research programmes, with shared responsibilities and resources, and fair distribution of scientific and economic revenues.

5.2.2. *Medium Term Action Plans*

Action I: Establish National Nanotechnology Laboratory

Action II: Enhance Nanotechnology Research in the Proposed Areas

Action III: Review Malaysia Regulatory Frameworks (Governance of Nanotechnology) to ensure that Innovation in Nanotechnology is regulated and controlled.

It is recommended that the NNC takes the following stand without jeopardising the research and development and application of nanotechnology:

Action IV: Create Public Awareness and Interest in Nanotechnology

It is recommended that NNC adopt the following approach in propagating the importance of nanotechnology:

(i) Educate the public through media, publication, journal and forum.
(ii) Emphasize on the heath, safety and environmental implications of nanotechnology to the public.

5.2.3. *Long Term Action Plan*

Action I: Inculcation of Nanotechnology Culture to the Society

Action II: Commercialization and Industrialization of Nanotechnology

Action III: Advanced Level Nanotechnology R&D, Integration of Technical and Human Resource Development

6. Nanotechnology Industry

Current local businesses in nanotechnology are mostly at a trading and distributing level. Companies involved acquired basic nanotechnology-based products mainly nanosilver from countries such as South Korea and Japan for redistribution. Although there are companies claiming their products contain nanotechnology elements, their claims cannot be verified. Furthermore, some companies are reluctant to provide current information or future plans citing the matter as their "trade secret".

There are a number of products of R&D in nanotechnology being developed for commercialization. The following is a list of nanotechnology-based companies:

(1) Nanopac (M) Pte. Ltd. - Nanosolution for airconditioning
(2) Usains Holdings Pte. Ltd. - Carbon Nanotubes
(3) ICE Pte. Ltd. Industriatech Corp. Ltd. - Maerogel insulator, mesoporous materials
(4) Unitechnologies Pte. Ltd. - Nanostructured Catalysts, nanomembranes
(5) UPM Holdings Pte. Ltd. - Nanocomposites, nanobiofertilizers
(6) Malaysian Agri HI-Tech Co. Ltd. - Nanobiofertilizers, Nanocomposites
(7) UKM Technologies Pte. Ltd. - Nanobiosensors
(8) Pakar Management Tech. Pte. Ltd. - BioFil for treatment of industrial and sewage effluent
(9) Silterra Malaysia Pte. Ltd. - Fabrication of nanoparticles
    (1) Sime Tyre Pte. Ltd. - Fabrication of nanocomposites
    (2) Malaysian Palm Oil Board - Nanocatalysts

7. Enhancement of Nanotechnology Research Activities

By end of 8 MP, Malaysia has spent more than RM160 million on various areas of nanotechnology related research projects. Currently, nanotechnology research is conducted by separate research groups in various local public universities and government research institutions. Major areas pursued by Malaysian scientists are molecular manipulation and generation of nanomaterials, nanoelectronics and living systems. One area of life-sciences in which local nanoscientists are active in is the design and synthesis of biologically functional nanostructures by genetically modified living systems. Research in nanotechnology is governed by molecular static and dynamical properties at nanometer scales, where the disciplines of chemistry, physics, biology and computer simulation all now converge. Therefore, the implementation and management of research in nanotechnology needs to be coordinated, organised and interdisciplinary. To be an effective technology watcher and obtain key information from the international research community in particular, the government needs to sponsor some nanotechnology research on topics of recognized significance.

A study on Business Opportunity and R&D in Nanotechnology in Malaysia; conducted recently, concluded that the areas of research in Nanotechnology need to be focused, enhanced and sustained. Based on the availability of resources, needs, infrastructure and technical strength, it is proposed that nanotechnology research in Malaysia, in the long term, should focus on the following areas:

(1) Solar Cells - Photovoltaic and Dye-sensitized
(2) Lithium ion batteries
(3) Plant Vaccines
(4) Nano Biodevices - lectin chips/glycol chips and environment management using DNA chip analysis
(5) Nano Biosensors - nanoagricultural diagnostics with nanomaterials
(6) Drug Delivery Systems - liposome technology and nano-encapsulation for food and drug delivery

8. Conclusion

Looking to the near future, it is necessary for Malaysia to introduce and create awareness among the public by organizing a programme in nanotechnology. Celebrating fifty years of Malaysia's independence, the first convention on nanotechnology called the Nanotech Malaysia 2007 is organised, supported by the Academy of Sciences Malaysia, Ministry of Science Technology and Innovation and Ministry of Defense. In conjunction with Nanotech Malaysia 2007, Malaysia is also hosting the fourth Asia Nano Forum Summit. This important event will strive to introduce, demonstrate and explain the worldwide investment in nanotechnology to the public and demonstrate the nation's commitment towards embracing the new paradigm of nanotechnology. Malaysia's active participation as a member of the Asia Nano Forum (ANF) is a way forward and a good effort of widening her networking, creating research collaborations and exploiting business opportunities at the international level.

Advances in nanoscience and nanotechnology promise to have major impacts on people in the coming decades. Among the expected breakthroughs are order of magnitude increase in device efficiency and the emergence of entirely new phenomena in physics, chemistry and biology. Thus, many exciting prospects for new frontier science knowledge, product's quality and technological innovations will soon be available and implemented.

For a matured nation celebrating her 50 years of independence, nanotechnology will almost certainly affect the lives of her growing population and have a major impact on health, environment and defense to name a few. Malaysia must not miss the vital opportunities nanotechnology has to offer nor be caught off-guard by new threats.

References

1. H. Hamdan, Nanotechnology in Malaysia: A New Paradigm, British Malaysian Chamber of Commerce Directory, pg 83 (2007)
2. MIGHT, Identification of Business and R&D Opportunities in the Application of Nanotechnology in Malaysia (2007)
3. S. Fritz (ed) Understanding Nanotechnology. From the editors of Scientific American, Warner Books, New York (2002)
4. M. Wilson, K. Kannagara, G. Smith, M. Simmons, B. Raguse, Nanotechnology basic science and emerging terchnologies, Chapman & Hall/CRC, New York (2004)
5. R. Jones, Soft Machines Nanotechnology and Life, Oxford University Press, UK (2004)

Author Biography

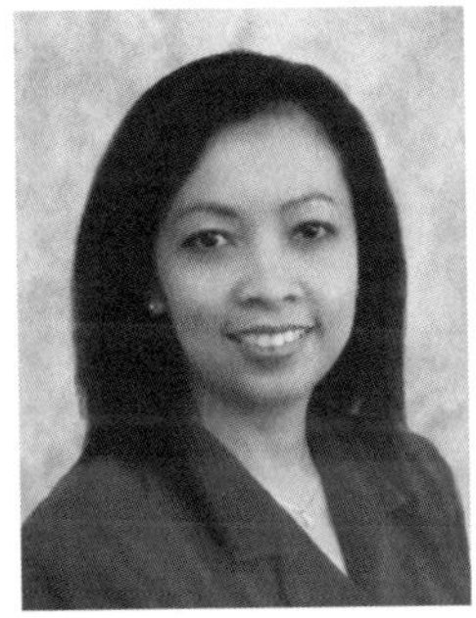

Halimaton Hamdan is a Professor at Universiti Teknologi Malaysia with a PhD from University of Cambridge, UK. Dr Halimaton (Hali) pioneered the Zeolite and Nanostructured Materials Research Group and synthesized Maerogel; the lightest nanomaterial from rice husk waste. Her current research is on new generation of hybrid, chiral, bifunctional and functionalised heterogeneous catalysts, drug delivery system and nanostructured materials. Hali is a Fellow of ASM and a committee member of the National Nanotechnology Initiatives and Asia Nano Forum (ANF). She has authored over a hundred of journal publications, six patents and won several research awards. Maerogel has been nominated as product of the year by International Clean Energy Circle UK. She received the TechnoFund Award of RM6.1 million from the Ministry of Science Technology and Innovation to build an exploratory pilot plant for the production of Maerogel. She won the IFIA Cup Award for women inventor 2008 from the International Federation Inventors Association and her Biphased Nanostructured Catalyst (BNC) was coveted The World Invention Award at the British Invention Show (BIS2007).

CHAPTER 9

THE MACDIARMID INSTITUTE AND NANOTECHNOLOGY RESEARCH IN NEW ZEALAND

Paul T. Callaghan

MacDiarmid Institute for Advanced Materials and Nanotechnology
School of Chemical and Physical Sciences
Victoria University of Wellington, Wellington, New Zealand
paul.callaghan@vuw.ac.nz

Richard J. Blaikie

MacDiarmid Institute for Advanced Materials and Nanotechnology
Department of Electrical and Computer Engineering
University of Canterbury, Christchurch, New Zealand
r.blaikie@elec.canterbury.ac.nz

An overview of nanotechnology research in New Zealand is given, with particular emphasis on the work of the MacDiarmid Institute, a multi-university and government laboratory partnership.

1. Introduction

In 2001 two apparently unrelated, but ultimately connected, events took place. First, Alan MacDiarmid made a nationwide tour of New Zealand, subsequent to his 2000 award of the Nobel Prize in Chemistry, an award shared with Alan Heeger and Hideki Shirakawa for their co-discovery of conducting polymers (see Fig. 1). Alan greatly raised the enthusiasm of New Zealanders, and in particular the Government and its advisors, for excellence in research in general and excellence in basic science in particular. He further drew attention to the remarkable developments taking place in advanced materials research worldwide. Second, the New

Zealand Government announced that it intended to fund a small number of Centres of Research Excellence in New Zealand universities. These Centres were not necessarily to be located in a single place. Indeed universities were encouraged to consider partnerships that would group together critical masses of excellence in specific fields of research.

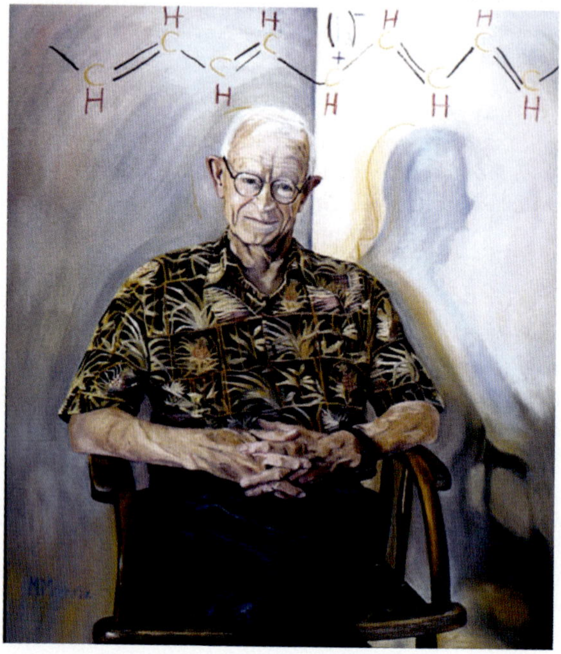

Fig. 1. Portrait of Nobel Laureate, Alan MacDiarmid, by New Zealand painter Marianne Muggeridge, and commissioned by Royal Society of New Zealand.

New Zealand has always had a vibrant condensed matter physics and materials chemistry community and it was natural that applications would emerge from this group. A coordinated bid took the name, *The MacDiarmid Institute for Advanced Materials and Nanotechnology*. We were successful, being awarded an operating grant of NZ$4.7 million per annum over six years, along with a capital start up grant of around NZ$10 million. The operating grant effectively doubled the external income of the group.

In February 2007 Alan died at his home in Philadelphia, while preparing to board a plane to New Zealand to attend the 3rd International Conference on Advanced Materials and Nanotechnology, to be held in Wellington. He did not live to see the next stage of development of the Institute that was to take place in June 2007, the awarding of continued funding at NZ$6.5 million *per annum* through to 2014, and the award of a second injection of NZ$10 million in capital funds. To celebrate Alan's achievements, the Institute has produced a 42 minute movie entitled "Super Plastics Man", the story of Alan's life. This can be viewed at, or downloaded from, the website[(1)].

This chapter gives an overview of nanotechnology research in New Zealand, but with particular emphasis on the MacDiarmid. We do not claim to present a complete picture of New Zealand nanotechnology research, indeed there are many researchers scattered around the country, who may classify their research as nanotechnology or nanoscience. But it is fair to say that the Institute is the only large and coherent grouping of nanotechnology and nanoscience researchers who have a critical mass of quality capital equipment, and who operate in a spirit of partnership across institutions, sharing equipment and ideas.

The MacDiarmid Institute is a partnership between Victoria University of Wellington (VUW), The University of Canterbury (UC), Otago University (UO), Massey University (MU), Industrial Research Ltd (IRL) and GNS Science (GNS). It comprises over 40 Principal Investigators (PIs), 30 Postdoctoral Fellows, 80 PhD students and 20 Masters students, at 7 Institutions, 5 universities (the above group along with The University of Auckland), and the 2 Crown Research Institutes (CRIs). Leadership is provided through the Founding Director (Paul Callaghan at VUW), the Director (Richard Blaikie at UC) and the Deputy Director (Shaun Hendy at IRL). The research is grouped into five Themes, Nanoengineered Materials and Devices, Novel Electronic, Electro-optic and Superconducting Materials, Conducting Polymers, Soft Materials and Advanced Materials. These will soon be supplemented by a sixth Theme in the area of bio-nanotechnology. Details about the research objectives and our PIs can be found on the Website: www.macdiarmid.ac.nz.

2. Overview Profile of New Zealand Nanotechnology Research

The major nanotechnology groupings in New Zealand are, in addition to the MacDiarmid Institute and its CRI partners, the industrial materials research centres at the University of Auckland, and the Biopolymer Network (a joint venture between research institutes at Ag Research, Crop and Food Research, and Scion). Much of the research is basic, or investigator led, and there is growing engagement from biological and agricultural research teams. There is a strong emphasis on the synthesis of nanostructured materials, and the development of nanoscale devices.

Focusing on "non-core" MacDiarmid activities, IRL also has nanoscience research that is not formally part of the MacDiarmid Institute, such as on functionalised surfaces and multi-scale modelling. The industrial materials research centres at the University of Auckland (the Centre for Advanced Composite materials, the Polymer Electronics Research Centre (recently partnered with the MacDiarmid Institute), the Research Centre for Surface and Materials Science, Advanced Materials Research Centre and the Light Metals Research Centre (also teamed with MacDiarmid), have long been active in areas closely related to nanoscience. The Biopolymer Network grouping seeks develop bio-based materials for industrial use, with emphasis on the use of fibres and other materials derived from plants and wool.

There has been some commercial spin-out from these activities, most notably the MacDiarmid Institute related companies, Nano Cluster Devices Ltd with self-assembling assembling nanowires technology developed out of research from the University of Canterbury, BioImprint (University of Canterbury) which is developing novel lithographic and imaging techniques and Anzode Ltd, with new battery electrode technology, from Massey University. Also from Massey University is PolyBatics Ltd which is seeking to develop a nanoparticle platform technology based on bacterial storage granules while Australo Ltd is a spin-out from the University of Otago, which is developing analytical equipment that incorporates nanotechnology.

The location of these major nanotechnology research centres is illustrated in Fig. 2.

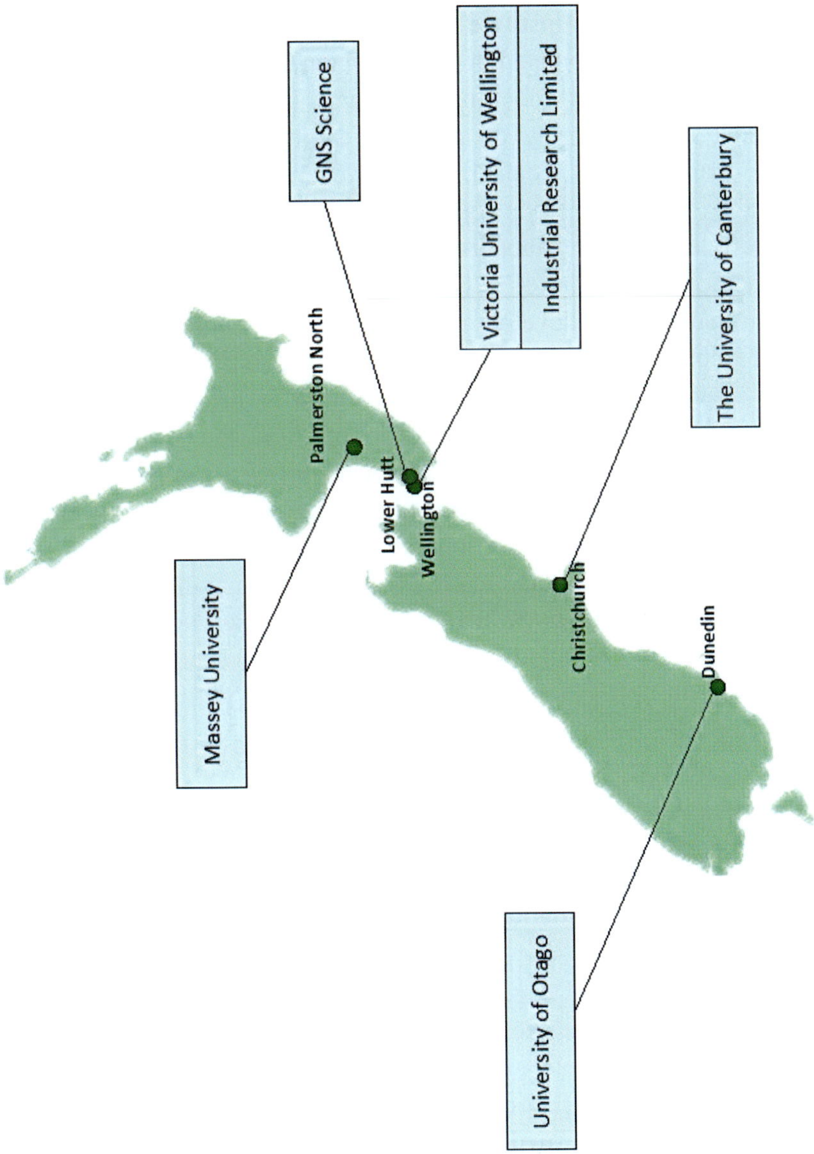

Fig. 2. Location of the main nanotechnology research institutes in New Zealand.

The MacDiarmid Institute has also commissioned social research in New Zealand associated with nanoscience and nanotechnologies at The Agribusiness and Economics Research Unit at Lincoln. This group has undertaken a survey of public at attitudes to nanotechnology[(2)], finding that the New Zealand public generally views nanotechnology favourably, but that there is some aversion to products where people can be directly exposed to nanoparticles. Nanotechnology was not far behind biotechnology in terms of general assessments of risk, and there was concern that medical developments from nanotechnology would be used for self improvement or human enhancement rather than for treatment to alleviate suffering. A major recommendation is that the public needs to be informed about areas of concern, for both risks and benefits.

In 2006, the New Zealand Ministry of Research, Science and Technology (MoRST) commissioned a "Road Map" on Nanotechnology research in the New Zealand context[(3)]. This document summarises the state of research today, and where New Zealand may have opportunities in the future. It needs to be noted immediately that New Zealand's *per capita* investment (population 4.2 million) in nanotechnology research is small, with an upper estimate of this "difficult to define" activity being given by MoRST as NZ$18.0 million per annum, but with a more likely figure at around NZ$8.0 million per annum. World-wide advanced country (e.g. OECD) investment levels are roughly US $3.0 *per capita* of population. This means that the investment level in New Zealand is running at around 1/3 to 1/2 that *per capita* rate. Again, in that context, the MacDiarmid Institute represents a significant component of the total New Zealand investment.

The MoRST RoadMap document gives a useful, if government-oriented, perspective on New Zealand nanotechnologv research (Fig. 3). In many ways it presents a confusing picture by failing to identify those components that form part of a coherent partnership, preferring to break the picture down into small institutional parts. Furthermore, the report has not clearly advanced an effective way forward for nanotechnology research, either through suggesting new funding initiatives or strengthened partnership. Instead it has provided actions and directives for other agencies (including funding agencies) to consider its findings, such as a directive that from 2010 the nanoscience and nanotechnology communities should change their focus towards being more relevant to New Zealand's *existing* industries; these are largely agricultural and biologically-based, so this directive will arguably reduce opportunities for new, disruptive technology developments to take place here. In addition, there are recommendations for future research to include studies of societal or environmental impacts, without clearly identifying mechanisms for funding this additional work.

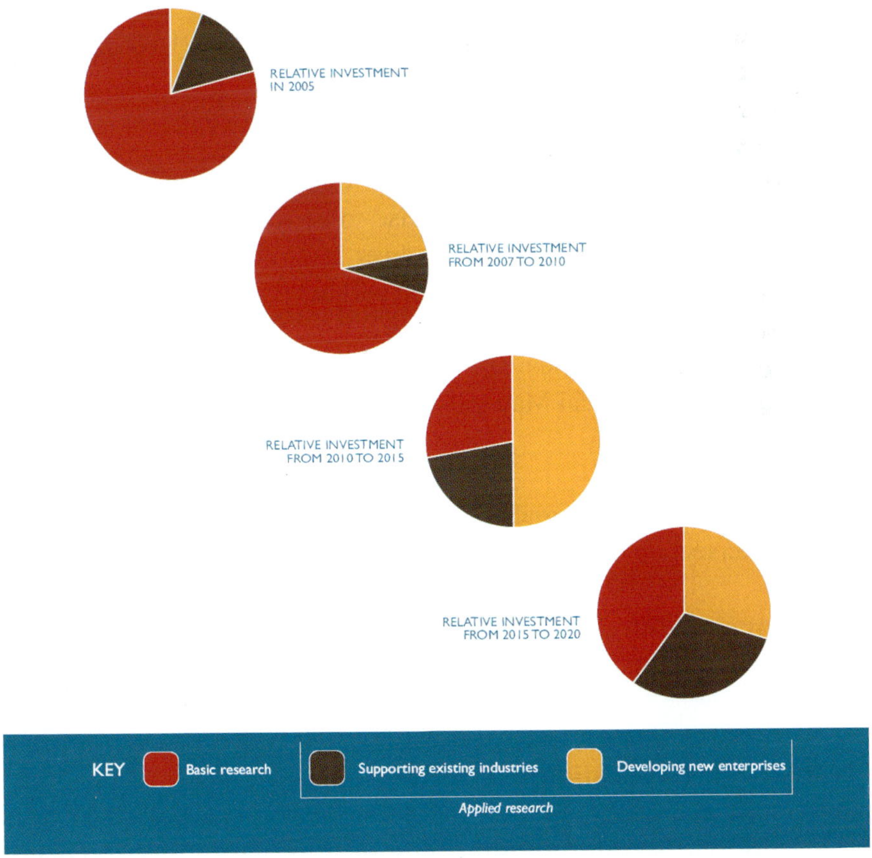

Fig. 3. Changes in future nanotechnology investment strategy proposed in the 2007 MoRST Nanosciences and Nanotechnology Roadmap[3].

The first funding rounds are currently underway where these roadmap messages will be considered, so we are in the midst of an interesting experiment here. Ironically, the agency that has been most successful in funding nanotechnology research, and in assisting nanotechnology partnership, has not been MoRST, but the New Zealand Ministry of Education and its Centres of Research Excellence funding agency, the Tertiary Education Commission. It is also ironic that much of the recent Ministry of Research Science and Technology interest in nanotechnology has been around ethical issues and public perceptions, while the Ministry of Education has made possible the funding of the actual science.

3. The MacDiarmid Institute Themes

A detailed description of our existing objectives can be found at http://macdiarmid.ac.nz/themes.php. Also to be found are a complete set of publications coming from our work over 2002-2006[4]. From 2008, the six Themes of the MacDiarmid Institute will be:

(1) Nanofabrication and Devices
(2) Electronic and Optical Materials
(3) Molecular Materials
(4) Soft Materials
(5) Inorganic Hybrid Materials
(6) The Intersection of Nanoscience and Biology

Just those activities specifically related to nanotechnology and nanoscience are detailed here, with the emphasis being on future directions.

Theme 1: Nanofabrication and Devices

Whilst nanotechnology is tremendously diverse, nanofabrication is the key for engineering nanostructures since it allows precise control over device dimensions and properties. In this theme we explore new approaches in areas of optical nanolithography and cluster-based self-assembly.

(i) Optical Nanopatterning: controlling light on the nano-scale using near-field and/or surface plasmon (SP) effects is of great interest, and we have contributed to the development of new new-field optical nanolithography techniques. Nanometer-scale structures showing phenomena such as 'negative refraction' and/or 'superlensing' are now under intense scrutiny both theoretically and experimentally; our research will extend understanding of superlensing in layered silver films that are predicted to show enhanced resolution, and explore further practical implications of plasmon-assisted near-field lithography alongside collaborators at MIT.

(ii) Cluster-based Assembly and Devices: The above lithography-based approach to nanofabrication is complemented by "bottom up" assembly of nano-devices using metal clusters as building blocks. Percolation, template, stencil, and no-lift-off lithography based cluster assembly techniques have been developed and led to the formation of NZ's first nanotechnology company, Nano Cluster Devices Ltd. The focus of ongoing work will be the production of reliable transistor structures, and the formation of cluster-assembled devices that are predicted to exhibit fractal conductance fluctuations.

(iii) New Devices and Applications: Our fabrication expertise gives us the ability to branch out into device and applications areas, many of which have strong links into other Themes. For example, combining our fabrication capability with expertise in laser physics and ultra-sensitive measurement will allow us to construct optical biosensors based on precision high-Q microcavities. Such systems have applications in environmental monitoring and health-related industries as well as providing platforms for fundamental science.

Field-emission devices will also be investigated using the ion-beam capabilities at GNS Science. In addition, the materials effort in Theme 2 will provide impetus for new nanoelectronic devices. We will study the design of low-dimensional semiconductor structures with the goal of simultaneously reducing thermal conductivity and electrical resistivity, building on our previous experience with photovoltaic energy conversion, and fabrication and analysis of quantum-confined structures.

(iv) Theory and Modeling: Molecular dynamics simulations have provided critical support to our cluster-based device development and will focus on complex behaviours and instabilities near the melting point. Our theoretical insight will be applied to design devices that utilise unconventional properties of charge carriers arising from size quantisation, correlations, and spin-dependent effects.

Theme 2: Electronic and Optical Materials

Researchers in this theme are recognised for their contributions to research in high-T_c superconductors, strongly correlated electron systems, semiconducting and metallic nitrides, glass ceramics, conducting polymers and surface enhanced Raman scattering. This work extends the classic independent-electron solutions, so useful in discussing crystalline semiconductors and metals, to the electronic and optical properties of (i) more complex electron systems, where electron correlation is a critical facet, and (ii) nanometer-scaled macro structures, including superlattices, fibres, carbon nanotubes, low-dimensional substructures, and quantum dots. Following a demonstration of vibrational pumping in Surface Enhanced Raman Scattering (SERS), we will expand on this by investigating SERS for more complex multiple-particle nanostructure geometries. In other optical-materials work we will extend our study of glass-ceramics to ordered 2d crystal–in-glass arrays with photonic band-gaps, produced by photo-induced generation of metallic nanoparticle nuclei.

Theme 3: Molecular Materials

Molecular materials find diverse applications and researchers in this theme have expertise in many of these, including: molecular magnets, solar energy and electroluminescent materials, functional surfaces and supramolecular assemblies. Our challenge now is to further develop molecules and use our knowledge in the construction of larger (>100 nm) assemblies whether on a surface or as a three-dimensional structure. These larger ordered systems are critical to success in a number of applications including solar cells, organic light emitting diodes (OLEDs), sensors and magnets.

(i) Functional Materials: We have made a strong contribution to the design and synthesis of new compounds including functionalized porphyrins and thiophene oligomers, self-assembly and templating methods to create single molecule magnets, and coordination chemistry to improve charge flow in electronic materials such as OLEDs. In making further progress design challenges need to be met. We will investigate the problem of systems photodegradation in solar cells and OLED systems. In single molecule magnets we will undertake a systematic study of structure-magnetic behaviour as a function of changes to the ligand employed.

(ii) Functional Surfaces: The success of devices fabricated from molecular materials lies in an understanding of electrode surfaces and control of substrate positioning. We have considerable experience in these areas and will study of the electrochemically-controlled nucleation and growth of conducting phases.

(iii) Supramolecular Arrays: The conduction of charges in a spatially controlled fashion is a desirable property in electroactive materials. We aim to form conducting wires that are microns long and nanometres in diameter (3–4 nm) by using discotic liquid crystal forming materials that will self-organise within polymer nanotubes.

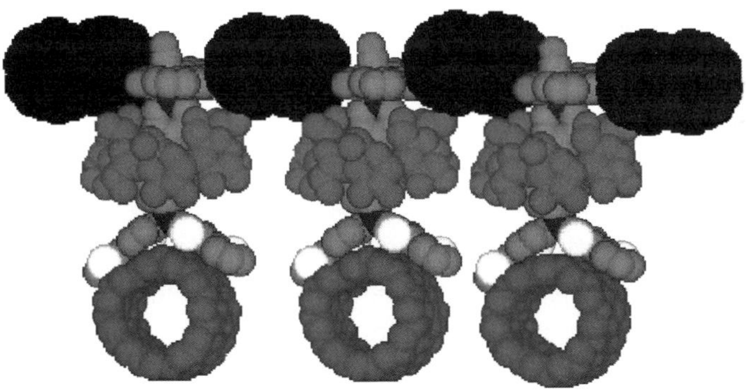

Fig. 4. Part of the nanolayer assembly of an organic solar cell with intercalated fullerene acceptor.

Theme 4: Soft Materials

Soft matter is where physics meets chemistry, and where physics and chemistry meet biology. Soft materials characteristically exhibit hierarchical structures organized on multiple length-scales, which emerge from molecular and supra-molecular self-assembly. We have particular strengths in novel rheo-NMR and rheo-optical methods, as well as small-angle X-ray scattering (SAXS), microscopy, laser diffraction, rheology, microrheology, AFM and ellipsometric techniques. Our work is focused on a range of problems concerning dynamic processes in emulsions' lyotropic liquid crystals and polymers. A theoretical modelling capability complements the extensive experimental suite available in the Theme. There is also programme of research in flow and diffusion, with flow in micro and nanochannels being examined by NMR and using our theoretical and computational expertise.

Theme 5: Inorganic Hybrid Materials

Theme 5 research will develop new inorganic hybrid materials whose hierarchy of nano-, micro- and macro-scaled features gives rise to new chemical, physical and biological functionality. We take two complementary approaches to the design and fabrication of these materials: These include chemical manipulation of the surface functionality of nanostructured fibres, arrays or templates — a '*bottom up*' fabrication strategy, and incorporation or grafting of nanoscaled particles, fibres or tubes into or onto micro- and macro-scaled substrates — a '*top down*' fabrication strategy. Other research areas include, Hybrid Nanostructured Arrays and Conformal Coating of Micro and Nano-Porous Materials, Novel Bio-Nanofibres from Cellulose Precursors, and Inorganic Polymer Hybrid Nanocomposites.

Theme 6: The Intersection of Nanoscience and Biology

Understanding and exploiting the promise of biological systems is a complex and challenging task best addressed through a multidisciplinary approach. This theme combines a group of researchers who together, and with their collaborators, have a unique set of skills and facilities to tackle research problems at the intersection of nanoscience and biology. Three threads run through this research:

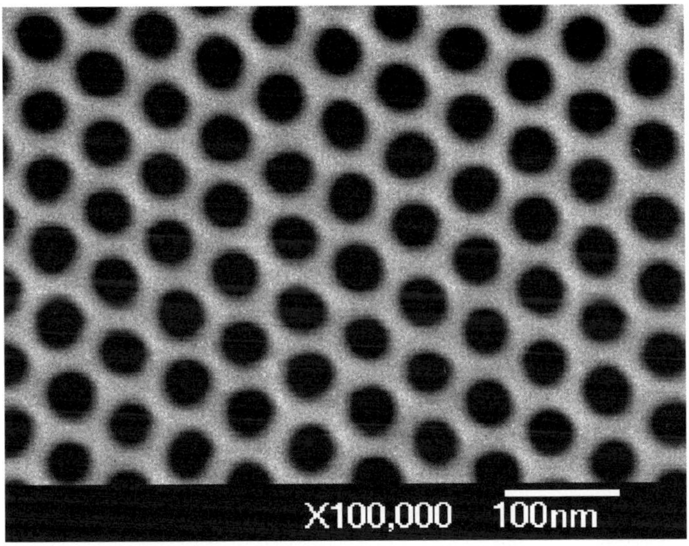

Fig. 5. Nanostructured ceramic array with 30 nm pores.

(i) Advanced physical techniques and methodologies for the study of biological systems: Investigating the dynamics of conformation change of membrane proteins, *in situ*, using sum frequency generation spectroscopy, and, spectroscopic measurements (IR, Raman and CD) and modeling studies of biomolecules, including amino acids, nucleotides and RNA, at high temperatures and pressures In addition, Exocytosis in cancer cells will be studied using the biochip and BioImprint technologies we have developed while targeted delivery of carbon nanotubes and ion channel proteins to supported bilayer membranes will also be studied using related techniques.

(ii) Nanomaterials for medical applications: Biomolecules, such as antibodies, will be attached to the Quantum Dot surfaces with the aim of targeted delivery of drugs to disease sites. Computer simulation and modeling of QD properties will accompany the synthetic work. We also propose to expand a new research effort on the preparation of nanostructured electrodes to the use of carbon nanotube arrays for direct electron transfer to redox proteins.

(iii) Biomolecules as templates and building blocks for materials: Two projects take different approaches to the study of how biomolecules control the growth of inorganic materials. The first will use antifreeze proteins as novel modifiers of the nucleation and growth of metal oxide nanocrystals. The second will be a continuation of a study into how Nature uses soft interactions controlled by biomolecules to form nanostructured hierarchical materials. Focusing on two marine organisms, the goal is to establish the structure/function relationships of the biomolecule in determining the material characteristics.

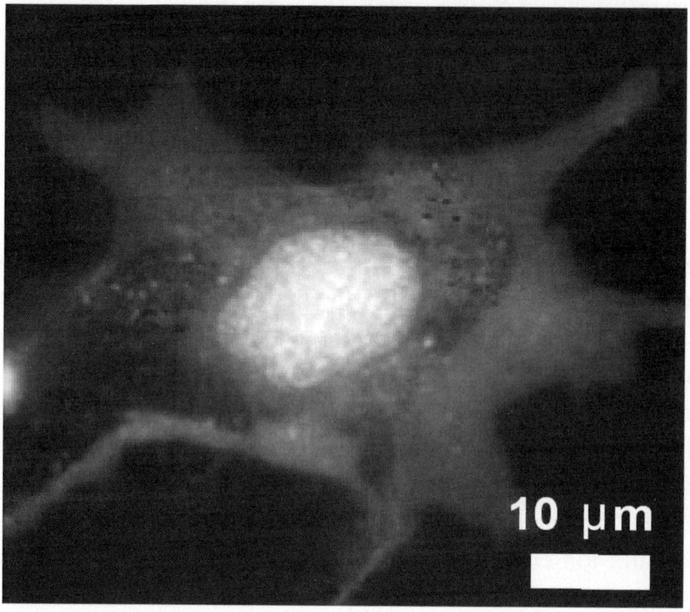

Fig. 6. AFM of a pituitary cell imaged using a BioImprint technique developed from bio-nanotechnology (10 micron scale bar).

4. The Capital Equipment

The formation of the Institute resulted in a unique capacity grouping in equipment resources. The capital injection (NZ$9.8 million) in 2002 was a key component in building the network. Our pre-existing capital included Molecular Beam Epitaxy (MBE), Ultra High Vacuum (UHV) cluster deposition apparatus, plasma etching and electron beam lithography, broad-band laser spectroscopy, NMR systems including solid state NMR, micro-imaging and Rheo-NMR, X-ray diffraction, UV, visible and IR spectroscopy, Raman spectroscopy, variable field magnet facilities, vacuum and cryogenic capabilities, thin film preparation facilities, photoelectron spectroscopy and sputtering ion mass spectrometry, and a nuclear accelerator with Rutherford Backscattering and Nuclear Reaction Analysis capability. The capital injection has given us world class Transmission and Scanning Electron Microscopy (TEM and SEM), Electron Beam lithography, new confocal laser Raman systems, excimer, Kr-ion, Cd, and UV lasers, RF sputtering facilities, SQUID magnetometry, new NMR capability, a supercomputer cluster, new surface probe microscopes, dynamic light scattering, and both stress-controlled and strain controlled rheometers. Now with a new injection of NZ$9.8 million in 2007, we will gain a further enhancement.

Institute policy has been to give all Investigators free access to all our equipment, provided that the work undertaken contributes to Institute objectives. At the same time, the local schools or departments hosting these capital items may charge external users for machine time, thus assisting the recovery of local costs. Depreciation is managed through a partnership in which the Institute pays 45% and the local university 55%. The result has been a win for all concerned. Most importantly, in each case the PIs responsible for managing these facilities have become nodal points for collaboration, building new connections across the Institute.

5. Beyond the Science

At our inception in 2002 we established a set of guiding principles that would help determine not only the governance and management structure but also our future course. We believe that these initial principles have served us well and provide an excellent basis for the future. First we were determined that the Institute would be based on partnership in which no one institution would dominate. Second we believed that the Institute could work most effectively by enhancing the capacity of its partnering institutions, rather than acting as a rival or alternative employing or owning entity. In consequence we determined that we would be a "virtual institute", employing no-one and owning nothing. Of course this second principle created a very special challenge, for it carried the risk that the Institute would be so virtual that it barely existed except as a funding agency for groups of researchers. Hence, we were challenged to build a strong sense of identity on the part of those who belonged. That identity needed to be more than a commitment to excellence. Identity had to spring from operational effectiveness and shared values, clearly understood and expressed.

What are the values of the MacDiarmid Institute, beyond excellence? They are in essence cooperation, collegiality, fairness, effectiveness and pride. *Cooperation,* in building new research groups across institution boundaries, in sharing equipment, in sharing ideas. *Collegiality,* in the celebration of each other's successes, in which we do not see fellow PIs as rivals, but as partners in building the reputation of the Institute to the benefit of all. *Fairness,* in the distribution of resources and in the telling of the stories of the Institute. *Effectiveness,* in that we are committed to tangible outcomes from our vision. *Pride*, in our sense of heritage (Rutherford and MacDiarmid stand behind us) and in the difference we can make to New Zealand science.

Many operational aspects followed from those principles. We needed a binding partnership agreement that would provide a legal framework for our operation, and a Board structure that affirmed the cooperation embodied in that agreement. We needed a newsletter of sufficient quality and regularity that we could tell our stories, a newsletter which gave everyone their spot in the limelight and in which, by our choice of stories on science quality, on outreach, on business, we could emphasise our values. We needed to get together as much as possible, through video seminars, through regular meetings. We needed a major conference, New Zealand-based, which would connect us to the best in the world and provide a highly visible local presence. We needed a participative management structure. We needed independent research on the effectiveness of our governance and management arrangements[(5)]. And finally, we needed a bottom line guide, for us perhaps best expressed as "let the science decide". Hence, we had to be an open structure, allowing new players to join and existing players to depart. This latter aspect of commitment to excellence is the most difficult to manage. We think we have been able to achieve this through strong governance based on fairness.

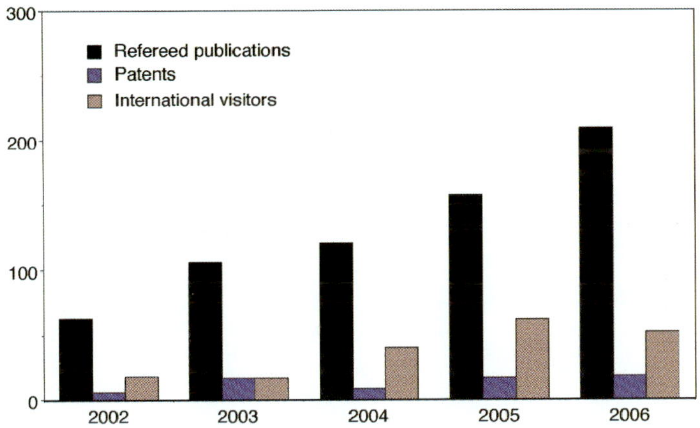

Fig. 7. Growth in refereed science publication per annum, as well as patents and numbers of international visitors.

Of course quality science research underpins our existence and the large number of publications, conference presentations and patents indicates our ability to reach a broad international audience through traditional science channels. However, as explained above, the MacDiarmid Institute sees its role see our role as much more than providing high quality science research. Our ambitious vision concerns providing leadership in educating a new generation of physical scientists, who are excellent, entrepreneurial, communicative and socially aware. We want to build a new role for physical science in New Zealand, one where physical science is seen as wealth generating, internationally connected, and a science focal point for the society, especially the young.

We are committed to commercialisation and knowledge transfer, having "spun-off" several new companies (Nanocluster Devices, Anzode, Magritek). These companies are now employing our graduates. The Institute has organized "Industry Fora" in which New Zealand companies sent representatives to learn about or research work, and to tell us about their own interests and research problems. The MacDiarmid

Institute has run three highly successful international conferences, AMN-1 (Wellington 2003, 300 participants), AMN-2 (Queenstown 2005, 350 participants) and AMN-3 (Wellington, 2007, 450 participants). Each has attracted several Nobel laureates and each was dominated by international participants. The Institute has hosted a number of foreign dignitaries, including, *inter alia*, the Prime Minister of Singapore, successive Directors of the US National Science Foundation, and various delegations of visiting official groups.

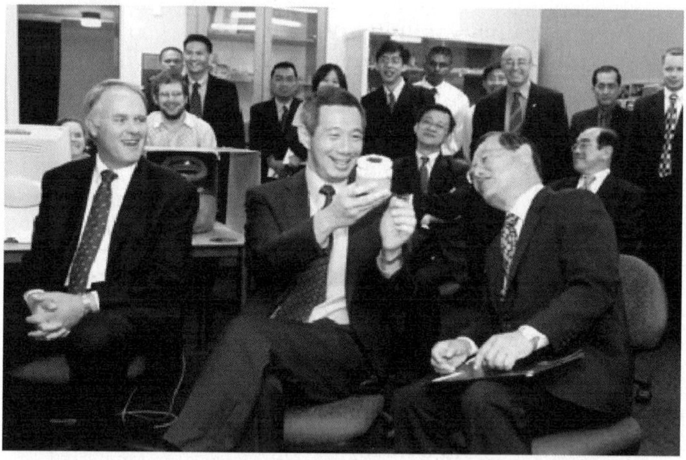

Fig. 8. 2004 visit of then Deputy Prime Minister, now Prime Minister Lee of Singapore to the MacDiarmid Institute. Mr Lee is examining a solution of wormlike micelles.

We have continued to raise the international standing of New Zealand science in the minds of New Zealand officials and policy makers. The director has had a monthly radio show, speaking about a wide range of science issues. We have worked hard to develop a good relationship with the media, and this has resulted in many favourable stories about the Institute and about science in general. Our communication of science to the public at large comprises a number of vehicles including a high quality six-monthly "Interface" newsletter, an annual report "Profile", a very good website, and participation by the director, deputy director and our other staff in numerous community and media speaking engagements.

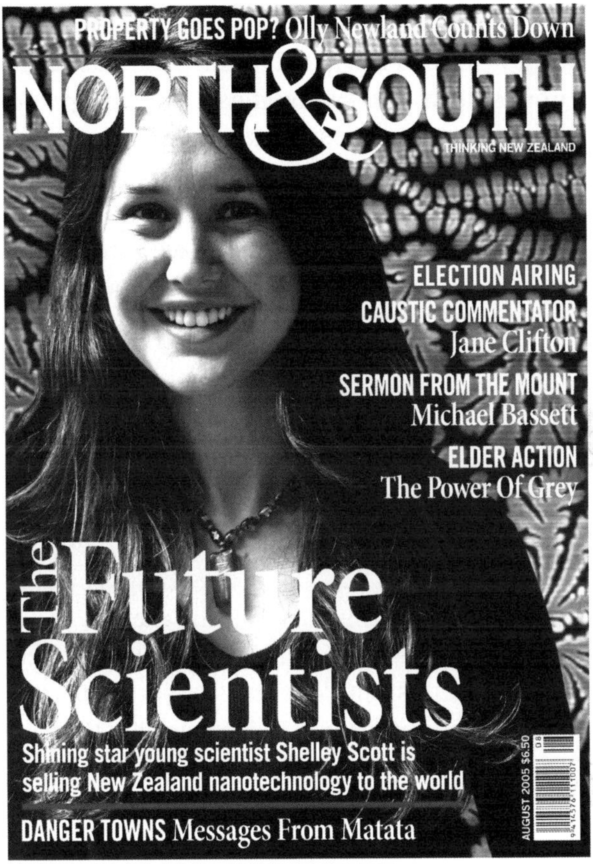

Fig. 9. A monthly magazine featuring a story about the MacDiarmid Institute and its work, with PhD student Shelley Scott on the front cover.

6. Conclusions

The MacDiarmid Institute has been an experiment in building partnership across a range of New Zealand science institutions, and across a range of science disciplines. It is a virtual institute, owning nothing and employing no-one, but applying its funding through each of the partners. A key factor in the success of the Institute has been the purchase of a wide range of high quality capital equipment that is made available at no cost to all Principal Investigators and their teams. This type of partnership is ideal in the area of nanotechnology, where different disciplines meet and where access to good experimental tools is vital. An irony of the MacDiarmid Institute is that its funding impetus came from educational sources, causing us to place a high value on educational goals, developing people, encouraging the young and reaching out to the public at large. This has reaped us rewards of goodwill and increased interest in science generally, reflected also by increased student enrolments in physical sciences courses associated with our activities. What we have discovered, almost by accident, provides an interesting model for the way science research partnerships might be better organised in the future.

Acknowledgements

The work of the MacDiarmid Institute is supported by the New Zealand Tertiary Education Commission Centres of Research Excellence Fund. Many researchers in the Institute are also supported by grants from the New Zealand Foundation for Research, Science and Technology, and the Royal Society of New Zealand Marsden Fund.

References

1. http://www.macdiarmid.ac.nz/news/stories/super.php.
2. http://www.lincoln.ac.nz/story_images/2679_RR289_s7764.pdf
3. www.morst.govt.nz/Documents/work/roadmaps/MoRST-Nanotechnology-Roadmap.pdf
4. http://macdiarmid.ac.nz/publications.php
5. http://macdiarmid.ac.nz/news/stories/collaboration.php

Author Biographies

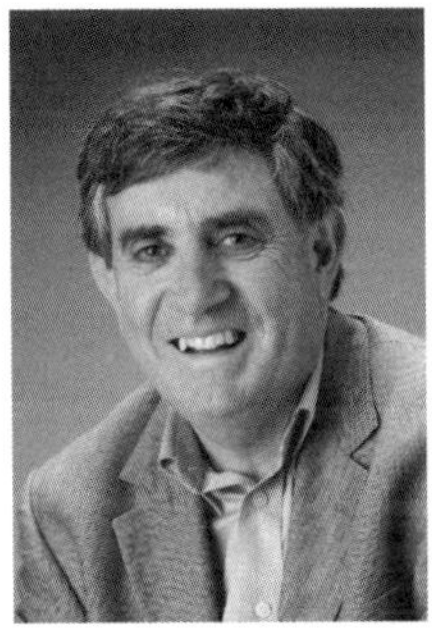

Paul Callaghan was born in Wanganui, New Zealand and took his first degree in Physics at Victoria University of Wellington. He then did a DPhil degree at Oxford University, working in low temperature physics. On his return to New Zealand in 1974 he took up a lecturing position at Massey University where he began researching the applications of magnetic resonance to the study of soft matter. He was made Professor of Physics in 1984, and in 2001 was appointed Alan MacDiarmid Professor of Physical Sciences at Victoria University of Wellington. He is the founding director of the MacDiarmid Institute for Advanced Materials and Nanotechnology. Paul is Past President of the Academy Council of the Royal Society of New Zealand. He has published around 240 articles in scientific journals as well as a book on magnetic resonance. He is also a founding director of "magritek", a small spin-off company based in Wellington, which exports NMR instruments. In 2001 he became the 36th New Zealander to be made a Fellow of the Royal Society of London. Paul was awarded the Ampere Prize in 2004, the Rutherford medal in 2005, and in 2006 was appointed a Principal Companion of the New Zealand Order of Merit. In 2007 he was awarded the Sir Peter Blake Medal.

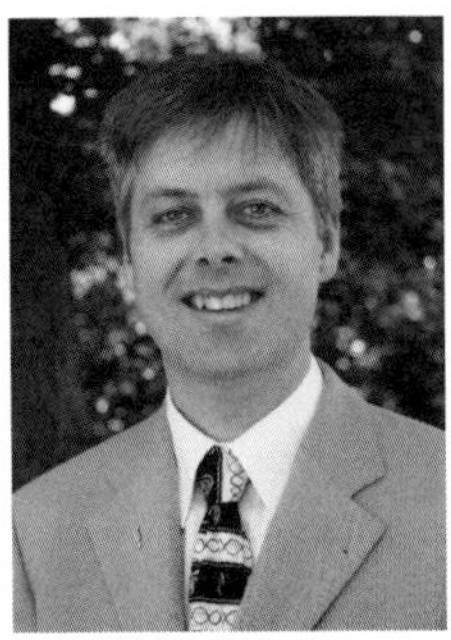

Richard Blaikie received the B.Sc. (Hons) degree from the University of Otago, New Zealand, in 1988 and the Ph.D. degree in physics from the University of Cambridge, U.K., in 1992. For one year, he was a visiting scientist at the Hitachi Cambridge Laboratory, investigating single-electron transport effects in semiconductor nanostructures. He returned to New Zealand in 1993, taking up a position in the Department of Electrical and Computer Engineering at the University of Canterbury, where he is currently a Professor. Blaikie is a founding member of the University's Nanostructure

Engineering, Science and Technology (NEST) group, and is currently the Director of the MacDiarmid Institute for Advanced Materials and Nanotechnology. In 2005 was appointed by the Minister of Research, Science and Technology to serve on the Marsden Fund Council, which administers New Zealand's most prestigious investigator-initiated research fund; he has just completed his term as the chair of the Physical Sciences and Engineering panel. His principal research interests are the development of low cost nanolithography techniques using near field illumination, and the utilisation of sub-wavelength-structures at sub-mm and visible wavelengths. This applied electromagnetics research led to the award of the 2001 T.K. Sidey Medal of the Royal Society in New Zealand. His research interests also include polarisation modulation in optical communications systems, modelling of semiconductor device structures, and the application of nanofabrication techniques to new electronic, optical, chemical and biological devices.

CHAPTER 10

NANOTECHNOLOGY IN SINGAPORE

Hiranmayee Vedam

NanoConsulting Pte. Ltd.
20 Maxwell Road #09-17, Maxwell House, Singapore 069113
hiranvedam@nanoconsulting.com.sg

Hsu Ann Goy

Science and Engineering Research Council
Agency for Science, Technology and Research
1 Fusionopolis Way, #18-10, Connexis North, Singapore 138632
goy_hsu_ann@a-star.edu.sg

Khiang Wee Lim

Science and Engineering Research Council
Agency for Science, Technology and Research
1 Fusionopolis Way, #18-10, Connexis North, Singapore 138632
kwlim@imre.a-star.edu.sg

Singapore is transforming itself into an innovation-driven economy, competing on knowledge and talent, and in so doing, sets itself out on the path towards sustained economic growth. Nanotechnology is recognized as a key enabler to nurture Singapore's capabilities in new growth areas and deepen its capabilities in existing industries such as electronics, chemicals, engineering and biomedical sciences. Substantial investment has gone into nanotechnology-related R&D and manpower development to reflect this significance of nanotechnology to the Singapore economy. In this chapter, the overall investment climate for new technology development and its commercialization, and key developments in nanotechnology research in Singapore are discussed.

1. The Little Red Dot

Singapore is the smallest country in South East Asia with an area of 707 sq. km (and growing with land reclamation) located about 1° north of equator. It has been ranked as the easiest place[1] and the most cost-competitive place[2] to do business in Asia. Singapore is a strong logistics hub within six hours reach of most major cities in the Asia Pacific. It is also home to companies from all parts of the value chain and has strong intellectual property (IP) protection laws. In addition, Singapore is ranked 1st for having the best labor force[3] and 2nd for the most attractive environment[4] for highly-skilled foreigners. Its low tax rates and high standard of living combined with high political stability and low crime makes it an ideal place for families. In fact, foreigners formed 30.9% of the city-state's total employment pool as of December 2006.

To maintain its place as an attractive destination for businesses of the future, Singapore has shifted its focus from cost-effectiveness and efficiency towards innovation and knowledge driven economy in 2006. As a part of this transformation, the Singapore government increased its R&D budget to S$13.6 billion and established the National Research Foundation (NRF) with a budget of S$5 billion in its Science and Technology 2010 plan. Out of the S$13.6 billion, another S$5.4 million was allocated to Agency of Science, Technology and Research (A*STAR)[1] to boost local R&D capabilities, and $2.1 billion to Economic Development Board (EDB) to promote private sector R&D in Singapore. This plan supports key programs to develop research talent, to strengthen public sector research capabilities in niche areas of environment and water, life sciences and, information and digital media, to promote private sector R&D, to strengthen technology innovation capabilities in small and medium enterprises (SMEs) and to increase commercialization of public research. While nanotechnology is not identified as a strategic trust area in the Singapore Science and Technology 2010 plan, it has been recognized as a key enabler for

---

[1] Business-readiness indicators for the 21st century, PricewaterhouseCoopers, 2007
[2] KPMG Cost Competitive Alternatives Study, 2006
[3] BERI Labor Force Ranking 2005-2007
[4] IMD World Competitiveness Yearbook 2005

healthcare (e.g. treatment, monitoring and diagnostics), lifestyle (e.g. performance materials for improved for equipment systems that are easy to use, adaptable, efficient, of low cost, and are portable), smart devices and systems (e.g. intelligent devices and systems that can sense and response), minimal maintenance infrastructures (e.g. materials with self-cleaning properties), green energy (e.g. novel materials for enabling energy production); and biostructures (e.g. biodegradable and biocompatible materials for minimal surgical intervention), and hence one of the 12 priority research areas by A*STAR.

1.1. *Increasing Commercialization of Public Research*

Singapore has over 3200 granted US patents[5] between 2000 and 2007. Of these, a large portion comes from institutes of higher learning (IHLs) and public research institutions (RIs). NRF launched the National Framework for Innovation and Enterprise (NFIE)[2] to develop innovation and enterprise in Singapore. Under this framework, NRF provides proof-of-concept grants of up to S$250K to IHLs to demonstrate initial commercialization potential of the IP generated by them. There are also translational R&D grants available to support polytechnics to perform translational research on the R&D output from universities and research institutes (RIs) to bring research breakthroughs to the marketplace. NRF is also introducing innovation vouchers to SMEs to procure R&D and other services from IHLs and research institutions. A*STAR has several schemes to help accelerate commercialization of technologies coming out of its 14 RIs. It has a Commercialization of Technology (COT)[5] grant of S$1M that the RIs can use work with SMEs and make the technology market ready. A*STAR has a larger "Flagship program" that provides up to S$3M to identify commercially valuable technologies earlier in the research value chain for larger and more impactful outcome through interaction with industry. The Prime Minister's Office offers The Enterprise Challenge (TEC)[6] grants to develop innovative product/service for pilot/trial to improve Singapore's public service.

[5] See http://www.uspto.gov/go/oeip/taf/cst_all.htm

Unlike other grants, TEC does not have a limit on the amount of funding received and covers 90% of the costs.

1.2. *Developing the Singapore Enterprise Eco System*[6]

The Singapore Economic Development Board (EDB) and SPRING Singapore are two agencies under Ministry of Trade and Industry (MTI) tasked to develop the enterprise eco system in Singapore. EDB plans and executes economic strategies that enhance Singapore's position as a global hub for business, investment and talent. Within the EDB, the New Businesses Group is responsible for developing the Singapore nanotechnology eco system. The EDB offers several incentives[3] such as preferential corporate tax rates, approved royalties incentive to offset royalties and technical fees payable to non-residents, further deduction for R&D expenses, development schemes such as Innovation Development (IDS) Scheme, Research Incentive Scheme (RISC), Initiatives in New Technology (INTECH) and Strategic Attachment and Training Program (STRAT) to attract multinationals and foreign companies to expand and perform R&D in Singapore.

SPRING Singapore is tasked with growing innovative companies and fostering a competitive SME sector and offers different program to meet the needs of enterprises at different stages of development. In addition to various financing schemes to support capital investments and working capital, SPRING Singapore offers funding for companies to map and execute their IP strategy, hire external experts and increase in-house technical innovation capability through its LETAS program and Technology Innovation program. SPRING also administers the

---

[6] A business eco system is defined as "*An economic community supported by a foundation of interacting organizations and individuals — the organisms of the business world. This economic community produces goods and services of value to customers, who are themselves members of the ecosystem. The member organizations also include suppliers, lead producers, competitors, and other stakeholders. Over time, they co-evolve their capabilities and roles, and tend to align themselves with the directions set by one or more central companies. Those companies holding leadership roles may change over time, but the function of ecosystem leader is valued by the community because it enables members to move toward shared visions to align their investments and to find mutually supportive roles*" *See Harvard Business Review May/June 1993 Vol 71, Issue 3, 75-86.*

Technology Enterprise Commercialisation Scheme (TECS) which was launched as part of the NFIE. Under the TECS scheme, technology based SMEs can get proof-of-concept grants of up to S$250K which covers 100% of qualifying costs. They can apply for further proof-of-value grants of up to S$500K which covers 85% of qualifying costs to develop a working prototype. Nanotechnology covers three of the four strategic areas identified for projects under this scheme. SPRING Singapore also offers tax deductions for investment losses of up to S$3M (Singapore does not have capital gains tax consequently capital losses are generally non-deductible) in startups to both individuals and institutions.

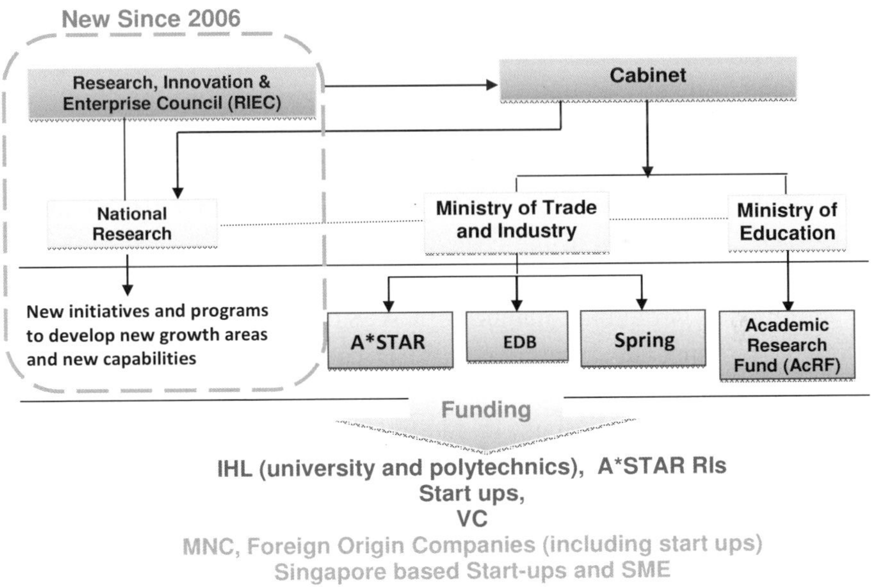

Fig. 1. Singapore National R&D Framework (Adapted from Singapore Science and Technology 2010 plan[4]).

As part of the NFIE framework, NRF plans to provide 85% co-funding to companies funded by approved technology incubators with an option to buy out NRF's share at the next round of funding. The EDB's Fast-Tech Incubator scheme for Cleantech companies operates very similarly. These schemes are meant to provide an environment for systematic nurturing of young companies before they are ready for venture funding. NRF also matched 1:1 the funds raised (up to S$10M) by five early stage venture funds. The venture funds are required to invest these funds in Singapore-based technology start-ups. The key incentives and grants from all agencies most relevant to nanotechnology companies are summarized in Table 1.

A*STAR and its 14 research institutes develop and sustain a substantial pipeline of research talent to meet industry needs. These research institutes are clustered into the biomedical cluster (managed by the Biomedical Research Council (BMRC)) and the physical science cluster (managed by the Science and Engineering Research Council (SERC)). All the research institutes under BMRC are in Biopolis which is a purpose-built biomedical research hub where researchers from the public and private sectors are co-located. Similarly, the research institutes under SERC are co-located with private sector companies in Fusionopolis which was launched in late 2008. Under EDB's Technology for Enterprise Capability Upgrade (T-Up) Scheme, companies can get 70% of the costs incurred in hiring most of the more than 2200 A*STAR researchers on a temporary basis for up to two years. In addition to providing manpower, A*STAR also provides shared facilities to companies establishing R&D centers in Singapore. So far, these partnerships with industry have resulted in 750 joint projects and consortia in the last five years. In addition, about 122 research scientists and engineers (RSEs) have been seconded to 93 companies under the Technology for Enterprise Capability Upgrading (T-Up) secondment scheme, and 51 technical advisors have been appointed to 48 companies. A*STAR has also developed 85 operations and technology roadmaps for 72 companies under such support programs.

Singapore is also home to three science parks set up to provide infrastructure for commercial R&D to flourish in Singapore. These are located near the universities, hospitals, Biopolis and Fusionopolis. Office space ranges from 400 sq ft to more than 15,000 sq ft. Thus far these science parks have attracted nearly 400 companies.

Table 1. Incentives and Grants available for Technology Commercialization in Singapore

Funding Agency	Grant/Fund Name	Grant Amount	Grant Criteria
NRF	Early Stage Venture Fund (ESVF)	Up to SGD10M	Need to raise or investment minimum SGD10M
	Technology Incubation Funding Scheme	SGD50M	Investors/Angels/Entrepreneur in partnership with NUS, NTU and SMU; 15% co-investment
	Disruptive Innovation Incubation Fund	SGD10M	
	Proof of Concept (POC)	SGD250K (program total SGD50M)	Researchers based in University and Polytechnics
	Technology Incubation Scheme (TIS)	Equity investment up to SGD500K, 85% support	Seed stage funding for incubators in univ. and polytechnics
SPRING[7]	Equity Financing Scheme-SEED	Up to SGD1M	3rd party commitment
	Business Angel Scheme	Up to SGD10M	For business angel group
	Technology Enterprise Commercialization Scheme (TECS)	POC-SGD250K, POV-85% qualified cost support up to SGD500K	Singapore based SMEs
	Technology Innovation Program (TIP)	50% support on qualified cost	
NUS	NUS Venture	Up to SGD300K	NUS Spin-offs
	NUS CleanTech Incubator	Up to SGD500K	

Table 1 (continued)

Funding Agency	Grant/Fund Name	Grant Amount	Grant Criteria
NTU	NTU Nanofrontier Research Partnership Equity Funding	Flexible	Need to identify a joint research project with NanoFrontier
EDB	Initiatives in New Technology (INTECH)	SGD250/Day up to two years	R&D training
	Strategic Attachment and Training Program (STRAT)	SGD4000~4500/month for up to two years	
	Fast-tech Fund	SGD500K per incubatee	
A*STAR	Labs in Research Institute (Labs in RI)	Share facilities and manpower	
	Technology for Enterprise Capability Upgrade (T-UP)	70% support on manpower	

2. Nanotechnology in Singapore

Nanotechnology is recognized as a key enabler to sustain future development of the Singapore economy and Singapore agencies have put more and more emphasis on it since the late 1990s in response to growing awareness of nanotechnology worldwide. The Singapore government spent about US$300 million between 2003 and 2007 in nanotechnology-related R&D and manpower development. It is estimated that the number of researchers and engineers working in nanotechnology-related fields in the Republic, in both the public and private sectors, totals almost 1,000. Singapore is an active member of the Asia Nano Forum (ANF, www.asia-anf.org)[8] and chairs the standardization working group in the ANF. The Institute of Materials Research and Engineering (IMRE) of A*STAR is currently hosting the ANF secretariat. Singapore is also a participating member of the International Organization for Standardization (ISO) and International Electrotechnical Commission (IEC) technical committees on nanotechnology.

2.1. *Research Infrastructure*

Research in nanotechnology mainly takes place in the two technological universities in Singapore namely, National University of Singapore (NUS) and Nanyang Technological University (NTU) and A*STAR RIs.

At NUS, the National University of Singapore Nanoscience and Nanotechnology Initiative (NUSNNI)[9] provides the focal point for nanotechnology related activities in six strategic areas namely: nanobiotechnology, nanomagnetics and spintronics, nano/microfabrication, nanophotonics, sustainable energy and health and environmental impacts of nanomaterials. It consists of a loose network of more than 120 academic staff spanning 14 research labs and 5 research centers across science and engineering faculties. One of these research centers is the Singapore Synchrotron Light Source (SSLS) which provides synchrotron radiation based fabrication and characterization services to academia and industry around the world. NUS is also home to the Center for Ion Beam Applications (CIBA) which consists of a proton beam radiation source which can be used to fabricate high aspect ratio 3D nanostructures and to image deep tissues. Some state-of-the-art labs at NUS include the Organic Nano Devices Laboratory (ONDL), Silicon Nano Devices Laboratory (SNDL) and Surface Science Laboratory (SSL). NUS launched NanoCore - an embedded entity within NUSNNI - in 2008 to build focused group of interdisciplinary labs of excellence in selected areas where NUS can build world class teams and make global impact. These areas include oxide electronics, graphene electronics, active plasmonics, spintronic materials, optoelectronics, ion-beam imaging and fabrication and bionanotechnology. NanoCore[10] has state-of-the-art infrastructures like the Orion He-Ion sub-nm microscope and e-beam writer. The Solar Energy Research Institute of Singapore (SERIS) located at NUS conducts industry-oriented research and technology development and use-inspired basic research in the field of solar energy conversion, in close collaboration with industrial partners. NUS is also home to Centre for Quantum Technologies - a S$150 Million Research Center of Excellence setup by NRF. This center focuses on the development of reliable quantum components, their interoperability and integration to construct quantum structures on chips and run quantum algorithms and quantum simulations.

NanoCluster[11] is an NTU-wide network of research centers with shared facilities for nanofabrication, nanocharacterization, and exploitation of nanotechnology applications in nanodevices, energy and catalysis, nanobiotechnology, nanomaterial synthesis, organic, molecular electronics and nanomagnetics and photonics. It consists of a loose network of 90 academic staff across five research centers. Major facilities include three cleanrooms (total net 1,000 sq m, net) for semiconductor processing, MEMs, Bio/Organic/Glass/Metal fabrication/processing, advanced characterization (TEMs/FIB surface analysis) and advanced materials synthesis. NTU also houses Computational Nanoelectronics Initiative which coordinates nanoelectronics modeling work among the IHLs and RIs.

Among the 14 A*STAR RIs, nanotechnology related research is concentrated in Institute of Materials Research and Engineering (IMRE), Institute of Microelectronics (IME), Singapore Institute of Manufacturing Technology (SIMTech), Institute of Bioengineering and Nanotechnology (IBN) and Data Storage Institute (DSI). A*STAR has setup the SERC's Nanofabrication and Characterization Facility (SNFC) to provide access all the fabrication and characterization facilities available under its SERC to Singapore nanotech community. This facility - managed by IMRE - consists of fabrication facilities such as inkjet printing system, metalorganic CVD systems and e-beam lithography equipment. It also has characterization equipment such as microRaman system, UHV TEM, TOF-SIMS and X-ray photoelectron spectroscopy.

IMRE also manages a wafer scale nanoimprint lithography facility that provides commercial prototyping and foundry services. IME's Silicon Photonics MPW (Multi-Projects Wafer) Prototyping facility provides access to 8-inch wafer-scale CMOS fabrication facilities to academic and industrial R&D groups to perform research and prototyping of photonic integrated circuits. SIMTech has setup a Microfluidics Manufacturing Programme to accelerate prototyping and pilot run base for microfluidic development. A*STAR also has a Center for Nanometrology Excellence, a 200mm Si-based wafer processing and a MEMS prototyping facility that are available for use by the wider nanotech community in Singapore. A*STAR's DSI and NUS have setup Information Storage and Materials Laboratory (ISML) to investigate advanced materials for ultra-high density storage, such as spintronics and nanomagnetics research.

In addition to world-class research infrastructure, Singapore has a wide variety of programs to train manpower required for nanotech companies at different levels of expertise as well. The Institute of Bioengineering and Nanotechnology (IBN) - an A*STAR RI - recently launched Nano-Bio kits to educate school children and these kits have been incorporated into secondary school curriculum in Singapore. The Nanyang Polytechnic and Republic Polytechnic offer diploma level courses in nanotechnology. The National University of Singapore offers double degree B.S. in Physics and Material Science which emphasizes nanotechnology and an Engineering Science program in Nanotechnology. Both NUS and NTU offer a variety of research programs in nanotechnology at masters and Ph.D level through their traditional departments. There is no specific M.Sc or Ph.D in Nanotechnology in either university at present although NUS' NanoCore has plans to launch a joint engineering and business Ph.D level program.

2.2. *Commercialization Infrastructure*

Technology transfer is well developed at IHLs and A*STAR RIs in Singapore. At NUS, technology transfer and collaborations are managed by the Industry Liaison Office (ILO) which is part of NUS Enterprise[12] - NUS Enterprise also has a division called the NUS Entrepreneurship Center (NEC) which administers the proof-of-concept grants, assists in the formation of spin-offs and selectively provides seed stage funding for them. NEC also manages a cleantech incubator to mentor cleantech startups under EDB's Fast-tech program. NanoCore supports the commercialization and entrepreneurship activities specifically in nanotechnology through NanoSpark.

The Innovation and Technology Transfer Office (ITTO)[13] at NTU is the key point of contact for technology transfer and collaborations at NTU. In 2005, NTU with support from EDB setup NanoFrontier Pte. Ltd.[14] - a nanotechnology incubator - to provide companies in various industries a platform to expand their research and development efforts in nanotechnology related areas. NanoFrontier also provides start-ups and technopreneurs a safe and strategic place to develop their nanotech inventions both technically and commercially.

All IP generated by the A*STAR institutions are commercialized by Exploit Technologies Pte. Ltd. Exploit Technologies also manages collaborations with A*STAR RIs and provides seed funding for their spin-offs companies.

In addition to the grants schemes discussed in Section 1 and the commercialization infrastructures here, Singapore also has a number of

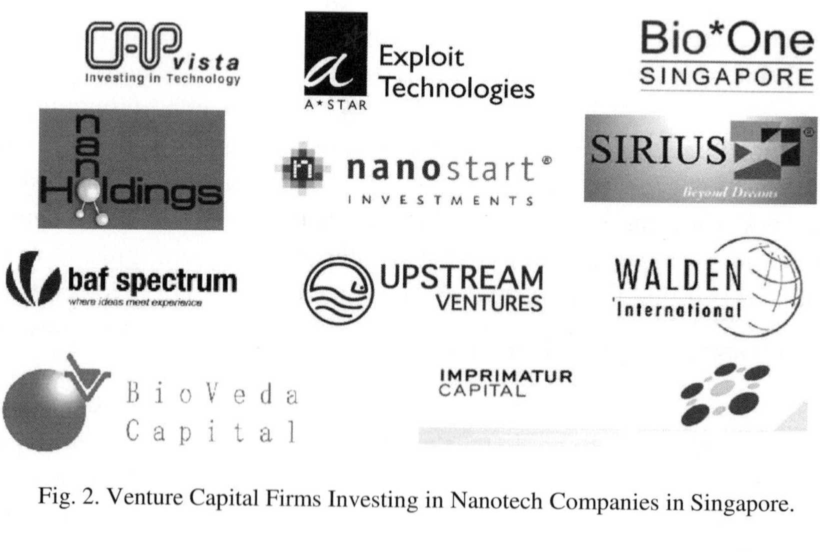

Fig. 2. Venture Capital Firms Investing in Nanotech Companies in Singapore.

private sector venture capital firms actively investing in nanotech companies. These companies are shown in Fig. 2.

NanoGlobe[15] and NanoConsulting[16] manage a nanotechnology network platform SingNano (www.singnano.org)[17] which serves as a one stop for Singapore nanotechnology database and activities, bringing together policy makers, researchers, entrepreneurs, industrialists, investors and the public to:

- Promote Nanotechnology policy, R&D and business
- Foster Nanotechnology Industry growth in Singapore
- Facilitate Nanotechnology education to the general public as well business community
- Provide information services and business matching to its members
- Represent Singapore in global nanotechnology business leadership forums

SingNano organizes monthly seminars to promote nanotech companies in Singapore and to promote their R&D and commercialization efforts by providing a platform to enhance their visibility to partners and investors.

2.3. *Application Specific Strengths*

Singapore's investment in nanotechnology has already resulted in contribution of nanotechnology related products and services growing at about 8-21% and the number of nanotechnology companies in Singapore has grown from 10 in 2004 to the present 58. Multi-national companies such as BASF, Bayer, 3M, BOSCH, ST Microelectronics and Zyvex have decided to setup substantial R&D facilities in Singapore to take advantage and contribute to the growing nanotechnology ecosystem in Singapore. Here, we highlight a few application-specific strengths of Singapore and the ecosystem to capitalize on them.

2.3.1. *Biomedical applications of Nanotechnology*

Singapore has world class research in drug delivery systems, tissue engineering and bio-imaging applications of nanotechnology. NUS is ranked #[7] in drug release and drug delivery systems using nanoparticles, #2 (after MIT) in tissue engineering and bio-imaging application of nanoparticles. NUS also contributed six of the top ten researchers in this area in Singapore, while NTU and A*STAR's IBN contributed two researchers each to round up the top ten list.

2.3.1.1. *Drug delivery systems*

Researchers at NUS have developed novel biodegradable polymers for sustained and controlled release of anti-cancer and AIDS drugs increasing the oral bioavailability to 91% while decreasing side effects. They are also using these polymers for developing vaccines and gene therapies with minimal side effects. Another group at NUS is studying the role of nano-biomechanics to understand the patho-physiology of malaria and cancer to develop highly sensitive and accurate high-throughput assays that can help detect diseases in early stages. Researchers at NTU are developing biodegradable stents with capacity to deliver multiple drugs and to reduce restenosis and thrombosis. IMRE is developing thermal and pH responsive degradable polymers via graft of NIPAAm (N-isopropylacrylamide) that can be used in hydrogels and micelles for drug delivery applications.

2.3.1.2. *Tissue Engineering*

Researchers at NUS are developing a novel intra-guidance channel using electrospun biodegradable PLGA fiber that can be used with commercially available nerve conduit. Other tissue engineering applications of nano-fibers being investigated include biomineralized scaffolds for bone and cartilage repair, skin grafts for wound dressing and nanofiber covered stents to reduce restenosis. Another group at NUS is developing methods to combat bio-material centered infection using surface functionalization techniques especially in Ti implants.

---

[7] Based on citations from 2004-2009

Researchers at IBN have developed novel fibrous scaffolds where the porosity can be manipulated by hydroentanglement and use of microparticles. These scaffolds can deliver growth factors customized to the cells and incorporate extracellular matrix thus attaining effective tissue regeneration.

2.3.1.3. *Bioimaging*

Bioimaging is an area of intense research by various groups in Singapore. Most notable among them is the research work done at IBN to develop water soluble silica-coated magnetic quantum dot nanocomposites. At NUS, researchers are developing water-dispersible magnetic nanoparticles for MRI applications and thermotherapy. Other groups at NUS are developing water soluble upconversion nanoparticles for bio-imaging as they reduce photo-damage and for phototherapy as they can be excited using infrared light to generate heat at targeted sites.

Spin-Offs

Amaranth Medical (www.amaranthmedical.com)
Amaranth Medical Pte. Ltd. is a spinoff from NTU to commercialize the bioresorbable stent technology developed for treatment of peripheral vascular disease. The company is developing implantable bioresorbable stent technologies (drug eluting and non-drug eluting) for treatment of vascular disease. Amaranth Medical has two operational sites, one in Singapore and the other in Silicon Valley of California. It is incubated by NanoFrontier Pte. Ltd.

Curiox BioSystems (www.curiox.com)
Curiox BioSystems Pte. Ltd. is a spinoff from IBN to commercialize the technology for miniaturization of heterogeneous bioassay. Curiox's patent-pending miniaturization platform, DropArray™, provides up to 1,000 times savings in sample and reagent consumption, and up to ten times reduction in assay time. It is backed by NanoStart AG and Exploit Technologies.

Biomers (www.biomerbraces.com)
Biomers Pte. Ltd. is a spinoff from NUS to commercialize fabrication technology to produce fiber-reinforced polymer composite materials. The company is engaged in developing novel polymer composite products for numerous biomedical applications. While orthodontics is the immediate application area, dentistry and the medical field are also application targets for this innovative and patented technology. The company's first products are the patented translucent arch wires and retainer wires for the orthodontic and general dentistry treatment market. It is incubated by NUS Enterprise and received additional funding from private investors.

Other Companies in the Eco System

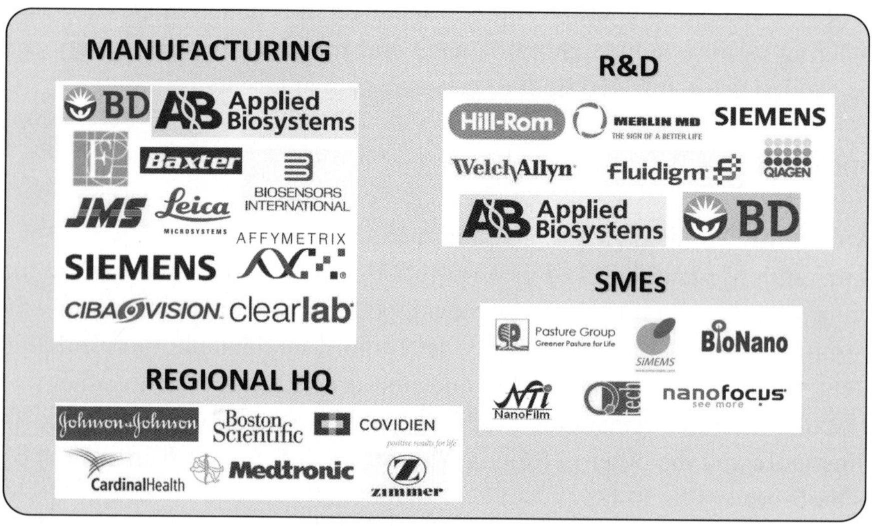

Fig. 3. Medtech Eco System in Singapore.

2.3.2. *Energy Applications of Nanotechnology*

Energy was recognized as one of the 12 priority research areas by A*STAR as part of a tech scan exercise it carried out prior to the launch of Singapore's Science and Technology 2010 plan. A significant portion of energy related nanotechnology research in Singapore is focused on third generation solar cells – specifically in organic and dye sensitized solar cells. Another significant area of research related to energy is the development of OLEDs.

2.3.2.1. *Dye Sensitized Solar Cells*

Researchers at NUS are investigating the use of diameter controlled anatase TiO_2 nanofibers in dye sensitized solar cells. They are also researching the impact of electrospinning and hot pressing 1D metal oxide nanorods on to substrates as guides for electron transport. Initial results indicate that they can produce dye sensitized solar cells with conversion efficiencies of ~6%. Another group at NUS is investigating mesoscopic metal oxide electrodes (TiO_2, Al_2O_3, etc.) and their assemblies with functional molecules to produce high efficiency, low cost dye sensitized and 3D solar cells. Other groups are investigating the use of conjugate polymers and nanocrystalline inorganic materials for solid state dye-sensitized solar cells. At NTU, researchers have developed dye sensitized solar cells based on ZnO nanoflowers with a conversion efficiency of 1.9%.

2.3.2.2. *Organic Solar Cells*

A*STAR's IMRE has developed translucent organic solar cells that can be easily printed on flexible substrates as part of a larger program on the synthesis of new materials and the fabrication of new nanostructures for organic/polymer field effect transistors and photovoltaics. These solar cells can not only be produced cheaply but also can have wider variety of applications from window panes to portable electronics. At NTU, researchers are incorporating silver nanoprisms as median layer between the electron and hole to increase the amount of light absorbed and enhance charge transport. At NUS, platform technologies such as advanced nanometal inks, DUV and i-line crosslinkable formulations for producing organic transistors and novel columnar hetrostructures for highly efficiency photovoltaics are being developed. Researchers at NUS are also investigating production of low-cost, high quality graphene by chemical exfoliation and its use in transistors and solar cells.

2.3.2.3. *OLED*

IMRE and NUS researchers have developed a technology to fabricate all-in-one white LEDs by growing multiple quantum wells using InGaN/GaN on sapphire substrate. This is an important milestone in obtaining while light LEDs that are cheaper, stable and less complex without using phosphors. Another research group at IMRE is developing top emitting OLEDs on flexible substrates and has developed robust plastic substrates with effective barrier against oxygen and moisture to increase their life time. They have also achieved significant improvement in electroluminescent efficiency in top emitting OLED by overlaying an optical coupling layer on a semitransparent cathode. Researchers at IMRE have also developed blue emitters with increased lifetime and efficiency that is solution-processible, making them cheap to produce.

Spin-Offs

NanoBright Technologies (www.nanobright.com.sg)
NanoBright Technologies Pte. Ltd. is a spin-off from NUS to develop applications using luminescent materials. Some of the applications that NanoBright is capable of developing include - security ink that can be used in defence or anti-fraud detection; long afterglow paint that will reduce the need for lighting in dark areas; increasing the efficiency of solar cells using our up and down conversion materials and bio-imaging probes with the capability to replace quantum dots, being some of them.

Other Companies in the Eco System

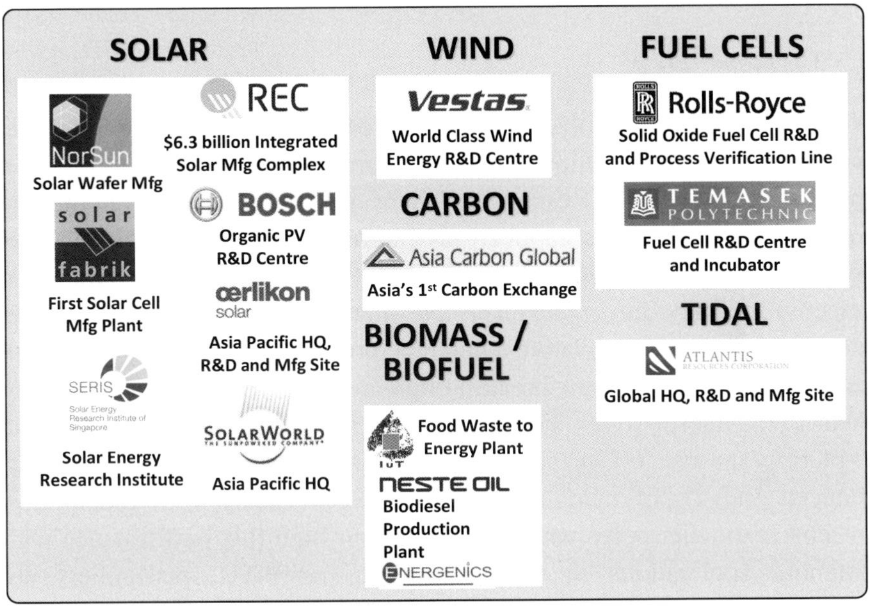

Fig. 4. Energy Eco System in Singapore.

2.3.3. *Environment and Water Related Applications of Nanotechnology*

Singapore is an Environmental Technology (ET) hub in Asia. Together with local companies and major global firms that provide ET solutions. Singapore has become an extensive user of environmental technologies in specialized applications such as water resources. The goals of sustainable development have been recently defined in the recent Singapore 2012 Green Plan[18], which is both the catalyst and driver for focused research and development efforts including the application of nanotechnology in sustainable development and environmental technologies. Environment and water sector was also recognized as one of two new strategic sectors for growth in Singapore in the Singapore Science and Technology 2010 plan. A significant part of environment and water related applications of nanotechnology is in the area of membranes with NUS being recognized as the world leader. Environmental sensors are another area of increasing research.

2.3.3.1. *Membranes*

NUS is the first to publish a study of forward osmosis by fabricating nanofiltration (NF) hollow fiber membranes for osmotically driven membrane processes. Researchers there are studying novel membrane materials for natural gas applications and have synthesized several novel and promising flourinated polyimides with high gas permeability and selectivity. They have enhanced anti-plasticization properties of the membranes using innovative chemical cross-linking technologies and have achieved significant breakthrough in membrane distillation using dual-layer microporous hollow fibers. Another group at NUS is exploring the use of surface modification techniques such as plasma-induced graft copolymerization on electrospun nanofibrous membranes to achieve smaller pores while retaining their high flux performance with potential applications in water filtration. At NTU, researchers are investigating the use of TiO_2 nanofiber membranes for concurrent photocatalytic oxidation and filtration to reduce membrane fouling problems in large scale water treatment applications. They are also developing bimetallic zerovalent metal particles for catalytic reduction of contaminants in water and nanostructured materials such as layered double hydroxides to adsorb trace inorganic contaminants and biomaterials.

2.3.3.2. *Environmental Sensors*

Researchers at NUS have developed an instrumentation free sensor system to detect the presence of mercury at room temperature using DNA modified gold nanoparticles. Another group there has developed liquid crystal based detection system for environmental applications. At IBN, researchers have developed an ultrasensitive electrochemical detection system for biomolecules using nanoparticle tagging. NTU's researchers, meanwhile, have synthesized semiconductor and ferroelectric nano-structured metal oxide and composite materials for use in gas sensors. Another group there has developed sensors based on modified gold electrodes using ZnO nanocombs.

Companies in the Eco System

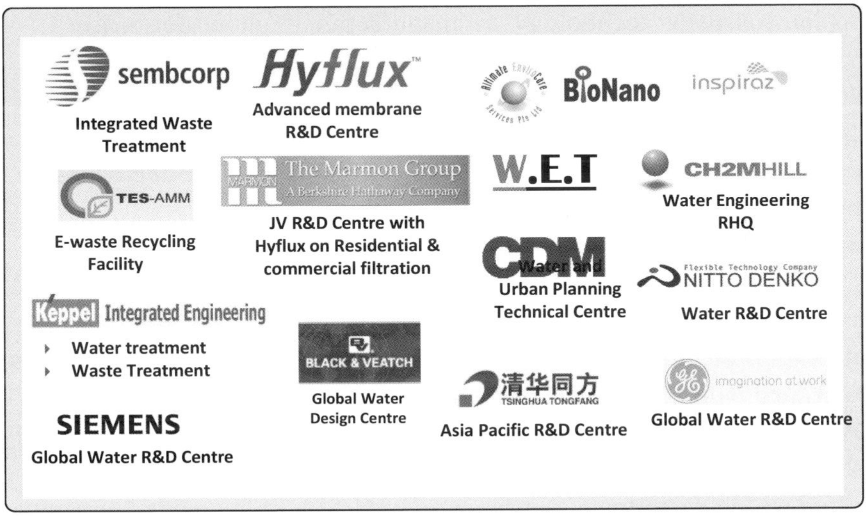

Fig. 5. Environment and Water Eco System in Singapore

2.3.4. *Electronic Applications of Nanotechnology*

Electronics is one of the main industry clusters in the Singapore manufacturing sector and data storage and semiconductors are the two most important sectors within it. Hence there is a critical mass of research in this area and infrastructure to support it in Singapore.

2.3.4.1. *NanoDevices*

NUS is first in the world to demonstrate the use of a new material comprising silicon and carbon in transistors to enhance the speed of electronics. NUS broke the world's sub-threshold record for semiconductor devices by developing a CMOS-compatible L-shaped impact-ionization MOS (IMOS) technology based on silicon-germanium (SiGe) to enable next generation of ultra-low power devices. It is also first in the world to demonstrate HfN/SiO_2 and HfN/HfO_2 gate stack technology and to develop a proton beam writing technology with unique ability to direct-write three-dimensional structures down to the 30nm level in both polymer and semi-conductor materials. NTU developed the world's first metamorphic indium phosphide double-heterojunction bipolar transistor technology to manufacture high performance III-V monolithic microwave integrated circuits cheaply. It has also developed filtered cathodic vacuum arc (FCVA) coating technology which overcomes many of the disadvantages in existing FCVA technologies. Potential applications of FCVA technology being explored include the deposition and characterization of ZrO_2 to replace SiO_2 as a gate dielectric and the deposition of ZnO on silicon substrates for optoelectronic applications.

A*STAR has an extensive program address a broad range of issues (e.g. architecture and design, transistor scaling, materials, processing and fabrication methodologies, integratable memory and embedded devices) associated with advanced 45-nm technologies and beyond for high speed, low power processor applications of the future. IMRE successfully fabricated high-quality high-k oxides on semiconductors and the improved the performance of Ge-based MOSFETS by using Gemanides such as NiGe as Schottky source/drain. Researchers there have also been successful in using silicide materials such as NiSi and $ErSi_2$ in sub-90nm nanodevices. IME successfully demonstrated the integration of Gate-All-Around (GAA) Si-nanowire transistors into CMOS inverters using top-down approach. They also fabricated P-channel omega-gated SiGe nanowire FETs with high-k/metal gate using a top-down approach, thus enabling the integration of nearly pure Ge nanowire transistors into CMOS logic circuits. Another key result achieved by researchers at IME is the development of Wafer Transfer Technology to effect wafer level transfer of circuit patterns from Si onto plastic, glass or rubber. Researchers at SIMTech have developed lead free nanocomposite solders by strengthening the solder with fine second phase particles to provide good mechanical, electrical and thermal properties.

2.3.4.2. *Photonics and Magnetics*

NUS researchers are developing novel spintronic materials such as oxide based ferro-magnets created by doping wide bandgap oxides such as TiO_2, HfO_2, Cu_2O, SnO_2, ZnO and using them to build multifunctional spintronic devices for information storage and processing. Another group at NUS is exploring heterostructure, supperlattices, or/and quantum wells between various epitaxial films with particular emphasis on materials with unconventional electronic, optical, magnetic and thermal properties. Other groups at NUS are examining nonlinear optical effects of intense laser pulses on novel materials with large higher order nonlinear optical susceptibility to identify materials for applications in optical switching, or optical limiting, or optical imaging. Researchers at NTU are developing mesococoscopic structures for routing and processing light that can be integrated with active devices such as lasers and detectors in a very compact form factor. They are exploring the integration of photonic devices based on III-V quantum dots grown using the molecular beam epitaxy process into a silicon-based substrate platform. Another group there has fabricated ZnO waveguides that achieve amplified spontaneous emission, pointing the way to directional ZnO lasers. That work is part of a larger program to fabricate of ZnO optoelectronics based on silicon substrate and integrate them with the silicon electronics using FCVA deposition technique. IMRE researchers have developed perfluorinated materials for use as barrier films and lubricants for hard disk drives. At IME, researchers established a full suite of passives components library based SOI nanowires and SiN waveguides with the final goal of achieving monolithic integration of all electronic and photonic devices in a single chip. They have also monolithically integrated Ge-Photodetector on Si-CMOS compatible photonics platform. Another group at IME has reported electron luminescence on an electrically pumped silicon light-emitting device with thin multi-layer stacked amorphous silicon/silicon nitride structure.

2.3.4.3. *Organic and Molecular Electronics*

NTU researchers developed a tunable computer-generated hologram stored in polymer-dispersed liquid crystals bringing them one step closer to rewritable holographic disks. Another group there is developing superior semiconductor single wall carbon nanotubes networks to build field effect transistors (FETs) with high mobility, on/off ratio and yield for printed electronics. As mentioned in Section 0, NUS researchers are developing a platform technology to develop organic electronic systems. Another group at NUS is developing new methods to make graphene related materials using chemical and physical means and integrating these materials into specially created device structure for investigating electron transport. NUS is also studying charge and spin transport in lithographically patterned graphene nanostructures and the manipulation of the magnetization of nanoscale ferromagnets by means of spin currents. As mentioned in Section 0, IMRE researchers have developed top emitting OLEDs on flexible substrates. They are developing nanocomposite dendrimers consisting of a rigid core with active organic peripheries for use in solar cells and transistors.

2.3.4.4. *Characterization*

NUS is home to the inventor of portable scanning electron microscope (SEM) concept and researchers there have developed several attachments to SEM to significantly enhance the performance of standard SEM. Another group there has developed an image collection and processing system to control, capture, store and process images from SEMs. This group developed a cathodoluminescence detection system capable of performing monochromatic and extended the capabilities of the SEM to deep, sub-surface analysis presently not detectable in the SEM using the detection of thermal-acoustically generated signals. There is also ongoing research at NUS to build photoelectron emission microscope (PEEM) to enable imaging of plasmonic devices with high special and temporal resolution. IMRE researchers are using photoemission spectroscopy to determine hetero-junction band offsets to optimize and tune devices to specific applications. Another group there has developed

Ballistic Electron Emission Microscopy (BEEM) – which is a modified version of Scanning Tunnel Microscope (STM) which allows local charge injection barrier measurements for a buried interface such as in electronic devices. Researchers there have also developed a high sensitivity oxygen and water vapor permeation measurement system for use in flexible OLED device lifetime and degradation studies.

Spin-Offs

Semicaps Corporation (www.semicaps.com)
Semicaps Corporation Pte Ltd is a spinoff from NUS to commercialize the scanning electron microscope image collection and processing system (SEMICAPS). They offer a range of optical fault localization microscopy for FA, Product Engineering and Yield Enhancement. Features of these microscopes include: Multi detector capabilities (SiCCD, InGaAs, MCT), visible and NIR spectroscopy for defect fingerprinting, Multiple laser stimulation techniques (TIVA, LIVA, OBIRCH, TBIP, OBIC, SCOBIC, SDL, RIL), ultra high precision stages (<0.5μm repeatability). They have more than 300 installations worldwide.

Nanofilm Technologies (www.nanofilm.com.sg)
Nanofilm Technologies International Pte Ltd is a spinoff from NTU to commercialize Filtered Cathodic Vacuum Arc (FCVA) technology to produce thin films. The company has expanded its patent portfolio and expanded the application of the technology multiple industry sectors from hard disk drive (HDD) industry to the glass lens mold industry.

Other Companies in the Eco System

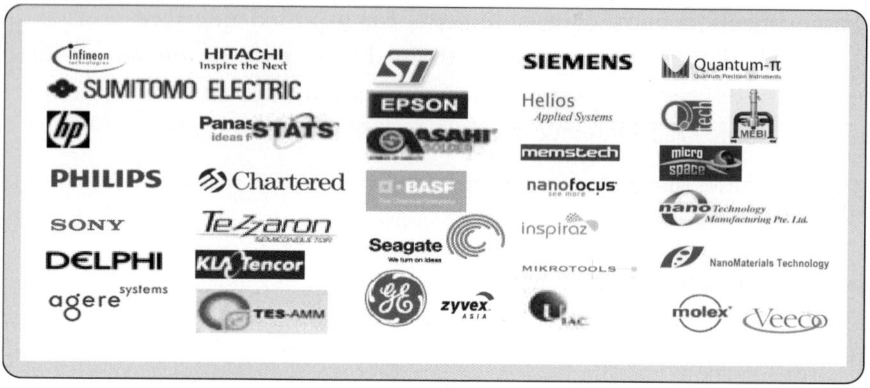

Fig. 6. Electronics Eco System in Singapore.

3. Conclusion

Nanotechnology is recognized as a key enabler for future economic growth and sustainability of the Singapore economy and this significance is reflected in the investment that has gone into nanotechnology-related R&D and manpower development here. This investment has already translated into world class technical achievements in different application areas such as nanobiotechnology and nanoelectronics with many exciting developments expected in areas as diverse as photonics and metamaterials to third generation energy systems in years to come. Many multinationals have setup their R&D facilities here and the local SMEs are slowly maturing. This makes for an exciting eco system for nanotechnology companies from around the world to plug in.

References

1. A*STAR - http://www.a-star.edu.sg/
2. NRF National Framework for Innovation and Enterprise - http://www.nrf.gov.sg/nrf/otherProgrammes.aspx?id=1206
3. EDB Incentives - http://www.edb.gov.sg/edb/sg/en_uk/index/why_singapore/Guide_to_Investing_in_Singapore/financial_assistance.html
4. Singapore Science & Technology 2010 - http://app.mti.gov.sg/default.asp?id=885
5. A*STAR Technology Commercialization - http://www.exploit-tech.com/For-Industry/A-STAR-Technology-Commercialisation.aspx
6. The Enterprise Challenge - http://www.tec.gov.sg/TEC%20Home/home1.htm
7. SPRING Programs - http://www.spring.gov.sg/Content/WebPageLeft.aspx?id=746dba13-abbc-4614-866b-fe2b29f98b8c
8. Asia Nano Forum - http://www.asia-anf.org/index.php
9. NUS Nanoscience and Nanotechnology Initative – http://www.nusnni.nus.edu.sg/
10. NUS NanoCore - http://www.nanocore.nus.edu.sg/
11. NTU NanoCluster - http://www.ntu.edu.sg/nanocluster
12. NUS Enterprise - http://www.nus.edu.sg/enterprise/
13. NTU Innovation and Technology Transfer Office - http://www.ntu.edu.sg/itto/Pages/default.aspx
14. NanoFrontier - http://www.nanofrontier.com.sg
15. NanoGlobe - www.nano-globe.biz
16. NanoConsulting - www.nanoconsulting.com.sg
17. SingNano - www.singnano.org
18. Singapore Green Plan 2012 - 2006 Edition - http://www.mewr.gov.sg/sgp2012

Author Biographies

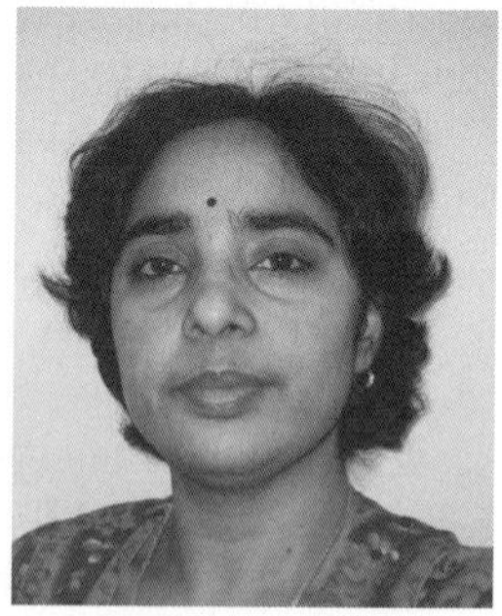

Dr Hiranmayee Vedam is the Managing Director of NanoConsulting Pte. Ltd. She is a commercialization expert specializing in technology based companies in materials, wireless and energy. She combines her experience in technology, business, IP and finance to provide venture capital companies and SMEs with deal sourcing, due diligence and business development services. She provides strategic policy support to governments and maintains the Singapore nanotechnology eco system map as part of the SingNano initiative. Prior to her current role, Hiran was a senior manager, Industry Liaison at the National University of Singapore (NUS). There, she was involved in spinning-off companies, licensing technologies to commercial companies and in business development activities to promote collaboration between NUS and industry. Prior to joining NUS, Hiran founded and managed MAtrika Pte. Ltd., a wireless application company and fostered the establishment of Honeywell's R&D center in Singapore. She graduated with a B.Tech in Chemical Engineering from IIT-Madras and completed her MBA at INSEAD and Ph.D. in Chemical Engineering at Purdue University.

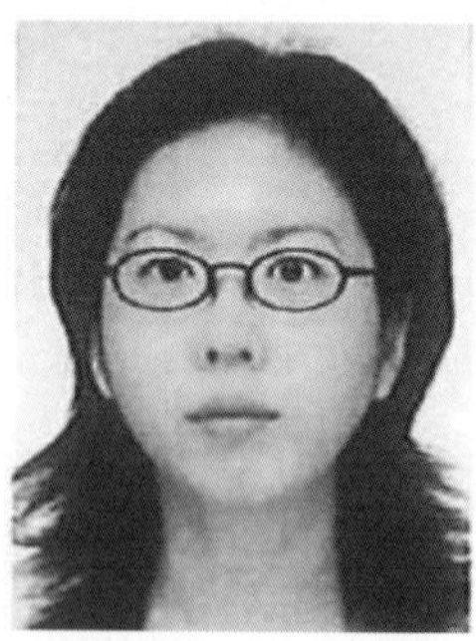

Hsu Ann Goy completed her BSc in Materials Science from the National University of Singapore and MSc(Eng) in Engineering Materials at the University of Sheffield, UK. She joined the Science and Engineering Research Council in 2003 where she assisted in the planning and execution of policies towards the smooth running of A*STAR's research institutes; and supporting the Economic Development Board and other agencies in R&D pertinent issues. She was recently posted to the Planning & International Relations Department where she will assist in the formulation, review and implementation of policies and ensure their alignment to the A*STAR's mission.

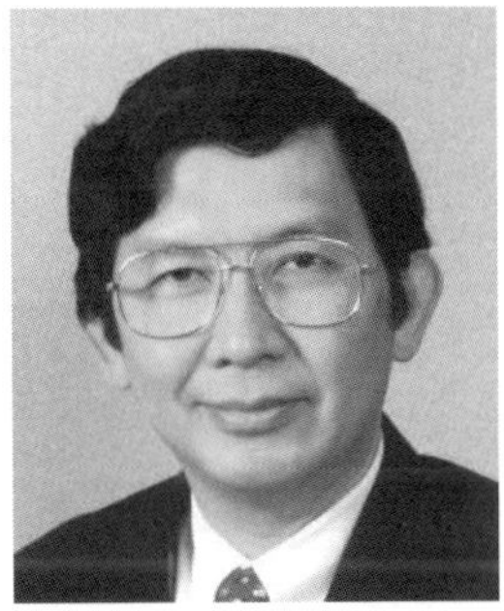

Dr Khiang Wee Lim is the Executive Director of the Institute for Materials Research and Engineering (IMRE) and concurrently the Programme Director for Research at Singapore's Science and Engineering Research Council, Agency for Science, Technology and Research (A*STAR). He graduated in Electrical Engineering from the University of Malaya and completed his DPhil in Engineering Science at Oxford University. Prior to A*STAR, he had academic appointments at the National University of Singapore and the University of New South Wales in Sydney. He is currently convening the Working Group for Standardisation, Risk Assessment and Safety of the Asian Nano Forum as well as leading the Singapore national ISO/TC 229 Working group that is part of the ISO effort to develop standards in nanotechnology.

CHAPTER 11

NATIONAL NANOTECHNOLOGY PROGRAM IN TAIWAN

Maw Kuen Wu[*,‡], Tsung Tsan Su[†,§] and Tsing Tang Song[†,¶]

[*]*Institute of Physics, Academia Sinica, Nankang, Taipei, Taiwan*
[†]*Nanotechnology Research Center, ITRI, Hsinchu, Taiwan*
[‡]*mkwu@phys.sinica.edu.tw*
[§]*tsung-tsan_su@itri.org.tw*
[¶]*ttsong@phys.sinica.edu.tw*

Taiwan government has lent long-term support to strengthen Taiwan's nanotechnology R&D capabilities in response to the socioeconomic needs. To this end, the government initiated the first phase of Taiwan National Nanotechnology Program (NNP) in 2003; the total budget of the program was approximately US$550 million over six years (2003~2008). Today, we have established one industrial common laboratory and nine academic core facilities centers. Acceleration strategies for nanotechnology commercialization are initiated and nanotechnology networking was set up as well. We have also developed talents cultivation program to train seed teachers and to enhance basic scientific knowledge and increase public awareness of nanotechnology. Since 2004, we have set up nanoMark certification system and established certification guidelines for nanoproducts. The nanoMark system aims to systematically manage nanotechnology products in the market and to let more public understanding and confidence in nanotechnology development. The second phase of NNP (2009~2014) has committed from the government, the budget of this six-year program is allocated over US$685 million (estimated in year 2008) from 2009 to 2014. The goals of the 2nd phase NNP are to strengthen the integration of Taiwan's competitive technologies for accelerating the commercialization of nanotechnology. We hope to derive hugely economical gains and induce a large number of private investments until year 2014.

1. Introduction

The Taiwanese government has pledged its long-term support to promote research and development (R&D) capabilities in various sectors, as a response to the country's socioeconomic needs. The government is committed to establish programs for training R&D personnel and implement a better mechanism integrating resources and personnel. To achieve these objectives, in addition to the regular R&D programs carried out or sponsored by various ministries, we have established the National Science and Technology Program, which all ministries have given their full support. Currently, there are seven National S&T Research Programs, which can be grouped in three categories:

(I) The development of digital learning, which includes the (1) Program for Digital Archives and E-learning;
(II) Promote innovative biotechnologies, which includes the (2) Program for Agricultural Biotechnology, (3) Program for Biotechnology and Pharmaceuticals, and (4) Program for Genomic Medicine; and
(III) Promote economic growth and develop new technologies, which includes (5) Program for Telecommunications, (6) Program for System-on-Chip, and (7) Program for Nanoscience and Nanotechnology.

The Taiwan National Science and Technology Program for Nanoscience and Nanotechnology was launched in 2003; the total budget of the program was set to around US$550 million spread over six years (2003~2008), and the second phase (2009~2014) of the program was recently approved with 30% budget increase from that of phase 1. The program aims to achieve academic excellence in basic research and to promote the commercialization of nanotechnology. The Ministry of Economic Affairs (MOEA) contributes 64% of the funding, while the National Science Councils (NSC) provides 28%. The remaining 8% consist of contributions from the Ministry of Education (MOE), Atomic Energy Council (AEC), Environmental Protection Administration (EPA), Department of Health (DOH) and Council of Labor Affairs (CLA) etc.

Prof. Maw-Kuen Wu from Academia Sinica and Dr Tsung-Tsan Su from Industrial Technology Research Institute (ITRI) serve as the Director-General and Co-Director of the National Program. There are four areas of focus in this initiative, which include excellence in academic research, industrialization of nanotechnology, the setting-up of core research facilities, and an extensive education program. The R&D funds are distributed as follows: 19% goes to the Academic Excellence Research Program, 64% for the Industrialization Nanotechnology Program, 15% for the Core Facilities Program, and 2% for the Education Program.

2. Scopes and Approaches

2.1. *Academic Excellence Research Program*

The goals of the academic excellence research program are to promote excellence in academic research, to support industrialization of nanotechnology, and to promote education and training for future scientists and engineers. We especially encourage innovation and interdisciplinary research from different disciplines to establish close interactions and to form a strong interdisciplinary research team. 89% of total research funding comes from the NSC and the other 11% from the DOH. Each project under this interdisciplinary program is comparatively much larger than most conventional research projects in Taiwan. The funding for each project approximates 0.5 million US and up to a period of three years. There are five research areas in the Academic Excellence Research Program for fundamental investigations, which include (1) basic research on physical, chemical and biological properties of nanostructures; (2) synthesis, assembly, and processing of nanomaterials; (3) development of manipulation techniques and fabrication of functional nanodevices; (4) energy and environment related nanotechnology; and (5) nano-biotechnology. We encourage collaboration between academic institutions and industrial laboratories, as well as other industrial sectors, to explore cutting-edge nanotechnology research and prototype-orientated studies. New programs promoting the collaboration between academia, industry, and university-research organizations collaboration were established in 2005. Each project lasts for three years, with funding of about 0.8 million US per year. The special fund derived from the NSC controlled foundation was allocated to speed up the industrialization of innovative research results by academia, and eventually to commercialization. Furthermore, we also strongly urge the participating groups to involve in further international collaboration.

2.2. *Nanotechnology Industrialization Program*

The goals of the nanotechnology industrialization program are to enhance core facilities and networks, to speed up the development of nanotechnology, to develop and apply the new features of nano-materials to leverage industrial knowledge and expand opportunities, and to integrate Taiwan's most competitive technologies and industries for a better chance of success in key areas. This program develops many emerging technologies, such as nanomaterials and processing techniques, nanoelectronics, advanced display, nanophotonics, electronic packaging, data storage, energy conversion and storage, traditional industry upgrade/transformation, and biotechnology. Many research institutes, including ITRI, the Chung-Shan Institute of Science and Technology (CSIST), the Taiwan Textile Research Institute (TTRI), and the Plastic Industry Development Center work under MOEA support. The Institute of Nuclear Energy Research (INER) of AEC also participates in this program. Besides, a joint research venture among three agencies – EPA, CLA and DOH initiated integrated environment, health and safety (EHS) projects that focus nanoparticle on nanotoxicology, occupational health and safety to promote the responsible development of nanotechnology. Research institutes participate in this program are listed in Table 1.

2.3. *The Core Facility Set-Up Program*

The core facility set-up program, which aims at setting up facilities for nanofabrication, precision measurements, and expertise, will provide

Table 1. Research institutes participate in Nanotechnology industrialization program

Research Institutes	Website	Introduction
Nano Tech-nology Research Center, ITRI	www.ntrc.itri.org.tw/eng/index.jsp	The goals of ITRI nanotechnology program are to build industrial applications and establish technical competencies under five major-thrust areas including information and communications, energy applications, process and equipment development, traditional Industry application, and biomedical applications. More than 10 research Labs/Centers in ITRI have involved in this program. NTRC performs several key functions; it is the window for strategic planning, coordination, and international contact.
Chung-Shan Institute of Science and Tech-nology	www.csistdup.org.tw	The CSIST is a research institute for national defense. For respond the needs of industry, it is assigned as the window of converting defense technology into industries.
Taiwan Textile Research Institute	www.ttri.org.tw	The goals of TTTR are to establish the manufacture and application technology of nanocomposite fibers or textiles for various functionalities such as antibacterial, fire resistant, electromagnctic wave, and absorbing and absorptive filter.
The Institute of Nuclear Energy Research	www.iner.gov.tw	The INER has applied its nuclear energy expertise on the development of nanotechnology, and its aim is to establish the key techniques for fuel cell utilizing the hydrogen energy and solar cell.

Table 1 (continued)

Research Institutes	Website	Introduction
Center for Nano-Medicine Research National Health Research Institutes (NHRI)	www.cnmr.nhri.org.tw	The goals of CNMR are to provide a platform for Nanotechnology research in bio-safety including extrapolation of animal results to human beings, interfacial chemistry between nanoparticles and biological tissues, and possible entry of nanoparticles into endogenous metabolic pool.
The Institute of Occupational Safety and Health (IOSH)	www.iosh.gov.tw	The IOSH is a research institute under the jurisdiction of the CLA. Its missions include application of scientific technology, surveys and analyses of various risk factors in the working environment, as well as development of countermeasures.
The Environmental Protection Administration (EPA)*	http://www.epa.gov.tw	The objectives of the EPA nanotechnology program are to promote environmental management in R/D labs and industries, to revisit existing regulations and management systems to ensure that nanotechnology specifications are taken into account and to study the impact of widely used nanoparticles on the environment.

* Call for proposals from universities and research institutes for nanotechnology research.

common support for research in nanotechnology. Common Laboratories are designed to support the academic community, R&D organizations, and industry. Until now, we have established one nano common laboratory which is located within ITRI and nine academic centers which are located throughout Taiwan at Academia Sinica (AS), National

Taiwan University (NTU), National Tsing-Hua University (NTHU), National Chiao-Tung University (NCTU), National Chung-Cheng University (CCU) distributed across a number of universities in central Taiwan, National Cheng-Kung University (NCKU), National Sun-Yet-Sen University (NSYSU), National Dung Hua University (NDHU), and National Synchrotron Radiation Research Center (NSRRC). Figure 1 shows the location of core facility set-up program in Taiwan.

2.3.1. *Nano Common Laboratory*

The Nano Common Laboratory center is located within NanoTechnology Research Center, ITRI. The aims of this laboratory are to offer world-class R&D environment including both software (service, technology) and hardware (equipment and instrument) to offer an excellent R&D environment for researchers. This lab has a floor space of 11,220 square meters and has analysis/measurement and fabrication equipment with extensive capabilities. It conducts microstructure/composition analysis, and detailed characterization for industrial applications.

2.3.2. *Core Facilities Center for Nanoscience and Nanotechnology at the AS (NanoCore)*

The NanoCore is located at the AS, its mission is to support and promote the advanced research, such as the studies on nanostructures, nanomaterials, nanodevices, and nanobiotechnology, and develop the state-of-the-art research tools for nano-science and technology. A novel facility in the realm of bio-imaging has built up for the ultrafast phase contrast wet-cell electron microscopy.

2.3.3. *Core Facility Center for Advanced Microscopy and Nano-Analysis*

The Center Core Facility Center for Microscopy and Nano-analysis is located in the Center for Condensed Matter Sciences of NTU. The goal of the center is to set up a frontier scanning transmission electron

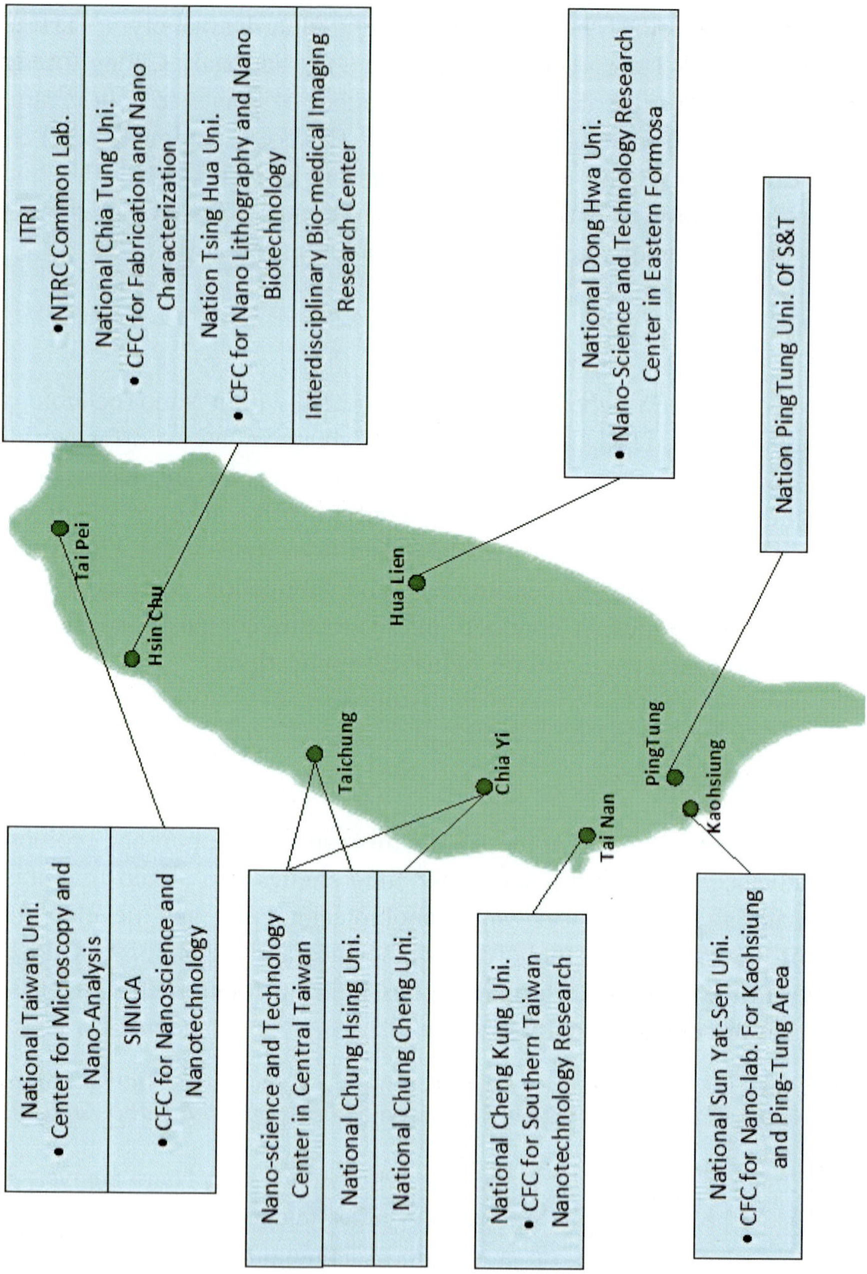

Fig. 1. The location of core facility set-up program in Taiwan.

microscope and other advanced microscopy tools for nano-science and technology applications, and to establish an expert-based system for the provision of expert consultations and promotion of innovative collaborations.

2.3.4. *Core Facilities for Southern Taiwan Nanotechnology Research Center*

The Core Facilities for Southern Taiwan Nanotechnology Research Center is located in the Micro/Nano Science and Technology Center at NCKU. The center combines experts in the field of micro-electro-mechanical-systems to design, construct, and optimize unique and high performance nanoimprint tools and technology for supporting future needs by the semiconductor and optoelectronic industries in Taiwan.

2.3.5. *Core Facility Center for Nano Fabrication and Nano Characterization*

The Core Facility Center for Nano Fabrication and Nano Characterization is located within Center for Nano Science and Technology in NCTU. They are three main topics for the center: (1) nano-structure growth and fabrication, (2) structure-analysis and (3) low temp and optical nano characterization. The center is working together with National Central University to form a regional equipment network.

2.3.6. *Core Facility Center for Nano Lithography and Nano Biotechnology*

The Core Facility Center for Nano Lithography and Nano Biotechnology establishes a nano lithography platform which includes a dedicated state-of-the-art SEM for e-beam writing and an advanced nanoimprint lithography system for nano device designs in NTHU, and an imaging, manipulation, and characterization core facility specifically suited for studying live cells, tissues, and biomolecules in National Yang-Ming University (NYMU).

2.3.7. *Nano-Science and Technology Center in Central Taiwan*

The Nano-science and Technology Center in Central Taiwan distributed across a number of universities to jointly establish core facilities by consolidating and enhancing their resources in nanotechnology. The universities include National Chung Cheng University (CCU), National Chung Hsing University (NCHU), National Changhua University of Education, National Chi Nan University, and National Yunlin University of Science and Technology. The Core Facility addresses the technology and human resources needs in nano science and technology of the region.

2.3.8. *Core Facilities Laboratory for Nano-Laboratory for Kaohsiung and Ping-Tung Area*

The Core Facilities Laboratory for Nano-laboratory for Kaohsiung and Ping-Tung Area is located in NSYSU. The goal of the center is to foster research teams that may become unique in Taiwan, such as the low-dimension semiconductor research and spintronics in oxide, and to educate abundant high quality research manpower for the academic research and industrial application.

2.3.9. *Nano-Science and Technology Research Center in Eastern Formosa*

The goal of this core facility center is to establish the infrastructure for a multidisciplinary Center of Nano Science and Technology in Eastern Taiwan with special emphasis on nano-biotechnology/medicine and nano-materials. This center integrated the existing complementary instruments and facilities at National Dong Hwa University and Tzu Chi University under a unified management.

2.3.10. *Interdisciplinary Bio-Medical Imaging Research Center*

The Interdisciplinary Bio-medical Imaging Research Center has been set up in National Synchrotron Radiation Research Center (NSRRC). The overall facility is closely linked with the life science research institutes including AS, NYMU, Mackay Memorial Hospital, NTHU and NHRI. This center aims to establish a very extensive biomedical imaging facility, by integrating with several cutting-edge biomedical microscopy modalities to perform multi-dimensional, functional, anatomical and in vivo microanalysis of live specimens.

2.4. *Education Program*

The main goals of the education program are to promote life-long learning in Nanotechnology Education, to provide higher quality professional education and K-12 education, and increase public awareness of nanotechnology. This program has established five regional personnel development centers, which serve as local resources within universities to form an education network for students and continuing education for professionals. We have also developed five regional K-12 personnel incubation centers to train seed teachers and to enhance basic scientific knowledge. To date, several books related to nanotechnology have been published. For example, a book entitled Nanotechnology Symphony, used for chemistry, physics and biology has been produced for high school students. A 3D comic book entitled Nano BlasterMan has been created for elementary and middle-school students. An animated cartoon video, Nana and Nono, has been created to help educate younger children. Other interactive multimedia disc, animated teaching disc and nanotechnology curricula were produced to help develop a comprehensive set of teaching materials for K-12 students. An e-learning networking system and online nanodictionary are established to bring nanotechnology understanding to the general public.

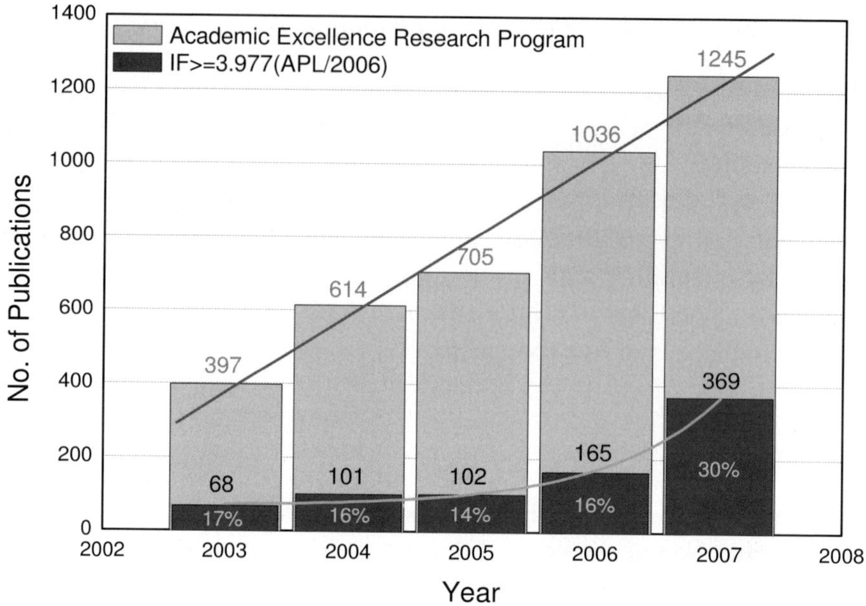

Fig. 2. Publications from 2003 to 2007 by scientists who receive support through the academic research program.

2.5. *Highlights of the Research Outcomes*

During the past five years, we have made substantial achievement in all areas of our programs. Figure 2 shows the publications from 2003 to 2007 by scientists who receive support through the academic research program. The overall number of publications increases linearly every year. More interestingly, the increase in high impact publications, here we use the impact factor of Applied Physics Letters as the lower bound of the criterion, can almost be fit to an exponential function. This set of data strongly suggest that people involved in nanoscience and nanotechnology research have gradually accumulated their expertise and are now applying that to generate new innovations. In Fig. 3, we highlight the nano-materials fabrication capability of the laboratory led by Prof. L.C. Chen and Prof. K.H. Chen. It is truly amazing to see a

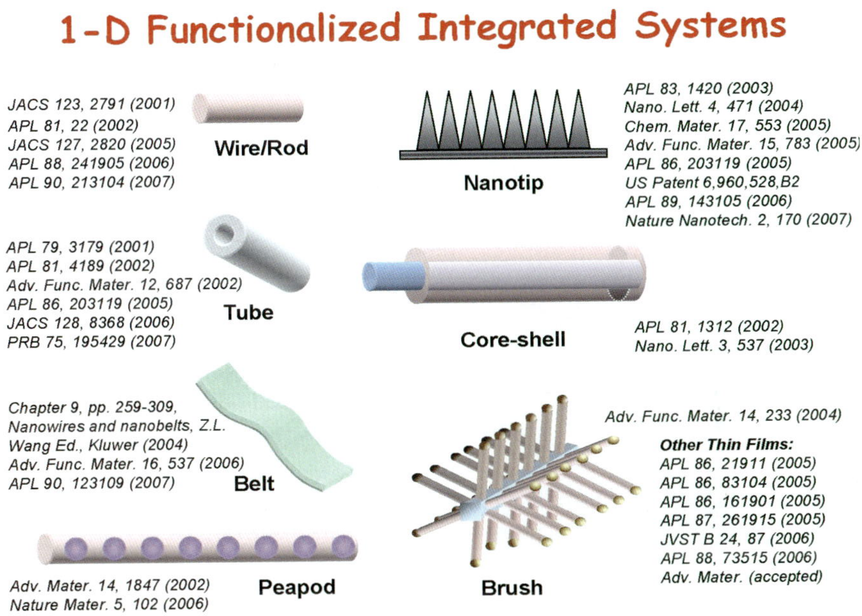

Fig. 3. The collection of research results on nanomaterial provided by Prof. K.H. Chen and Prof. L.C. Chen.

single laboratory produce so many novel nano-materials in such a short time.

A novel structure of polyimide (PI) with high Tg and high transparency have been developed in ITRI. The colorless PI/nanoparticle hybrid substrate has features of high Tg (~350°C), high light transmittance (~90%), excellent thermal tolerance, and dimensional stability. It is a promising substrate for the next generation active matrix flat panel display. A light, thin and flexible 7-inch active matrix TFT-LCD panel on a colorless PI substrate with an ultra low-temperature (200°C) a-Si:H TFT fabricated is successfully developed (Fig. 4). It has no glue

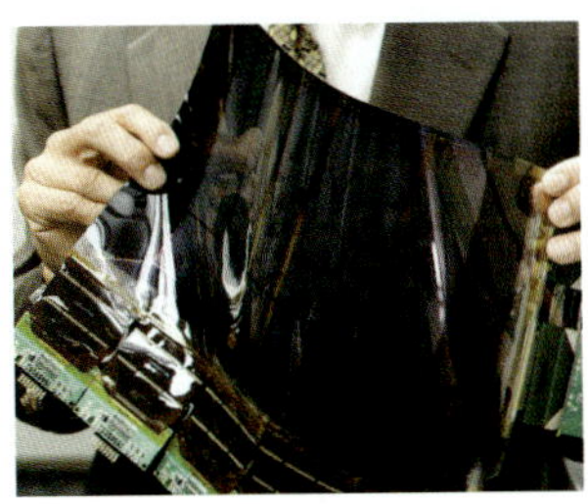

Fig. 4. Light, thin and flexible 7-inch active matrix TFT-LCD panel.

residue issues during the TFT backplane process. De-bonding easily is its great advantage. More innovated applications of flexible displays are in progress.

A novel amorphous functional fluoropolymer with transparent and low-dielectric constant has developed in ITRI for electret applications. The internal effective surface area increases substantially due to the nano-porous structure and average surface charge can reach as high as 400 V/μm. This light, thin and flexible nano-porous electret material has been used to make flexible speakers. Figure 5 shows an ultra-thin flexible electronic product consists of timing and information data processing functions, it can stick to the surface of wall, car, etc.

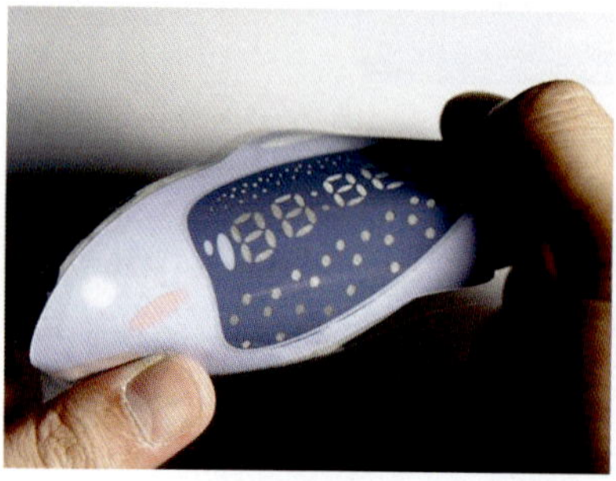

Fig. 5. Electret-based flexible speaker.

Table 2 summarizes the overall performances of the national program including academic publications, technical innovations (both patents pending and granted), and number of companies that invest in the development of technology. The total number of patents including both pending and granted are over 4000 in about 5 years, which means on average every US$24,000 investment will generate one patent. No other National Program sponsored by the government has such high productivity.

One other very important index to indicate the success of the industrialization program is the investment in R&D on nanotechnology by private sectors induced by the national program. The data, as also shown in Fig. 6, shows that the ratio between the contributions from private sector in R&D on nanotechnology to what the government invests to the industrialization program is close to 1 by year 2007. We expect this ratio to increase further in the future.

Table 2. Performance of National Sci-Tech Program for Nanoscience and Nanotechnology

Performance Index	2003	2004	2005	2006	2007	2008(~2. Quarter)	Total
Academic Credit							
Publications (Number)	668	1,074	1,165	1,514	1,997	599	7,017
International	518	873	1,040	1,314	1,693	546	5,984
Domestic	150	201	125	200	304	53	1,033
Technical Innovation							
Patents Pending (Number)	280	351	518	717	814	331	3,011
International	100	136	256	432	463	122	1,509
Domestic	180	215	262	285	351	209	1,502
Patents (Number)	205	215	167	220	254	94	1,155
International	50	128	55	77	148	30	488
Domestic	155	87	112	143	106	64	667
Technology Transfer (Number)	50	76	64	42	109	25	366
Early Participation Project (Number)	31	55	46	52	47	14	245
Economic Benefits							
Induced Investment (Number)	20	43	48	64	98	6	279
Amount of Investment (TWD Thousand)	187k	1205k	1210k	1357k	2174k	13k	6149k

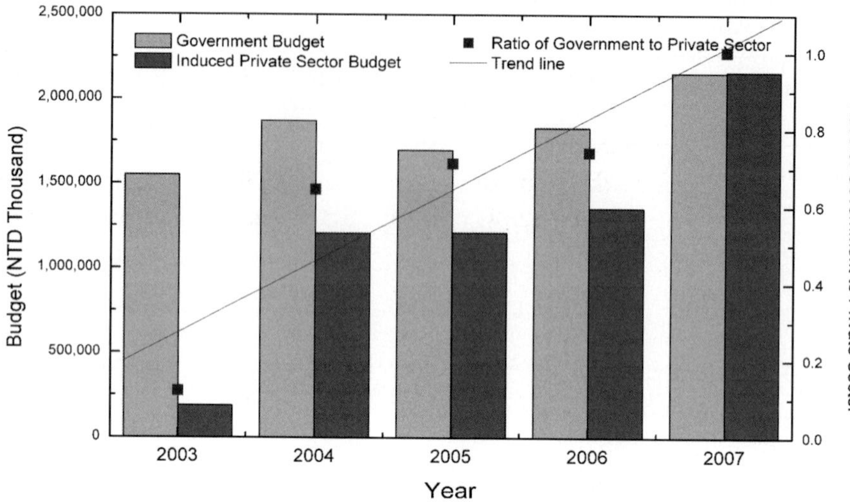

Fig. 6. Government and Private sector R&D budget for Nanotechnology Industrialization.

3. Rapid Industrialization Strategy

Since the beginning of the national nanotechnology program, several strategies to accelerate nanotechnology commercialization have been initiated. In order to realize the potential benefits of nanotechnology, however, there are still many challenges for the commercialization of nanotechnology. The challenges on the technology side include:

(a) A multi-sector market system is recognized in the field of nanotechnology. Nano product manufacturing must integrate the knowledge bases of different sectors, meaning that small enterprises need make lot of efforts involved;
(b) Many competing nanotechnologies have been developed rapidly at either commercial or pre-commercial stages at around the same time. The opportunities seemed huge, but it also reveals enormous challenges;
(c) Require to set up more test sites for users to develop new products.

On the nano business side, the challenges include:

(a) Only "Nano" cannot solve all the problems of technology, many other factors need to be considered when developing a process plan for an entire product.
(b) For profitability, the approach of pain-killers, for example, is more helpful than vitamin pills to achieve superior differentiation. Similarly, provided sustainability of series products and continuous improvement capability will enable sustained profitable growth.
(c) Sales channels are relatively limited.
(d) Public acceptance of nanotechnology is still lacking. A dialogue with the general public is required to reduce doubt in nanotechnology. More public understanding and support is crucial for facilitating the responsible development of nanotechnology.

In order to overcome the challenges mentioned above and take advantage of nanotechnology's opportunities, many approaches are being implemented in Taiwan. These approaches can be seen below:

(1) Technology-driven approach
For a multi-sector market, how to collaborate with various companies are important considerations to ensure the success of business. For example, a non-profit industrial alliance named Carbon Nano Capsule Industrial Alliance was founded in November 2002 with 30 group members. Within the alliance, they can share knowledge and information to accelerate carbon nanocapsule industrialization, to foster strategic partnerships among members, and to bridge government R&D resources and industrial needs for looking the potential opportunities. Currently, under the integral structure of using carbon nano capsules, many achievements are obtained, which include an anti-inflammatory conducting polymer for electronic devices, magnetic carbon nanocapsule for catalyst support, and cisplatin separation, automobile radiator, and heat pipe etc. This strategy will dramatically help improve product success in the market.

(2) Need-driven approach

This approach involves rapidly monitoring the changes of customer demand, and to identify a painkiller approach towards product development. It is necessary to perform further scenario and alternative planning to see how different combinations of requirements can best help create a potentially new feature product. The requirement will be identified as most significantly affecting market trends. For example, in order to produce a highly competitive flexible substrate for flexible devices, a low cost and new structure of high Tg, colorless polymer is being synthesized in ITRI. A hybrid of nanoparticles and polymers will retain transparency, decrease CTE, increase chemical resistance, and reduce birefringence of the flexible substrate respectively. This approach will completely solve the issues of the plastic substrate and provide significant benefit to the customers.

(3) Application innovation-driven approach

The application Innovation-driven approach focuses on one unique property that will conduct the different needs of the market to create potential opportunities. We generated the concept through brainstorming to motivate new applications ideas. Actions include conversion of knowledge bases in areas such as new phenomena, new features and new theories of nanotechnology into structured databases, and integration of structured knowledge into a patentable application concept. For example, an Electret Films trademark was named ELECTEN® from ITRI (See Fig. 5). It uses novel partial F-polymer to form nanoporous films to make a higher interior surface area, which can help keep long-term retention and stability of charges. Furthermore, based on this technology platform, more application Innovations have been created to enlarge application areas, such as touch panels, actuator microphone flexible speakers, flexible thin devices (combine speaker, display, battery etc.), hearing aids, and portable and flexible asthma detectors.

(4) Sustainability-driven approach

The sustainability-driven approach is focusing on one market orientation through collaboration with downstream users to offer a wide variety of products and services for customers. For example, we have applied the concept of Lotus Effect to build car wax by adding micro-/nano-powder to create rough surfaces. A series of products for car polishing, e.g. hard wax, soft wax, cream wax (liquid wax), and plastic grinding stone/clay were developed. Companies for Car Beauty and Care Services will enter partnership for understanding future needs. We believe that only one product cannot easily sustain a company, or service a market for sustainable operation.

The quick commercialization of new technologies has always been Taiwan's forte, and this is the strength we hope to employ for the development of nanotechnology. It is important to efficiently convert nanotechnology-related knowledge to application ideas, and to completely integrate the supply chain between upstream and downstream users to ensure the success of businesses. To this end, ITRI, under the support of the MOEA, has established the "Prototyping service project" to accelerate the realization of nanotechnology since 2005. Through the integration of worldwide nanotechnology development and brainstorming with potential users, practical product specifications would be defined to identify commercialized application at a faster pace. The practical criteria include: the supply of nanomaterials, the rapid prototyping of nanoproducts and function evaluation, the collection of user opinions for further improvement.

Within the past five years in Taiwan, many nanotechnology networking and R&D consortia have been established to promote nanotechnology industrialization. For example, the Taiwan Nanotechnology Industry Development Association (TANIDA) was founded in July 2004 to serve the nascent industry by 57 distinguished members from public associations, industries, universities and research institutions. There are five technology committees: the technology development and publication committee, the industry promotion committee, the standards committee, the international business committee, and the member recruitment and service committee. The aims are to facilitate the integration of R&D efforts at universities, research institutions and industry to accelerate the commercialization of nanotechnology and to meet industries' needs of developing high valued-added products and technology. Other nanotechnology networking groups, such as the Nanotechnology Community, which was set up in

July 2003, connects R&D institutes to industry; this alliance integrates upstream nanomaterials and downstream application users to share research results and information. The Nanotechnology Common Facility Alliance was set up in April 2003. Its members join the ITRI core facility center to share information and technology. The Taiwan Photocatalyst Industry Association was set up in June 2004 focusing on photocatalyst development and related product applications. In its entirety, 75% of nano-related companies joined the networks in Taiwan.

4. NanoMark System

Since 2004, we set up a nanoMark certification mechanism and established certification guidelines for nano products. The motivations of the nanoMark System are mainly to protect consumers who buy certified nano products, to protect good companies from unfair competitions, to enhance public trust for facilitating healthy development of nanotechnology, and to stimulate economic growth in the field of nanotechnology. In order to assess risk and systematically manage nanotechnology products in the market, the structure of certification system includes the Technical Committee and nanoMark Promotion Committee to promote nanoMark. The nanoMark Committee will certify companies' products with nanoMark, if it contains a material of nano dimensions less than 100nm, and new useful nanotechnology characteristics are identified. There are many working groups with different tasks operating under the technical committee to draft certification guidelines and testing methods and identify testing laboratories to support the required infrastructure of the nanoMark System. Applications for certification of nano products including anti-bacterial light tubes, air-cleaning light tubes, anti-bacterial ceramic tiles, anti-smudgy ceramic tiles, anti-smudgy paints, deodorized paints, anti-smudgy sanitary facilities, anti-abrasive PU synthetic leathers and resins, air purifier and filter, anti-bacterial household textiles, anti-smudgy household exhaust hood have begun in 2004.

Table 3. Summarized nanoMark companies in Taiwan

Company	Nanoproduct	Website
Taiwan Fluorescent Lamp Co., Ltd.	Anti-bacterial Light Tubes	www.tfc.com.tw
China Electric Mfg. Corp.	Anti-bacterial Light Tubes	www.chinaelectric.com.tw
Champion Building Materials Co., Ltd.	Anti-bacterial Ceramic Tiles	www.champion.com.tw
Hsin Mei Hua Paint Factory Co., Ltd.	Deodorized Paints, Anti-smudgy Paints	www.starcoating.com.tw
Tatung Fine Chemicals Co., Ltd.	Deodorized Paints, Anti-smudgy Paints	www.twtfc.com
Tien Kuang Enterprise Co., Ltd.	Anti-smudgy Sanitary Facilities	www.tenco.com.tw
NeoGila Co., Ltd.	Anti-abrasive PU Synthetic Leathers	www.neogila.com
United Ceramics Co., Ltd.	Anti-smudgy Ceramic Tiles	www.roma.com.tw
Lidye Chemical Co., Ltd.	Anti-abrasive PU Resin	www.43671990.boss.com.tw
Headway Group	Anti-abrasive PU Resin	www.headway.com.tw
Everest Textile Co. Ltd.	Anti-bacterial Household Texile	www.everest.com.tw
Long Qiang Nano Technology Co., Ltd.	Anti-bacterial Household Texile	www.longq.com.tw
Vitallife Nano Technology Co., Ltd	Air-purifier and Filter	www.vitallife.com.tw
Hocheng Co., Ltd	Anti-smudgy Sanitary Facilities	www.hcg.com.tw
Sunny Win Light Ray Nano Technology Co., Ltd.	Anti-bacterial Marble	www.sunnywin.com.tw

Table 3 (continued)

Company	Nanoproduct	Website
ABA Nano-tech Company	Anti-bacterial Household Textile	www.phyllotex.com.tw
Chin Hong Enterprise Co., Ltd	Anti-smudgy Metal Partition	www.chinhong.com.tw
Modern Alloy Plating Co., Ltd	Anti-rusting Metal Fastener	www.mapt.com.tw/ environment.htm
Kinglin Technical Industrial Co., Ltd.	Deodorized Paints	

As of Dec. 2008, 19 suppliers with more than hundred of nano products had passed the nanoMark certification. The certified companies for nanoMark are summarized in Table 3.

5. Conclusions

Since the implementation of the National Program, some significant outcomes have increased S&T efficiency and have positively impacted industrial output. The volume of nanotechnology literature has increased tremendously, and many patents were also generated. Hundreds of newly developed technologies have been transferred to domestic companies and induce over US$140 million investment. A survey by the TANIDA found that private investment has reached about US$3.4 billion in the field of nanotechnology by 2007, and around 1.8 billion will be invested continuously in the three years. We believe that the upcoming novel research results will present many more opportunities for novel industrial applications. Accelerated strategies for nanotechnology commercialization have been carefully planned, and nanotechnology networking has been set up as well. Some achievements of academic researches ready for further application studies have been identified and provided resources for more integrated collaboration efforts among the universities, industrial laboratories, and companies. We will continually strive to engage the university-industry partnerships in nanotechnology research. We are also seeking partners to build international networking and multi-lateral collaboration to share resources, experimental facilities, and complementary strength to promote nanotechnology study.

The second phase of Taiwan National Nanotechnology Program (2009~2014) was approved by the "Council Meeting" of the NSC in April 2008. The program encompasses six main themes including advanced research, electronics and optoeletronics, energy and environment, instrument development, nano-biotechnology, and traditional industry application. We will also continue to promote key strategic projects which include EHS issue, education program, nano standardization, nanoMark, university-industry partnerships, international collaboration etc. We believe that this National Program will contribute to substantial economic gains by encouraging major private investments in nanotechnology development by 2014. We anticipate that the success of this program will not only create innovative R&D for academia and industries, but also allow us to create a new "*Taiwan Miracle*".

Author Biographies

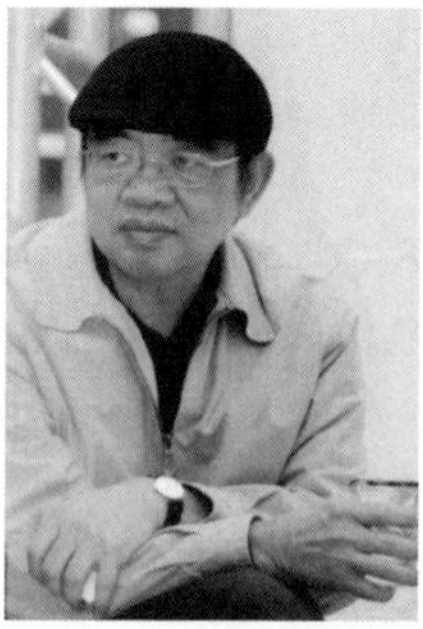

Prof. Maw Kuen Wu currently serves as the Director of the Institute of Physics, Academia Sinica. He is a solid state experimentalist specialized in magnetism and superconductivity. He received his bachelor and master degrees in physics from Tamkang University in Taiwan, and completed his Ph.D., also in Physics, at the University of Houston. He has been a professor of physics at several institutions in Taiwan and in the United States America, including University of Alabama (in Huntsville), Columbia University (New York City) and National Tsing Hua University (Taiwan). He has also served as Deputy Minister and then the Minister of the National Science Council of Taiwan prior to take up his current position at Academia Sinica. Because of his major accomplishment in the field of high temperature superconductivity, he has been awarded many honors including a special award from NASA (1988), the US National Academy of Sciences Comstock Prize (1988), the Bern Matthias Prize (1994), elected to the membership of Academia Sinica (1998), became foreign associate of the US National Academy of Sciences (2004) and member of the Third World Academy of Sciences (2004).

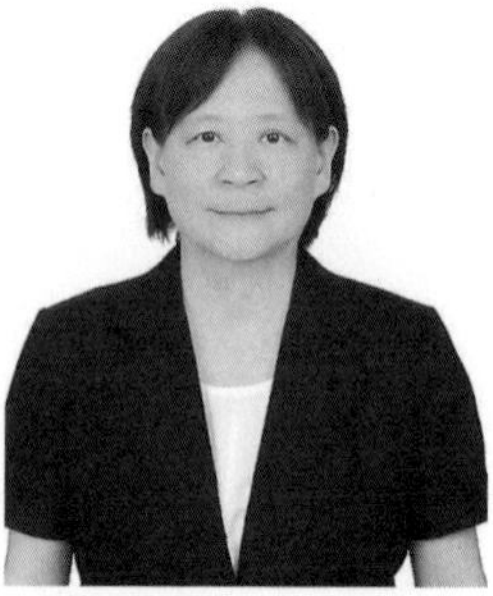

Dr Tsung Tsan Su is the general director of the NanoTechnology Research Center of Taiwan's Industrial Technology Research Institute (ITRI). In 2006, she was appointed as co-director of Taiwan National Science and Technology Program for Nanoscience and Nanotechnology, which consists of four R&D focuses, Academic Excellence, Industrialization, core facilities and Education. She received her BS degree (1972) in Chemistry from National Tsing-Hua University, Taiwan and Ph.D.

degree from Princeton University, USA in 1977. With over two decades of field experience in organic synthesis, analytical chemistry, pollution prevention, cleaner production and nanotechnology, she developed the ability to integrate interdisciplinary technologies. In 1994, she attended the International Senior Management Program at Harvard Business School to further refine her management skills. Her principal research interest is the development of nanotechnology, electronic chemicals as well as cleaner production.

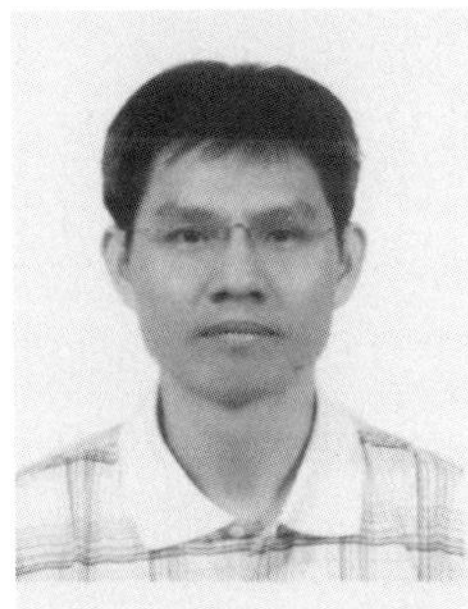

Dr Tsing Tang Song received his Ph.D. degree in chemistry from National Taiwan University. He is currently Chief Executive Officer of Taiwan National Program for Nanoscience and Nanotechnology, National Science Council. He is also Secretary General of the Taiwan Nanotechnology Industry Development Association, and Vice Chair of Technical Committee of nanoMark System, Ministry of Economic Affairs. In 1996, he joined Industrial Technology Research Institute (ITRI) and founded a research group in photosensitive materials working on new photoresist materials technology for semiconductor lithography application. The technology of his group was successfully transferred to industry in Taiwan. Dr Song is also a co-current research scientist in nanotechnology at NanoTechnology Research Center, ITRI.

CHAPTER 12

NANOTECHNOLOGY DEVELOPMENT AND OUTLOOK IN THAILAND

Wiwut Tanthapanichakoon*, Teerachai Pornsinsirirak†, Weeraya Pakawech‡, Manida Teeratananon§ and Sirasak Teparkum¶

National Nanotechnology Center, National S&T Development Agency
Ministry of Science and Technology 130, Thailand Science Park
Klong Luang Pathumthani, 12120, Thailand
**wiwutt@scg.co.th*
†teerachai@nanotec.or.th
‡weeraya@nanotec.or.th
§manida.prudhomme@ensiacet.fr
¶sirasak@nanotec.or.th

Nanotechnology is billed as the revolutionary manufacturing technology of the 21st century with immeasurable impact on the global economy and human society. Since the announcement of the U.S. National Nanotechnology Initiatives (NNI) in the year 2000, there have been tremendous movements in R&D support in nanotechnology all over the world. The Thai government also recognized the importance of this emerging cross-disciplinary technology. To keep up with the advancements in this field, the National Nanotechnology Center (NANOTEC) was established on August 13th, 2003, as an agency under the umbrella of the autonomous National Science and Technology Development Agency (NSTDA), Ministry of Science and Technology (MOST). The establishment of NANOTEC signifies the long-term commitment of the Thai government towards achieving excellence in this field. Approved by the Cabinet on June 12th, 2007, Thailand's first National Nanotechnology Strategic Plan spells out the direction in which the future of nanotechnology development in Thailand till 2013 would proceed in five key strategic areas:

- thrusting nanotechnology towards Thailand's key economic clusters
- developing and nurturing human resource
- establishing world-class infrastructure
- implementing rigorous R&D
- creating public understanding and appreciation

To be successful, it is crucial that the plan is supported at all levels, particularly the top level. Chaired by the prime minister, the National Nanotechnology Policy Committee (NNPC) was established in 2005. The current status in terms of R&D activities, human resource development, infrastructure development, technology transfer and business opportunities are described in this report.

1. Introduction

As a driving force for sustainable economic development, Science and Technology (S&T) is one also of the most important keys in any national development towards a knowledge-based society (KBS). In order to achieve harmonious, sustainable development, the developing countries must strive towards becoming a KBS and focus on comprehensive innovation — a complete cycle of innovations with big impact on the economy and society. The S&T system should not only produce scientific papers for "basic science" and "applied research" but also develop technology through patents and prototypes necessary for successful development of innovative products.

1.1. *Thailand's Big Step Towards Excellence in S&T and Knowledge-Based Society*

Back on September 17th, 1996, when the Cabinet approved the 1st National S&T Development Plan (1997-2006), four major areas were emphasized in Science and Technology Policy: (1) human resource development (HRD); (2) research and development (R&D); (3) technology transfer, and (4) infrastructure development. Though the aforementioned areas are vital, they are by no means complete. The Ministry of Science, Technology and Environment (MOSTE) initiated this 10-year plan and served as coordinator responsible for making related action plans[1].

As Bell (2002) pointed out, policy systems and processes had a wider scope than what the specific "S&T" focused upon. S&T should be embedded in perspectives on policy that addresses a broader spectrum of concerns about knowledge, innovation and competitiveness in the global economy[2]. In this national plan, the strategic goal is to develop the country's competitiveness based on knowledge in S&T, research and innovation. Core strategies must include building a favorable environment for useful creativity, learning, research and innovation as well as the development of unique S&T management systems at the policy and practical levels with objective evaluation.

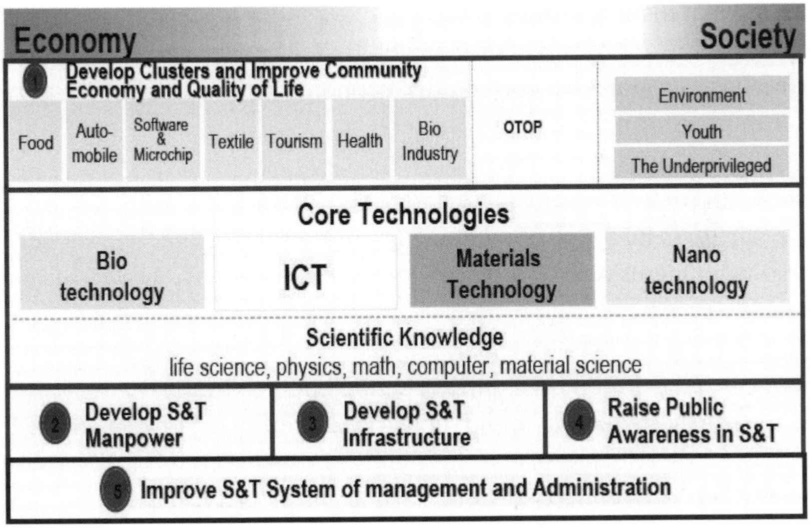

Fig. 1.1. The framework of National S&T Strategic Plan (NSTSP 2004-2013).

1.2. *National S&T Strategic Plan*

With the rapid relentless development of technology in today's world, the need to update the national S&T plan and keep up with changes has become crucial. In 2004, the government via the National Science and Technology Policy Committee (NSTC) issued Thailand's first 10-year

National S&T Strategic Plan (NSTSP 2004-2013). The main objectives of the Plan are to enhance Thailand's capability of responding to rapid changes in the era of globalization, thereby strengthening the country's long-term competitiveness. Using S&T as key propulsion tool, the NSTSP adopted a cluster development approach to reinforce the capabilities of the targeted economic sectors. The core concept focuses on systematic development, linkages among key sectors, namely manufacturers, suppliers, research and advanced academic institutes, financial institutions and other relevant organizations both public and private. The NSTSP describes five key development strategies as shown in Fig. 1.1.

Strategy 1: To develop industrial clustering and to improve community economy and quality of life
The objectives are to increase technological capacity and manufacturing productivity, to improve community economy and to upgrade quality of social services of the indicated clusters.

Strategy 2: To develop adequate manpower in S&T
There are two targeted numbers: (1) to reach 10 R&D personnel per 10,000 population and, (2) to reach 2 million-baht R&D expenses per researcher per year.

Strategy 3: To develop S&T infrastructure and institutions
The measures include a rapid construction of world-class centers of excellence where state-of-the-art facilities, specialized know-how and expertise are offered; expansion of science parks in Thailand's central and regional areas; amplification of technical services in the S&T fields to quantitatively and qualitatively serve the industrial demands.

Strategy 4: To raise public awareness and appreciation of S&T
To enhance awareness and appreciation of the vital importance of S&T among the target groups including the youth, the middle class and key public and private organizations, community learning centers will be developed along with S&T media. ICT infrastructure services will be

expanded to support future technologies such as Wi-Fi, thereby allowing convenient and efficient access to information and knowledge.

Strategy 5: To improve S&T system management and administration
To achieve better integration and higher efficiency, national agencies are to coordinate among themselves and collaborate with implementing parties in both the public and private sectors through the mechanism of the Chief Science Officer (CSO). The appointed CSO will be the policy and functional links between the NSTPC and their ministries or agencies. In addition, there will be personnel exchange and development of up-to-date monitoring and evaluation systems together with S&T indicators and database in line with international standards. The NSTPC also appointed a number of sub-committees to follow up the actions.

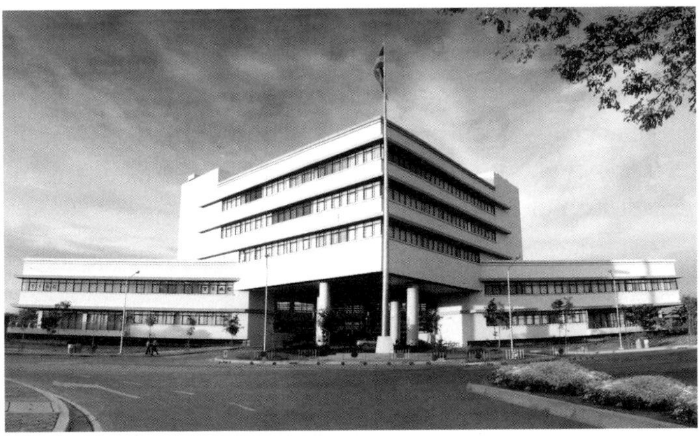

Fig. 1.2. National Science and Technology Development Agency (NSTDA).

1.3. *National Science and Technology Development Agency (NSTDA)*

Created by the Science and Technology Development Act of 1991 by bringing together four existing national units — the National Center for Genetic Engineering and Biotechnology (BIOTEC), the Science and

Technology Development Board (STDB), the National Metal and Materials Technology Center (MTEC) and the National Electronics and Computer Technology Center (NECTEC) — the National Science and Technology Development Agency (NSTDA) officially commenced its operation in 1992[3]. Unlike other public S&T organizations in Thailand, NSTDA was founded with the explicit aim to be an Agency with a high degree of autonomy and mobility unbound by the usual bureaucratic rules and regulations in order to conduct, support, coordinate, and promote efforts in S&T development and collaboration between the public and private sectors towards maximal benefit for national development. Since it was first launched, NSTDA has served as the hub where leading scientists and experts work to tackle scientific and technological issues of imminent concern to both national and international communities. Furthermore, it helps connect and create networking among experts in universities, and provides access to state-of-the-art facilities. NSTDA headquarter is located on an 80-acre of land in Thailand Science Park in Pathumthani province, shown in Fig. 1.2, on the outskirt of Bangkok metropolitan next to Thammasat University, Rangsit Campus, and Asian Institute of Technology. In 2003, the newest National Nanotechnology Center (NANOTEC) was inaugurated and started operations under NSTDA.

2. Nanotechnology Development in Thailand

Nanotechnology is the study, design, creation, synthesis, manipulation, and application of functional materials, devices, and systems through the control and the exploitation of novel phenomena and properties of matter at the nanometer scale (1–100 nanometers, one nanometer being equal to 1×10^{-9} of a meter), that is, at the atomic and molecular levels[4]. Nanotechnology has become a huge buzz word in today's S&T forefront since President Bill Clinton first announced the U.S. National Nanotechnology Initiatives (NNI) on January 21, 2000, standing in front of an image of the Western hemisphere written literally in gold atoms[5]. M.C. Roco reported in 2001 that the worldwide nanotechnology R&D investments announced by government organizations have increased by a factor of 3.5 between 1997 and 2001, and the highest growth rate of 90% was in 2001[6]. Since then the budget for nanotech R&D has increased remarkably not only in the U.S. but all over the world. Obviously S&T has prominently played an important role for shaping national economic and social development. In the U.S., the NNI budget for 2005 has increased almost four folds to US$982 million from $250 million in 2001[7]. The 2006 NNI budget request made by President Bush to the Congress called for a total of $1,054 million.

In Asia, Japan has invested an amount similar to its U.S. counterpart. There is a race among other Asian countries — Korea, China, and Taiwan — to position itself as the closet rival to the frontrunner Japan. In Southeast Asia, five countries are members of the Asia Nano Forum (ANF): Thailand, Singapore, Malaysia, Vietnam and Indonesia. Among them Singapore, Thailand and Malaysia have started national nanotechnology initiatives since 2002, 2003, and 2005, respectively, though the actual funding still lags behind those of the U.S., European Union and Japan.

The Thai government also recognized the necessity of this emerging field. When former Prime Minister Dr Thaksin Shinawatra visited NSTDA in Thailand Science Park in December 2002, he ordered then NSTDA President Dr Pairash Thajchayapong to investigate the feasibility of setting up a national nanotech center. As a consequence, the National Nanotechnology Center (NANOTEC) was born on August 13th, 2003, under NSTDA, Ministry of Science and Technology (MOST) with the following missions:

- Establish a national nanotechnology strategic plan and pursue policy research synergistic to the national S&T and National Economic and Social Development Plans
- Provide R&D funding in the field of nanotechnology and related areas
- Support HRD in the fields of nanotechnology and technology management
- Promote public awareness and appreciation of nanotechnology

The organizational structure of NSTDA with the four national centers is shown in Fig. 2.1. Currently, NSTDA is under the leadership of its 3rd president, Dr Sakarindr Bhumiratana national centers. The Technology Management Center (TMC) handles issues involving technology licensing and intellectual property rights, technology transfer to private sectors, and other special programs. The Corporate oversees Office of the President and handle all NSTDA's budgetary, administrative and personnel issues.

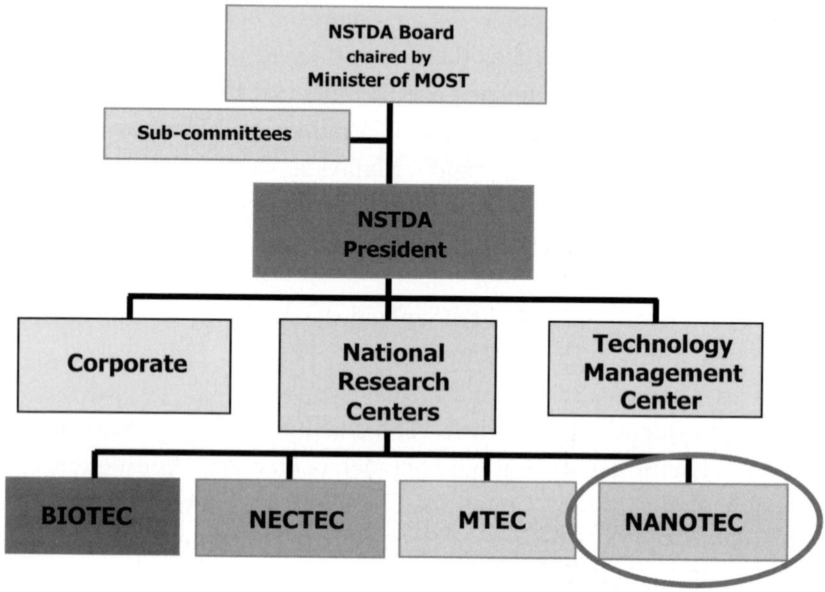

Fig. 2.1. NSTDA Organizational Structure.

2.1. *National Nanotechnology Strategic Plan (2007-2013)*

2.1.1. *Drafting Process*

Figure 2.2 shows the work flow diagram of the drafting of Thailand's national nanotechnology strategic plan (NNSP). Four working groups were formed in order to focus on four different but closely related targeted areas: Nanomaterials, Nanoelectronics, Nanobiotechnology, and Nano-Education/HRD. Each working group held numerous brainstorm sessions among scientists, experts and representatives from government agencies, universities and the private sector to formulate a roadmap for each area. Once all the information needed was completely gathered, final reports were written, integrated and subsequently rewritten into a draft of the NNSP. The draft was presented to the stakeholders and interested public at various public hearings for comments and was revised as necessary. The strategic plan was presented to NANOTEC and NSTDA executive boards and revision before being submitted to the National Nanotechnology Policy Committee, and the Thai Cabinet for final approval.

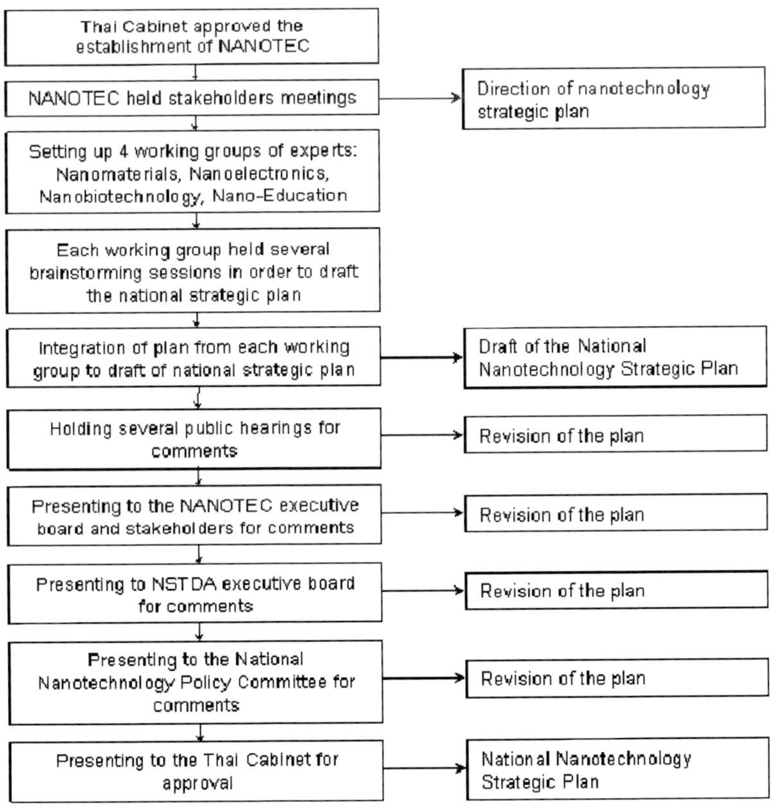

Fig. 2.2. Process flow of drafting the national nanotechnology strategic plan.

2.1.2. *National Nanotechnology Policy Framework (NNPF)*

As summarized in Fig. 2.3, Thailand's National Nanotechnology Policy Framework (2007-2013) is used as guideline to set the direction of nanotechnology development in National Nanotechnology Strategic Plan. It is based on the national social and economic visions and knowledge-based society (KBS) conceptual framework. The core technologies, such as nanomaterials, nanoelectronics, and nanobiotechnology as well as the goals in the year 2013 are identified.

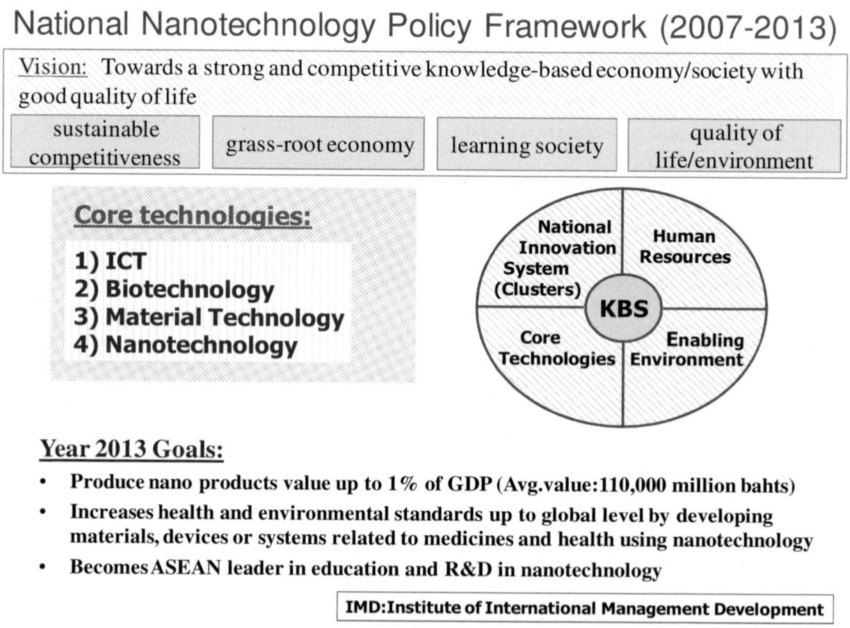

Fig. 2.3. Thailand's National Nanotechnology Policy Framework (2007-2013).

Under the current political administration, the national NNPF includes the creation of a strong and sustainable grass-root competitiveness, knowledge-based economy and learning society with good quality of life and better environment. Factors that enable a knowledge-based society consist of:

- Core technologies: nanomaterials, nanoelectronics, and nanobiotechnology
- Human resource development
- Enabling environment, such as tax systems, investment incentive and flexible regulations
- National innovation systems, such as, economic clusters

Thailand's national nanotechnology policy framework has the following goals to achieve in the year 2013:

(a) to generate the value of nano products in the year 2013 as much as 1% of Thailand's gross domestic product (GDP). This would yield an estimated value close to US$3 billion
(b) to raise the health and environmental standards up to the global level by developing nano materials, devices or systems related to medicines and health using nanotechnology
(c) to become ASEAN leader in education and R&D in nanotechnology

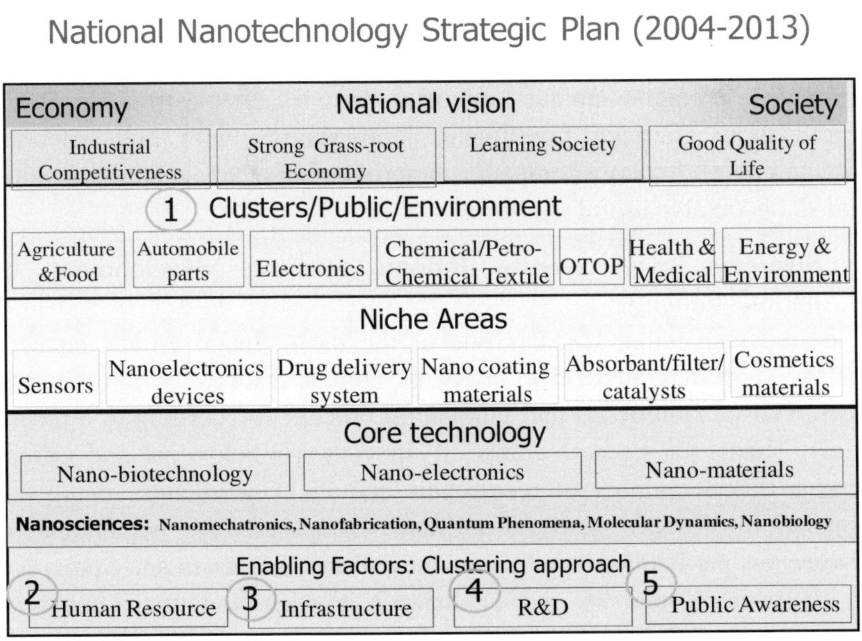

Fig. 2.4. National Nanotechnology Strategic Plan (2007-2013).

2.1.3. *National Nanotechnology Strategic Plan (2007-2013)*

It is important for any nation to have a clear picture of where it is heading towards in terms of technological development. For Thailand, the ultimate goal has been how nanotechnology will be able to drive forward the growth of the country's economy with the solid foundation of human resource and infrastructure development, rigorous R&D, and public understanding as backbones for success. This is summarized in Fig. 2.4. This plan was approved by the Thai Cabinet on June 12th, 2007.

Thailand's National Nanotechnology Strategic Plan focuses on 5 strategies:

- **Strategy 1: Drive forward nanotechnology to support strategic clusters**

The ultimate goal of R&D is to channel the results into the marketplace. As a key enabling technology, nanotech should be applied to create additional values to the seven strategic clusters indicated in Fig. 2.4. Its tremendous impact on the Thai economy can be effected by close complimentary collaboration and connection among key players in each cluster and across them. This includes the development of a "Nano Park" to incubate private companies and bring them together with public R&D organizations. Foreign investment, joint venture with high-potential foreign nanotechnology companies, improvement of financial regulations and so on are also useful measures.

- **Strategy 2: Accelerate human resource development in nanotechnology**

It is undeniable that HRD is one of the most crucial critical success factors. The lack of critical mass in human resource will definitely disrupt sustainable R&D and innovation process. According to a recent study, Thailand's research competitiveness is still below par and suffers from a dismal number of researchers and Ph.D. graduates among the population of more than 60 million. Comparing the population of researchers per 10000, Japan boasts more than 70, Taiwan and Singapore ca. 50 each, Korea 30, and Thailand 3. Furthermore, the number of home-grown PhDs is still much lower than those of the master and bachelor degrees. The ratio of Ph.D: M.S.: B.S. is 152: 7,000: 49,000, respectively[8].

This weakness can be alleviated by increasing the number of scholarships for students to study in nanotechnology-related fields either domestically or internationally. In addition, the government needs to promote nanotechnology education by creating a variety of new programs as well as the promotion of international collaboration in HRD or exchange program.

- **Strategy 3: Increase investments in nanotechnology R&D**

According to IMD Global Competitiveness Report 2003, the ratio of Thailand's gross expenditure in R&D to gross domestic product (GERD/GDP) is 0.26, compared to 0.49 for Malaysia, 0.84 for India, 2.16 for Taiwan, 2.80 for U.S.A., 2.92 for Korea, and 2.98 for Japan[9]. Historically, the turn-around point for sustainable effective S&T is around 1. In recent years Thailand has put its emphasis on becoming knowledge-based society and S&T-based economy. Thailand has to strategically invest in R&D areas that correspond with her industrial strength and niche applications in the six product groups shown in Fig. 2.4. One key performance indicator (KPI) is for the GERD/GDP ratio to exceed unity in the year 2013. In fact Thailand should use the government markets to drive forward her R&D capability instead of simply buying the technology from overseas. More specifically, Thailand has to set key success criteria, such as the annual numbers of patents filed with and approved by the U.S. Patent Office, and the number of papers published in academic journals with high impact factors.

- **Strategy 4: Develop basic infrastructure**

Currently, only a handful of Thai laboratories are equipped with state-of-the-art tools. Though NSTDA is definitely one of them, it is indispensable that R&D facilities in all regions of Thailand should receive sufficient funding to equip their laboratories with necessary instruments and tools to carry out basic and advanced researches. In addition to physical infrastructure, Thailand also needs to develop efficient policies, regulations and investment incentives. For examples, setting up and sharing core facilities and lab networks, training personnel regularly and continuously, and collaborating with foreign scientific peers.

- **Strategy 5: Create proper public awareness and appreciation of nanotechnology**

Before the establishment of NANOTEC, few Thais knew what, and how important, nanotechnology was. Therefore the government must provide mechanisms to create proper yet effective public awareness and appreciation of nanotech through the various forms of mass media. Moreover, nobody could overlook the issues of knowledge management, safety of nanomaterials, and ethical standards for scientists. The promotion of educational activities to uplift public awareness and achieve correct understanding is essential. Students should also be encouraged to think out of the box while encouraging them to participate in national competitions for nanotechnology inventions and innovations.

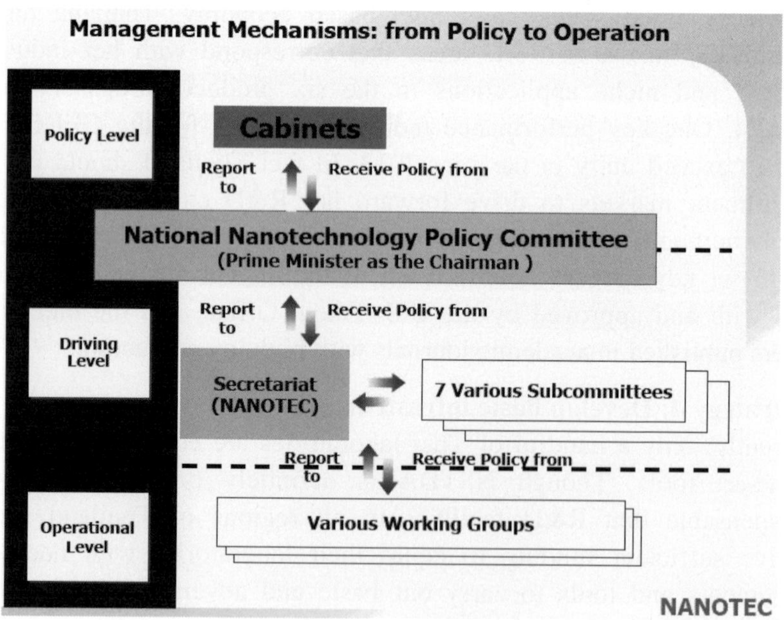

Fig. 2.5. Management Mechanisms.

2.1.4. *From Policy to Implementation*

The management mechanisms to deploy national nanotechnology strategic plan are shown in Fig. 2.5. At the driving level, NANOTEC acts as the secretariat to coordinate and report directly to the National Nanotechnology Policy Committee (NNPC) chaired by the Prime Minister. At the operational level, various working groups are formed to tackle specific tasks assigned by the relevant national subcommittees.

2.2. *Overview of Current Status*

2.2.1. *HRD in Nanotechnology*

Though the current number of nanotechnology researchers in Thailand was estimated to exceed 300, the actual number may be more, depending in the exact definition of nanotechnology and the individual scopes of research activities. To create synergism via networking, NSTDA (NANOTEC) has decided to proactively collaborate with universities in the creation and implementation of nanotechnology curricula. In 2003 the Cabinet approved a second-phase five-year plan of MOST to provide 1,500 full-expense scholarships for Thai students to advance their study in S&T abroad. The plan has also earmarked nearly 200 S&T scholarships for NSTDA, to recruit students in the various fields of S&T, including 48 for NANOTEC. These numbers are in addition to nanotechnology scholarships for students studying in national universities.

2.2.2. *Domestic Nano Education and Training*

The National Nanotechnology Strategic Plan aims to increase manpower in nanotechnology in the public sector up to 2,500 persons in 2013. In addition, the goal is to produce another 2,500 personnel in the private sector.

Currently, several national universities such as Chulalongkorn, Mahidol, Khonkaen, Prince of Songkla, Chiangmai, and King Mongkut's

Institute of Technology Lad Krabang offer nanotechnology courses and some of them even offer, or will soon offer, nanotech programs at master and/or doctoral levels. However, only Chulalongkorn Univ. offers a B. Eng. program in Nano-Engineering since June 2005 with a maximum annual enrolment of 100 students[10]. The current enrolment has about 5% foreign students. NANOTEC also provides five training courses designed for specific groups ranging from undergraduate students, teachers, technical officers, and entrepreneurs. The five courses are: Basic Nanotechnology, Nanomaterials, Nanobiotechnology, Nanoelectronics, and Nanotechnology and applications for industry

2.2.3. *R&D and Innovation in Nanotechnology*

Recently there are close to 40 laboratories nationwide working in nanotechnology related fields with more than half of them in nanomaterials. Examples of major research groups and their main interests are as follows:

- BIOTEC, NECTEC, MTEC and NANOTEC: biosensors, MEMS/NEMS, lab-on-a-chip, nanocoats, nanocapsules, nanoparticles, photocatalyst, drug delivery systems, solar cells and fuel cells
- Semiconductor Device Research Laboratory (SDRL), Department of Electrical Engineering, CU: multi-quantum well GaAs devices, including quantum wells, quantum wires, quantum dots and solar cells
- Center of Excellence (COE) in Catalysis and Catalytic Reaction Engineering, Faculty of Engineering, CU: nanosized single crystals of various metal oxides, such as titania, zirconia, and alumina, etc.
- COE in Particle Technology (CEPT), Faculty of Engineering, CU: synthesis and applications of nanoparticles
- Center of Nanoscience and Nanotechnology, Faculty of Science, Mahidol University (MU): nanoscale modeling and simulation, organic light-emitting diode (OLED), and nano electronics
- Center of Nanotechnology, Kasetsart University (KU): nanostructured zeolites and nanocatalysis

- Nanomaterials Research Laboratory, Physics Department, Faculty of Science, Chiangmai University: carbon nanotube and silicon carbide nano particles
- COE in Nanotechnology at AIT, Asian Institute of Technology (AIT): nano particle synthesis and applications, nano-sensors and nano-devices

In addition, there are other active labs in various universities, such as Prince of Songkla University (polymers, composite and drugs) and King Mongkut's Institute of Technology Lad Krabang (organic electronics and nanosensors). Some research groups or laboratories have strong links to the industry e.g., the KKU-Seagate Cooperation Research Laboratory at Khonkaen University, which focuses on the recording head technology for hard-disk drives. Since 2006, NANOTEC has set up a nation-wide nanotechnology network. More detail may be found in the section on NANOTEC.

The National Research Council of Thailand (NRCT) promotes a wide range of basic, applied and development research programs, including nanotechnology. Its legislated missions are national research policy making and planning, research promotion and coordination, and research information center.

Established by the Research Fund Act B.E. 2535 (1992) as an autonomous agency, Thailand Research Fund (TRF) focuses only on research management, particularly, funding projects and coordination of research network. In 2007 TRF also set up a directed research program in nanotechnology.

Another autonomous organization operating under MOST, the National Innovation Agency (NIA) adopts a broad-based, systematic approach towards enhancing the innovation system of Thailand. NIA plays a key role in organizing an innovation system that promotes and supports innovation and transfer technology to the industrial sector.

Numerous Thai enterprises have produced nanotechnology products developed jointly with NANOTEC. For examples, Siam Student Uniforms Ltd. offers anti-bacterial student uniforms coated with light-reactive titanium-dioxide. Grand Sport Company produces water-repellent and anti-microbial jackets and sports uniforms for the national sport teams at the 2006 Asian Games in Doha, Qatar.

3. National Nanotechnology Center (NANOTEC)

3.1. *Introduction*

Officially established in August 2003 under NSTDA, NANOTEC's mission is to enhance the competitiveness of Thailand via S&T development. With the rapid advancement in technology, Thailand cannot afford to miss the latest departing train called Nanotechnology. In addition to the preparation of the National Nanotechnology Strategic Plan, NANOTEC has focused on the urgent crucial issues of HRD and infrastructure development. It is vital for all stakeholders, including the government and industry, as well as the public to understand and appreciate nanotechnology and its potential contributions toward the prosperity of Thailand's current and future economy and sustainable social development. As NANOTEC works with the Federal Thai Industry to help educate its industrial members, basic nanotechnology-related products, such as neckties, scarves, shirts, cosmetics, odor absorbents and others, have started to show up in market places.

To accomplish its vision and mission, NANOTEC has recently completed a five-year Master Plan (2007-2011) which calls for its personnel to increase from 100 in 2007 to 230 and 350 by the year 2010 and 2013, respectively. With R&D as the topmost priority area, NANOTEC has put emphasis on recruiting strong dedicated researchers from all over the world.

3.2. *R&D Programs and Focuses*

As of 2007, NANOTEC R&D activities essentially belong to three main technology platforms: Nano-coating, Nano-encapsulation, and Functional Nanostructure. Figure 3.1 shows an overall picture of NANOTEC's technology platforms, supporting tools and target industries. In the Nano-coating platform, the current focuses are on binder and photocatalytic technologies, especially in textile and agriculture applications. Struggling with growing competition from China and India due to their lower wages, Thailand's textile industry needs to move up to functional and technical textile. Nanotechnology-enhanced surface finishes, such as coating with nano-silver, titanium dioxide, etc., are expected to provide product differentiation and add values to the synthetic and natural fabrics, especially traditional Thai silk and cotton fabrics. The coating of titanium dioxide on food packaging materials helps prolong the storage life of vegetables or agricultural products while suppressing microbial growth. In addition, coating glass and ceramic surfaces with TiO_2 makes them super-hydrophilic and easier to get rid of dust and dirt. Other research projects include fire-retardant and anti-static fabrics as well as anti-algae and anti-microbial ceramic stones for fish tanks. Figure 3.2 shows examples of product prototypes co-developed by NANOTEC: anti-bacteria jacket in (a) and fire retardant textile in (b).

In the Nano-encapsulation platform, the research activities focus on the capture of active ingredients or desired materials in nano-capsules for

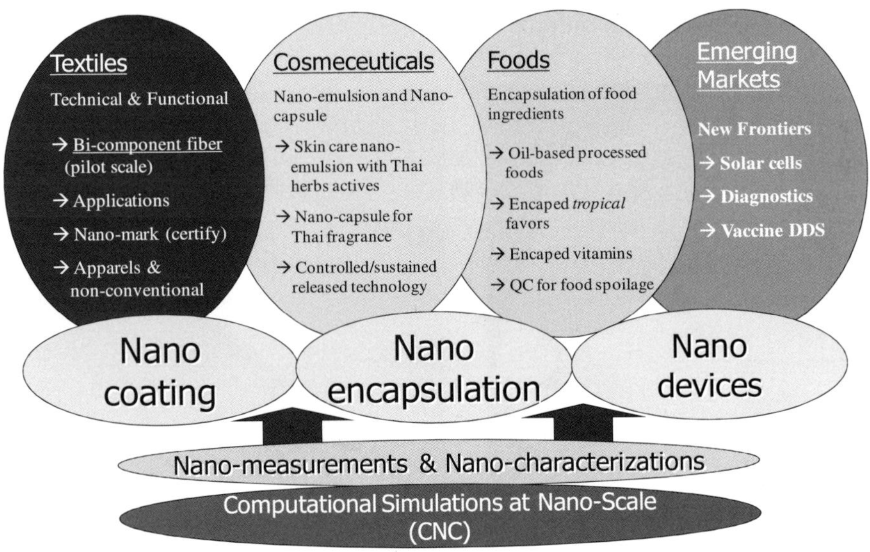

Fig. 3.1. Overall picture of NANOTEC's technology platforms and focuses.

drug delivery, food and cosmeceutical applications in order to respond to industrial needs for better and more effective methods to deliver drugs into bodies, make food taste better, and slowly deliver active ingredients into deeper layers of the skin. The platform investigates the core and shell materials of the capsules and also develops the encapsulation processes, such as nano-emulsion. Extracts from indigenous Thai herbal plants are popular as the core materials.

The Functional Nanostructure platform focuses mainly on the synthesis of organic and inorganic materials as well as the fabrication of nanostructured thin films for nano device applications such as solar cells and a variety of sensors, e.g. e-nose sensor and alcohol vapor sensor. As for solar cells, the focus has shifted from inorganic solar cells to newer and cheaper organic solar cell technology with great promise.

Last but not least, NANOTEC also cultivates researchers who specialize in computational nanoscience and safety of nanomaterials. In fact NANOTEC creates and runs a highly acclaimed nation-wide Computational Nanoscience Consortium. What follows are examples of NANOTEC's research outputs.

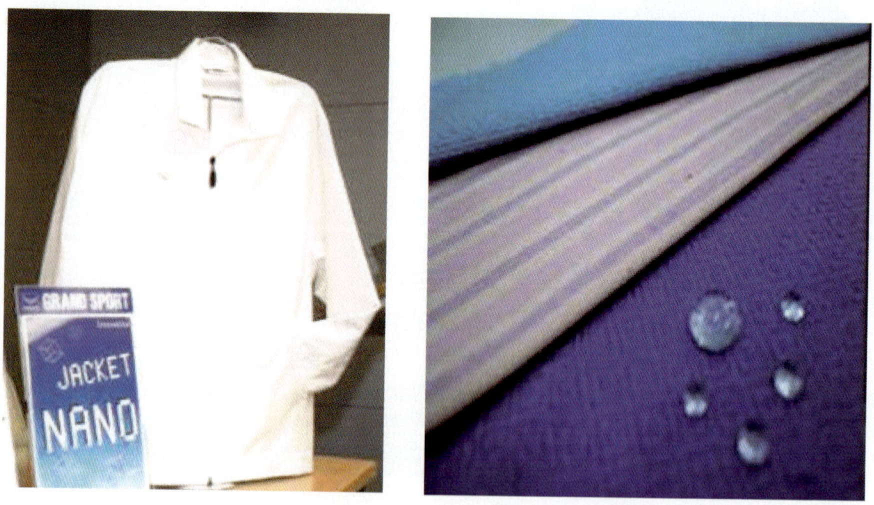

(a) **(b)**

Fig. 3.2. Examples of product prototypes: (a) ZnO-coated Nano Jacket (b) fire retardant textile.

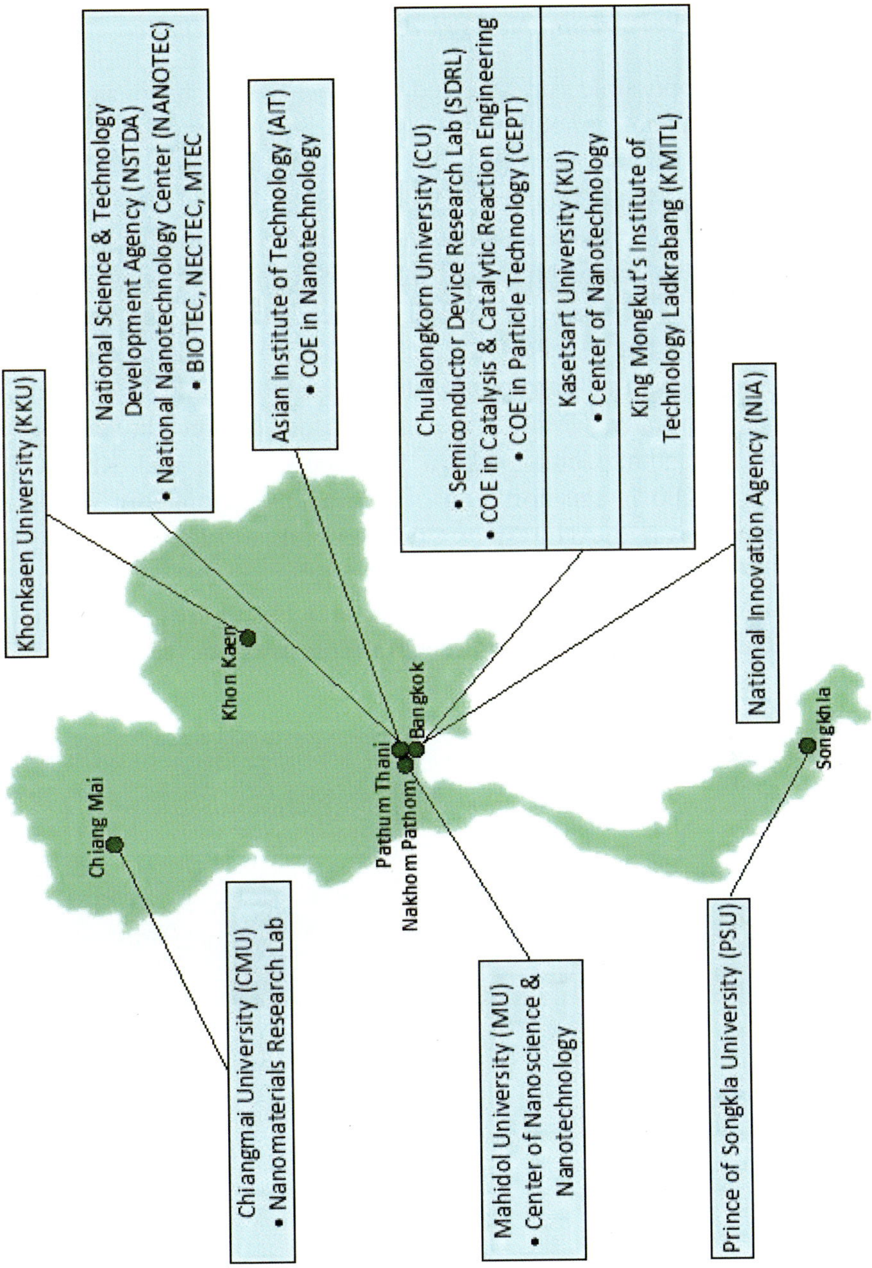

Fig. 3.3. Nanotec's national network of centers of excellence in Thailand.

3.3. *NANOTEC's National Network of Centers of Excellence*

It would be impossible for Thailand with her severe handicaps to be competitive in nanotechnology R&D without any strong complementary nationwide university network (see Fig. 3.3). In 2006, NANOTEC signed the first Memorandum of Understanding (MoU) with Kasetsart University (KU) to provide an annual funding of 5 million bahts for five years with the same matching fund from the counterpart university. As listed in Table 3.1, a succession of selected COEs in the following institutions have since followed suit: AIT, King Mongkut's Institute of Technology at Ladkrabang (KMITL), Mahidol University (MU), Chiangmai University (CMU) in the north, Prince of Songkla University (PSU) in the south, Chulalongkorn University (CU), and Khonkaen University (KKU) in the northeast of Thailand. Examples of research outputs produced by NANOTEC's COE Network are shown in Fig. 3.4. Figure 3.4(a) shows nano-crystal coated eyeglasses specially designed for forensic work developed by KMITL and Fig. 3.4(b) shows Self-Aligned Quantum Dots for Single Electron Transistors developed by CU.

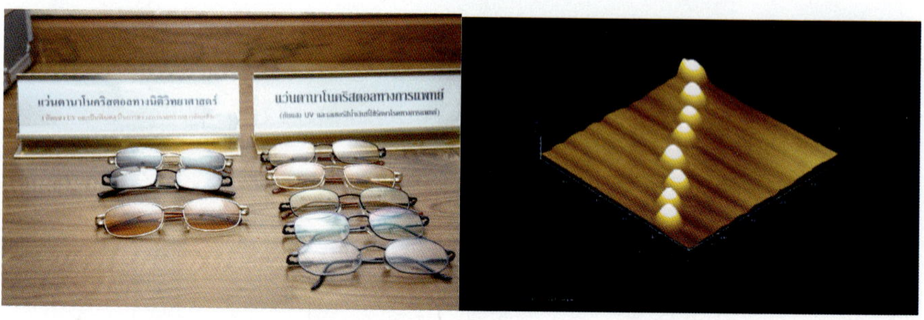

(a) (b)

Fig. 3.4. Example of research outputs: (a) new nanocrystal coated eyeglass lens (b) Self-Aligned Quantum Dots for Single Electron Transistors.

Table 3.1. NANOTEC's Network of Centers of Excellence and Their Focuses

Universities	Focuses
KU	Nanoscale Materials Design and Simulation
AIT	Synthesis and Application of Nanoparticles
KMITL	Organic Nanoelectronic Devices
MU	Nanodevice Engineering
CMU	Functional Nanomaterials
PSU	Nano-Biomaterials Synthesis, Nano-Material Engineering
CU	Nanoelectronics and Nanophotonics
KKU	Nanoparticles and Nanofibers for Medical Applications

3.4. *Key Infrastructures and Facilities*

No research center with decent laboratories can do without essential tools and characterization capability. NANOTEC has acquired several key pieces of equipment and the goal is to continue investments and expansion in key infrastructures and facilities to become one of the best in the world. The current 500-sq.m. lab area of NANOTEC will be tripled when the expansion work is finished by early 2008.

Key characterization and analytical instruments include ND-MDT's NTEGRA Spectra SPM instrumentation that analyzes chemical composition, structure and function of materials by combining confocal microscopy, scanning probe microscopy, and Raman spectroscopy in one platform, Environment Scanning Electron Microscope (E-SEM), Atomic Force Microscope (AFM), Gas Chromatography-Mass Spectrometer (GC-MS), Differential Scanning Calorimeter (DSC), Thermo Gravimetric Analyzer (TGA), UV-Visible Spectrophotometer, Fluorescence Spectrophotometer, High-Pressure Homogenizer, MTT

Assay, UV-Fluorescence Microplate Reader, Gel Documentation, Thermal Evaporator Vacuum System, Fourier Transform Infrared Spectrometer (FTIR), Surface Area and Pore Size Analyzer (BET), Particle Size Distribution Analyzers (Mastersizer and Nanosizer), Colony Counter, and High-Performance Liquid Chromatography (HPLC). Some examples of the instruments are shown in Fig. 3.5.

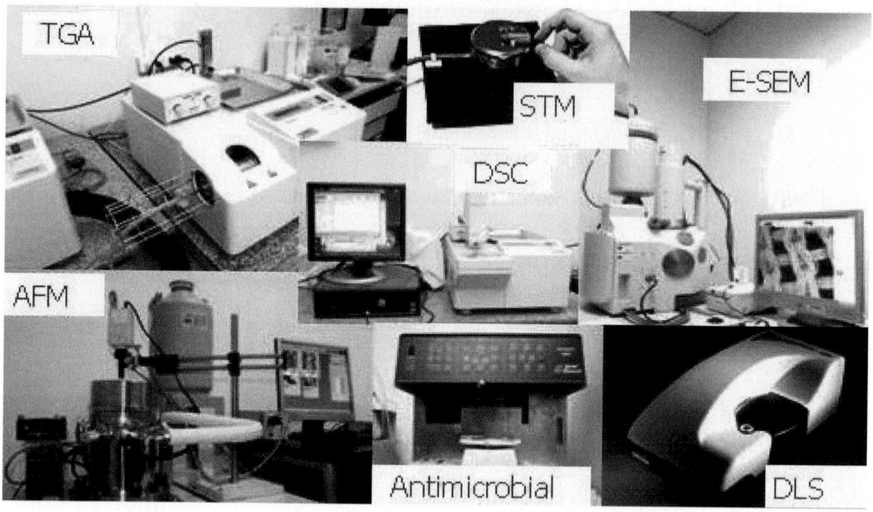

Fig. 3.5. Instruments used in nanotechnology research at NANOTEC.

3.5. *Technology Transfer and Communication*

Since 2001 various "nano-products" are prominent in the market; for example, nano-shirts, nano-filter, nano-TiO_2 spray, nano-cosmetics, etc. As a result the dissemination and transfer of nanotechnology knowledge and technology are crucial. NANOTEC focuses on three main categories of services: (1) business development (2) communications and (3) testing services. The scope of the business development encompasses technology licensing and collaborative researches between NANOTEC and private companies. The goals are to generate collaborative (or contract) research projects and enhance the competitiveness of Thai industries with nanotechnology-enhanced processes and products.

The collaborative research projects range from technical consulting projects to contracted research projects. Designed to fit the needs of individual customers, each project is flexible and the agreement is based on the IP policy of the NSTDA. The types of IP agreements NANOTEC has participated in are non-disclosure agreement, research agreement, collaboration agreement, and commercial licenses. As a national research center, NANOTEC prefers to limit the application fields of the licensed-out technology. Even when the grant is for an exclusive license, NANOTEC may retain certain rights, such as the right for further R&D study and the right for publication.

At present, projects successfully delivered to the clients include textile, cosmeceutical, paint and house-hold products.

The goal of communication is the nationwide dissemination of nanotechnology knowledge through training courses, publications, exhibitions, multi-media, and websites. Currently, five standard courses designed for teachers, students, and entrepreneurs are available: (1) Introduction of Nanotechnology, (2) Nanomaterials, (3) Nanoelectronics, (4) Nanobiotechnology, and (5) Nanotechnology for Industries. In addition to the creation of public awareness and appreciation of nanotechnology, one important goal is to have topics in nano science and technology interspersed in the standard educational curriculum of K-12, i.e. general sciences, physics, chemistry and biology. This endeavor of NANOTEC calls for a joint effort between the Ministry of Science and Technology and Ministry of Education.

In addition to in-house services, NANOTEC also provide technical testing services to the industry and other research institutions. Currently, physical and chemical tests include particle size distribution and nano-scale measurements, analysis of surface structure and chemical compounds. Biological tests cover anti-bacterial and cell cytotoxicity testing. Overall, more than 10,000 samples were tested in 2007.

4. Future Trend

It is estimated that the industrial and commercial markets generated by nanotechnology R&D will be worth one trillion US dollars per year by

2015[11]. To prepare itself for these new challenges, Thailand has reorganized its policy of public subsidies for R&D and innovation and introduced ambitious programs to encourage co-operation between manufacturers and public research establishments. Nanotechnology is expected to give a lasting boost to global economic growth over the next fifteen to twenty years. The Thai government has adopted the value creation approach, which combines the fruits of S&T and innovation to upgrade the quality and value of products and to manage proliferating non-tariff barriers such as food safety measures and so on.

The progress of nanobiotechnology worldwide and their applications to products related to human health have strengthened Thailand's intention to develop and utilize nanobiotechnology based upon her strength in the specific areas of food, medicine, pharmaceutics, agriculture, environmental protection and pollution control. Led by the National Biotechnology Policy Committee, Thailand has seen a rising growth in production and services that utilize indigenous R&D in the rice, plant, shrimp, animal, bioprocess, medical, and environmental industries.

Another area that will continue to enjoy tremendous growth in Thailand is the nanomaterials-related sectors. Several companies have ventured into and invested in the production of nanomaterials. Some companies prefer to shift away or migrate from their traditional industries by importing technology from overseas. Some seek out R&D organizations, like MTEC and NANOTEC, to be their R&D provider and keep them updated on current and future technologies. In fact several domestic and international companies have established their presence in Thai nanomaterial and chemical markets such as zinc oxide, nanosilver, titanium oxide, carbon nanotube, and so on. Nanotechnology applications in cosmetics are also one of the high-growth areas.

In nanoelectronics, the progress is inevitably slow due to lacks of domestic industrial strength. Nevertheless, the government has realized the importance of hi-tech push in this area and recently announced full support to facilitate future growth in the electronics sector, especially high-density storage disk drive.

To support new startup companies, foreign direct investments and joint ventures between Thai and foreign companies will be promoted through financial measures, fiscal measures, supportive regulations and infrastructure services. Furthermore, technology capacity of firms will be promoted via the cluster approach in the target industries. Regarding intellectual property issues, relevant Thai laws include copyright law, Thai trade secret law 2002, patents and Thai law on plant intellectual property based on the obligation to the World Trade Organization (WTO) and trade-related aspects of intellectual property rights (TRIPS) agreement.

Thailand's ninth five-year National Social and Economic Development Plan (2002-2006) sets a research investment goal at 0.40% of the gross domestic production (GDP). The GDP for 2006 is forecast to reach US$150 billion, which translates to a research investment of at least $0.6 billion. However, this value would still be significantly less than those of the advanced industrial countries. Therefore Thailand's limited budget must be spent wisely in the most appropriate and effective manners. Nevertheless, with full support from the Thai government, the future looks promising for S&T including nanotechnology development in Thailand.

Acknowledgements

The authors would like to thank the National Nanotechnology Center and her network of Centers of Excellence for the supply of information and support materials.

References

1. National S&T Development Plan (1997-2006); http://www.most.go.th, retrieved Oct 22, 2007.
2. Bell Martin (2002), Knowledge-Capabilities, Innovation and Competitiveness in Thailand: Transform the Policy Process, a report for the National Science and Technology Development agency (Thailand) sponsored by the World Bank.
3. National Science and Technology Development Agency, Ministry of Science and Technology, Thailand; http://www.nstda.or.th/english, retrieved Oct 22, 2007.
4. Fabio Salamanca-Buentello, Deepa L. Persad, Erin B. Court, Douglas K. Martin, Abdallah S. Daar, Peter A. Singer, 2005, "Nanotechnology and the Developing World", *PLoS Medicine,* Vol. 2, Issue 4, pp. 300-303.

5. The National Nanotechnology Initiative (2001); http://www.nano.gov, retrieved Nov. 2005.
6. M.C. Roco, 2001, "International Strategy for Nanotechnology Research and Development", *J. of Nanoparticle Research*, Kluwer Academic Publ., Vol. 3, No. 5-6, pp. 353-360.
7. The U.S. National Nanotechnology Initiative Funding; http://www.nano.gov/html/about/funding.html., retrieved Nov. 2005.
8. J. Svasti, Thailand Research Fund Young Researcher Seminar, 2004.
9. Thailand's National Science and Technology Strategic Plan (2004-2013) by NANOTEC, p. 35.
10. B. of Eng. Program in Nano-Engineering, Faculty of Eng., Chulalongkorn Univ.; http://ise.eng.chula.ac.th/~nano/index.php, retrieved Oct. 22, 2007.
11. M. Roco and W.S. Bainbridge, eds., "Societal Implications of Nanoscience and Nanotechnology," National Science Foundation, March 2001; http://www.wtec.org/loyola/nano/societalimpact/nanosi.pdf.

Author Biographies

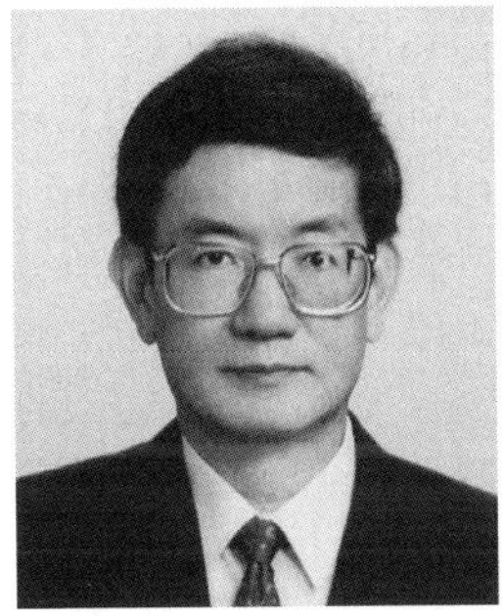

Dr Wiwut is Honorary Advisor of Thai Institute of Chemical Engineering and Applied Chemistry (TIChE), Chair Professor in Particle Technology, Chulalongkorn University (CU) and Associate Member of Thailand's Royal Institute (Petrochemical Technology discipline); Vice President for Special Affairs, Technology Promotion Association (Thailand-Japan) (TPA); Secretary General, Association of Thai Particle Industries (TAPI); Fellow of Chemical Engineering, Engineering Institute of Thailand (EIT).

Dr Wiwut received numerous research awards and honors worldwide, which include: Japanese Government Scholarship; Japan Society for the Promotion of Science (JSPS) Fellowship as Visiting Scientist to Kanazawa University, Kyoto University, Tokyo Institute of Technology; Association for Overseas Technical Scholarship: Program for Quality Management (PQM), Program for Solving Human and Organizational Problem (SHOP), AMEICC Training Program for Organizers of Centers of Excellence (AMOG); Senior Research Scholar/Team Research Award, The Thailand Research Fund, Prime Minister's Office; Outstanding Staff Award for Academic Excellence, Faculty of Engineering, Chulalongkorn University; Science and Technology Award, Thailand Toray Science Foundation; Outstanding Faculty Award in Science and Technology-University Level-Chulalongkorn University; Listing in *Marquis' Who's Who in the World*, Millennium Edition, 2000; *Marquis' Who's Who in Science and Engineering, 2003;* Award for the Most Excellent Paper in Terrestrial Applications Session, 6th International Heat Pipe Symposium; Distinguished Paper Award, 3rd Asia Pacific Drying Conference, etc.

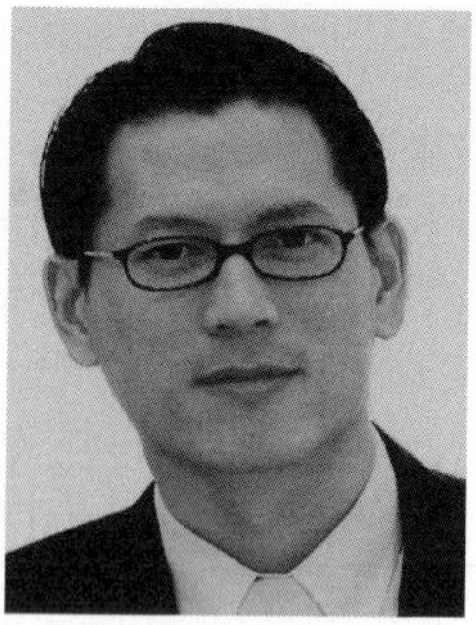

Dr Nick Teerachai Pornsin-Sirirak is the Deputy Executive Director at Thailand's National Nanotechnology Center (NANOTEC). He obtained his B.S. and Ph.D. degrees from California Institute of Technology and his M.S. degree from Stanford University. He helps pioneering nanotechnology development in Thailand from the beginning and was the key person to help draft Thailand's first 10-year nanotechnology strategic plan, which was approved by the Cabinet in June 2007. He has been integral in pushing nanotechnology for industrial use to create more innovative nanotechnology products and also helps increase public awareness about nanotechnology in Thailand.

Prior to joining NANOTEC, Dr Pornsin-Sirirak worked in semiconductor as well as satellite telecommunications industries in the United States for several years. He worked at Applied Materials as a researcher/technologist and global product manager for both front-end and back-end PVD Endura products. His research work included deep silicon etching, in which he holds a US patent, and low-resistance W gate materials for sub-100nm poly-metal gate applications. In addition, he was the Secretary-General of the Asia Nano Forum (ANF) that advocates the nanotechnology development in Asian countries.

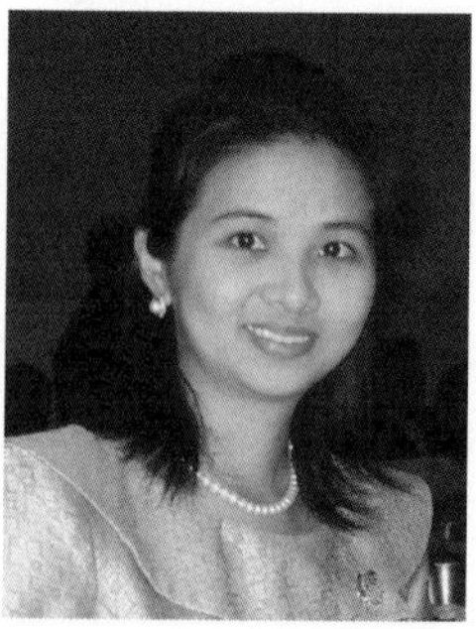

Dr Weeraya Pakawech is a Technical officer of National Nanotechnology Center (NANOTEC), Thailand. *She* takes full *responsibility* for organization strategy and evaluation section of NANOTEC. She led team in building and implementing strategy focused organizations with the Balanced Scorecard and successfully developed the first NANOTEC operation plan FY2008. In 2005, she got Ph.D. in Chemical Engineering from Chulalongkorn University. She studied as a research assistant with Prof. Wiwut Tanthapanicakoon under The Royal Golden Jubilee Ph.D. Program Scholarship. In 2003, she spent one year to conduct research with Prof. Chikao Kanaoka at Kanazawa University, in air filtration using electret filter supported by AIEJ (Association of International Education, Japan). *She* has always had a strong *interest* in nanotechnology *research* especially powder technology.

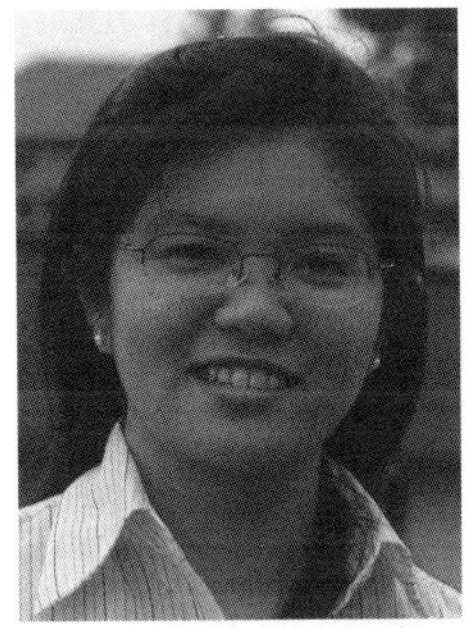

Dr Manida Teeratananon received her diploma of Analytical Chemistry (Second Honor award, 1997) and Bachelor of Science in chemical engineering from Chulalongkon University, Thailand (B.S., 2000). Afterward, she received a scholarship from Thailand Research Fund and pursued her doctorate in Process and environmental engineering from Chulalongkorn University and Institut National Polytechnique de Toulouse, France (Ph.D., 2004).

She has career experience in both private industry and public organization. Indeed, in 2004, she joined Western Digital Company, Thailand as a senior engineer in Head/Media Tribology section. She successful led team in identifying nano failure on media under harsh conditions. In 2006, she worked at National Nanotechnology Center (NANOTEC), National Science Technology Development Agency (NSTDA) as a technical officer of business development section. She was awarded Model Employee of the month in July 2007 and successful led team in closing business development deals via contract research and licensing agreement.

Regarding her marital situation, she has move to France and received post doc fellowship at Institut National Polytechnique de Toulouse. She is working on modeling of new electrochemical reactor with improved energetic yield, higher productivity and less environmental impact for nuclear application, under Sélection et Utilisation de Composites. Carbone-carbone pour l'Electrolyse Fluor (SUCCEF) project.

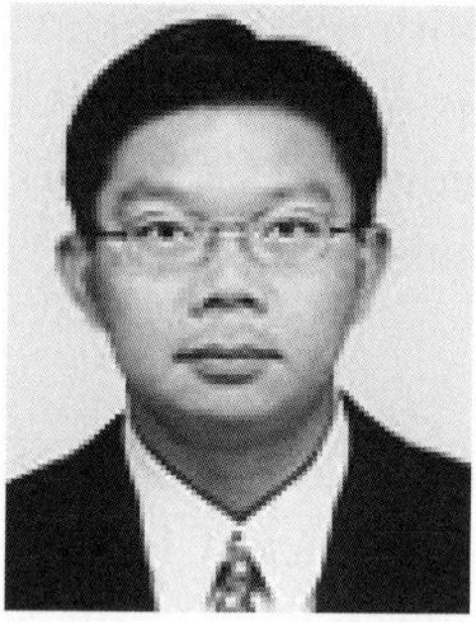

Dr Sirasak Teparkum, a director of technology transfer division, has long experience of various technology transfer project. Before joining NANOTEC, he worked at the National Center for Genetic Engineering and Biotechnology (BIOTEC), National Science Technology Development Agency (NSTDA) as a technical officer responsible for science communication, disseminating scientific information to the public thru writing articles, creating science media, giving scientific presentation for the public, and organizing science exhibitions. He wrote more than 30 published articles, edited 11 books, created science media so called, "Genomic Music" which DNA is decoded and transformed into a truly music of life. The story of Genomic Music was broadcasted and disseminate via national and international media i.e. The Manager (Thailand), The Nation (Thailand), Strait Time News (Singapore), Discovery Online (USA), and BBC News (England). In 2003, he was chosen to be a deputy director of "City of Science" in the 20th World Scout Jamboree which is hosted by Thai Scout Association at Suttahip, Thailand. In 2004, Dr Sirasak was involved in drafting a national strategic plan for nanotechnology and selected to be a secretariat of Nano-biotechnology working group for establishing the National Nanotechnology Strategic Plan. Afterwards, he was transferred to work with the National Nanotechnology Center (NANOTEC), NSTDA as a director of technology transfer division and has been working with NANOTEC since then.

Dr Sirasak Teparkum received the Bachelor degree of Horticulture at Kasetsart University and biology from Concord College, West Virginia. Afterwards, he pursued his graduate levels, Master and Ph.D, at Virginia Polytechnic Institute and State University in the field of plant tissue culture and plant-insect molecular interaction.

CHAPTER 13

INFRASTRUCTURE, RESEARCH AND DEVELOPMENT OF NANOTECHNOLOGY IN VIETNAM

Phan Hong Khoi

Institute of Materials Science, Vietnam Academy of Science and Technology
College of Technology, Vietnam National University, Hanoi
phkhoi@ims.vast.ac.vn

Phan Ngoc Minh

Institute of Materials Science, Vietnam Academy of Science and Technology
College of Technology, Vietnam National University, Hanoi
minhpn@ims.vast.ac.vn

An overview of the current status of infrastructure, research and development of nanoscience and nanotechnology in Vietnam is presented. Some results related to the research and development of Carbon Nanotubes and quantum dot materials are given as the illustration.

1. Introduction

After 20 years of transformation, Vietnam has gained important achievements in science and technology (S&T), industry and economic development. The continuous and high growth rates in economy, stable political and social situation, increasingly democratization and socialization trend have significantly improved people's life and improved international cooperation relations.

The socio-economic development strategy for the period from 2001 to 2010 affirms the overall objective: moving Vietnam out of the less developed situation, improving the people's spiritual and material life, creating the foundation so that by 2020 Vietnam will become a modern industrial country. In particular, strengthen human resources, S&T capacities, infrastructure, economic potentials, national defense and security; establish the socialist-oriented market economy and improve competitive position of Vietnam in the world. In terms of S&T, ensure the speed of increasing investment rates of S&T from the State budget should be higher than that of increasing expenditures from State budget, at the same time strength sources of non-State budget investments in S&T. It is striving that the whole society's total investment in S&T reaches 1% of GDP in 2005 and 1.5% of GDP by 2010[1].

Some research-development organizations and universities which are of the average advanced level in the region should be established in some key technology fields and some strong scientific branches of Vietnam. The first-stage construction of two high-technology zones of HoaLac High-tech park near Hanoi and Saigon High-tech park in HoChiMinh City should be completed; some approved National Key Laboratories will be put into operation and effectively exploited; technical infrastructure of some important S&T service providers in terms of S&T information, standards-measures-quality be upgraded. By 2010, our S&T will be able to acquire, master and use effectively imported modern technologies; capable of doing researches and using some modern technologies, especially information technology, biotechnology, advanced material technology, automation technology, mechanic–electronic technology; and catch up with the global level in some Vietnam's strong scientific fields.

Developing science and technology and growing high-tech industries in Vietnam is believed to be the key driving force for future and sustainable economic growth. During the period from now to 2010, our country needs to focus on the development of some key technologies include advanced technologies which have significant impacts on modernizing economic-technical branches and ensuring the national defense and security; create favorable conditions for forming and developing some new branches, and increase the competitiveness of the economy; new technologies can make use of our country's advantages of tropical agriculture and abundant labor force in rural areas and create products for export and jobs with high income for population classes.

In December 2003, Primer Minister of Vietnam has signed a National Strategy of S&T development of Vietnam until 2020, in which nanotechnology is formulated as one of high priority fields. The rapid global nanotechnology development and increasing strong awareness of its societal impact especially has inspired Vietnam R&D community, politicians, policy makers and industry to contemplate the role of Vietnam in the global nanotech arena. Recently, a new program on nanotechnology has been launched by the Ministry of Science and Technology. The key goal of the Vietnam nanotechnology program is the enhancement and extension of world-class nanotechnology research institutions, the rapid establishment of a robust education and training system on nanotechnology, a highly skilled, educated and diverse workforce, efficient infrastructure and integrated involvement in nanotechnology activities. The Vietnam nanotechnology program also emphasizes on international cooperation and global network. In addition the key issues such as infrastructure building, identifying key organizations, strategic research areas, building interdisciplinary research projects, education programs and management of the nano programs has been clearly identified and outlined.

2. Science and Technology Research, Development and Education Network in Vietnam

Over the past decade, a network of S&T organizations has been set up with more than 1,100 research and development organizations in every economic sector. There are nearly 500 are non-state organizations; 197 are universities and colleges, including 30 non-public schools. Infrastructure of institutes, research centers, laboratories, S&T information centers and libraries have been strengthened and upgraded. Associations among scientific research, technology development, production and business have been formed. The key scientific and technological R&D funding organizations in Vietnam are:

2.1. *Ministry of Science and Technology (MOST)*

The Ministry of Science and Technology (MOST) is a funding agency that fulfills state management functions in terms of science and technology activities, development of science and technology potentials. MOST is the main R&D policy making agency in Vietnam. Among many other functions, MOST is responsible for establishing the national plan of S&T development (in particular, the recent strategic guidelines for S&T of Vietnam until 2020). MOST allocates funding from national budget for national R&D priority programs. There are nine programs in the period from 2001 to 2005 with the budget about 300 billions VND, or approx. 20 millions USD. The funding is for developing national R&D key laboratories[1].

2.2. *Vietnamese Academy of Science and Technology (VAST)*

VAST (former name: National Center for Natural Science and Technology, NCST) is the largest S&T research institution in Vietnam with the mission to conduct the research and implementation of the natural science and technology in the key directions of the country. VAST was established in 1975 and is, in status, affiliated directly to the Vietnamese Government. VAST consists of more than 20 research institutes ranging from those of basic fields such as mathematics, physics, ecology, oceanography; to technological development such as biotechnology, information technology, materials science, and environmental technology employing more than 3000 researchers. There are more than 20 spin-off enterprises, employing totally more than 3000 researchers. The VAST is mostly funded by the national budget (in 2004: 205 billions VND or 12.5 millions USD, in 2006: 275 billions VND or 17 millions USD; in 2007: 362 billions VND or 22 millions USD)[2].

2.3. *Ministry of Education and Training (MOET)*

MOET is an education policy maker of Vietnamese Government. In the Education Development Strategy from 2001 to 2010, the goals of higher education in Vietnam have been clearly set out as “to provide high quality human resources in line with the socio-economic structure of the industrialization and modernization of the nation; enhance the competitiveness in fair cooperation for Vietnam in its international economic integration; to facilitate the expansion of post secondary education through diversification of educational programs on the basis of a path-way system that is suitable for the structure of development, careers and employment, local and regional human resource needs and the training capacities of education institutions; to increase the appropriateness of the training to the employment needs of the society, the ability to create jobs for oneself and for others".

In the academic years of 2002-2003, there are 111 universities and 119 colleges in the higher education system; of which 15 universities are private, two semi-public, and two private colleges. The total number of students reaches 1,020,670 and 64% of which are full-time students, 36% part-time. The student rate is 124.7 per 10 thousand citizens. The total number of lecturers are 32,205 of whom 5,476 lecturers have PhD degrees (17%), 9,543 have Master degrees (29.6%), and 17,186 have bachelor degrees (53.4%). Only 324 of these lecturers have been awarded with the title of professors (1%) and 1,124 associate professors (3.49%)[3]. According to the updated data by the National Professorship Committee of Vietnam, until 2008, January Vietnam has total 6600 Professors and Associate Professors. Among the 111 universities and 119 colleges in the higher education system, the strongest in R&D activities are: Vietnam National University-Hanoi (VNU-Hanoi)[4], Hanoi University of Technology (HUT)[5], Vietnam National University-HoChiMinh City (VNU-HoChiMinh)[6].

2.4. *Vietnam National High-Tech Parks*

To develop and grow leading industries and modernize processes of production and quality control via transfer of technology or joint ventures to produce high end technology intensive products, Vietnam Government decided to set up HoaLac National High-Tech Park in the North (40 km from Hanoi) and SaiGon National High-Tech Park in the South of Vietnam.

The schedule of HoaLac High-Tech Park's development is as follows: (1) The 1st stage - the 1st step: 1998-2003 on an area of 200 Ha by ODA of the Japanese Government; (2) The 1st stage - the 2nd step: until 2010 on an area of 800 Ha; and (3) The 2nd stage: will run until 2020 on the entire area of the zone.

The Saigon High-Tech Park (SHTP) was established in 2002. It is located in HoChiMinh City. The milestones of SHTP's development are as follows: (1) 2002 - established with the approval of the Prime Minister; (2) 2002 - Approval of Master Plan - Issuance by the government of Decree 99 on the code of conduct at hi-tech parks - Framework Agreement signed with Hewlett-Packard Corporation; (3) 2004 - Issuance by the government of Decision 53 on incentives in hi-tech parks and Establishment of the SHTP R&D Centre; (4) 2005 - Issuance of investment license to the Nidec Corporation and establishment of the SHTP Training Centre in 2006; (5) - Selected as one of the five focal economic projects in HoChiMinh City during 2006-2010 period and issuance of $1 billion investment certificate to Intel Corporation and establishment of SHTP Incubation Centre. The SHTP's vision "is to develop a technopolis that will greatly enhance the economic, technological, and intellectual base of HoChiMinh City and the Southern Economic Region of Vietnam and that will ultimately serve as a model for Vietnam technological innovation, intellectual capital development, and innovation economy".

Being one of Vietnam's only two national High-Tech Parks, SHTP has received exceptional supports from both the central and local governments, as well as from other relevant state agencies. As a result, the Park has been authorized to offer the highest tax incentives and one-stop investment application service to investors.

The priority of investment of Vietnam National High-Tech Parks is for the fields of micro-electronics, information technology, communications, automation, precision mechanics and nanotechnology.

3. Nanotechnology Research and Development

3.1. *General*

The research on nanotechnology in Vietnam officially started in 1993 at the Second National Conference on Solid State Physics, Director General of the NCST (the former name of the Vietnam Academy of Science and Technology, VAST). Prof. Academician Nguyen Van Hieu put forward a major initiative on nanotechnology research. To respond this initiative, many physicists, researchers in the whole country showed their interest in research on nanoscience and nanotechnology. Since then the National Program on Basic Research allocated funding to selected projects on nanoscience and nanotechnology.

In 2003 the Ministry of Science and Technology (MOST) launched an infrastructure-building program on Nanoscience and Nanotechnology in Vietnam. It is a new priority program for the period of 2004-2006 under umbrella of the National Program on Basic Research. The program on Nanoscience and Nanotechnology aims to identify researchers in various areas of Nanoscience with specific expertise; to nurture a nanoscience and nanotechnology research culture among researchers in Vietnam; to prepare a comprehensive human resource development program for training scientists and to gradually upgrade and equip the Nanoscience and nanotechnology laboratories with state-of-the-art facilities.

Besides funding the national programs on science and technology, the Vietnamese government has put in considerable effort to upgrade the R&D infrastructure and facilities in selected areas of science and technology. There are now 19 National Key Laboratories focusing on research in various areas such as Materials (three Labs.), Biotechnology (four Labs.), Information Technology (three Labs.) and other fields (nine Labs.). They are funded with about 50-60 billions VND (about 3.5-4.0 millions USD) per lab for a period of 3-4 years. For instance, a National Key Laboratory for Electronic Materials and Devices was set up at the Institute of Materials Science, VAST, a National Key Laboratory for Genome was already established at the Institute of Biotechnology, VAST with 3.5 Million $US each. The National Key Laboratories are

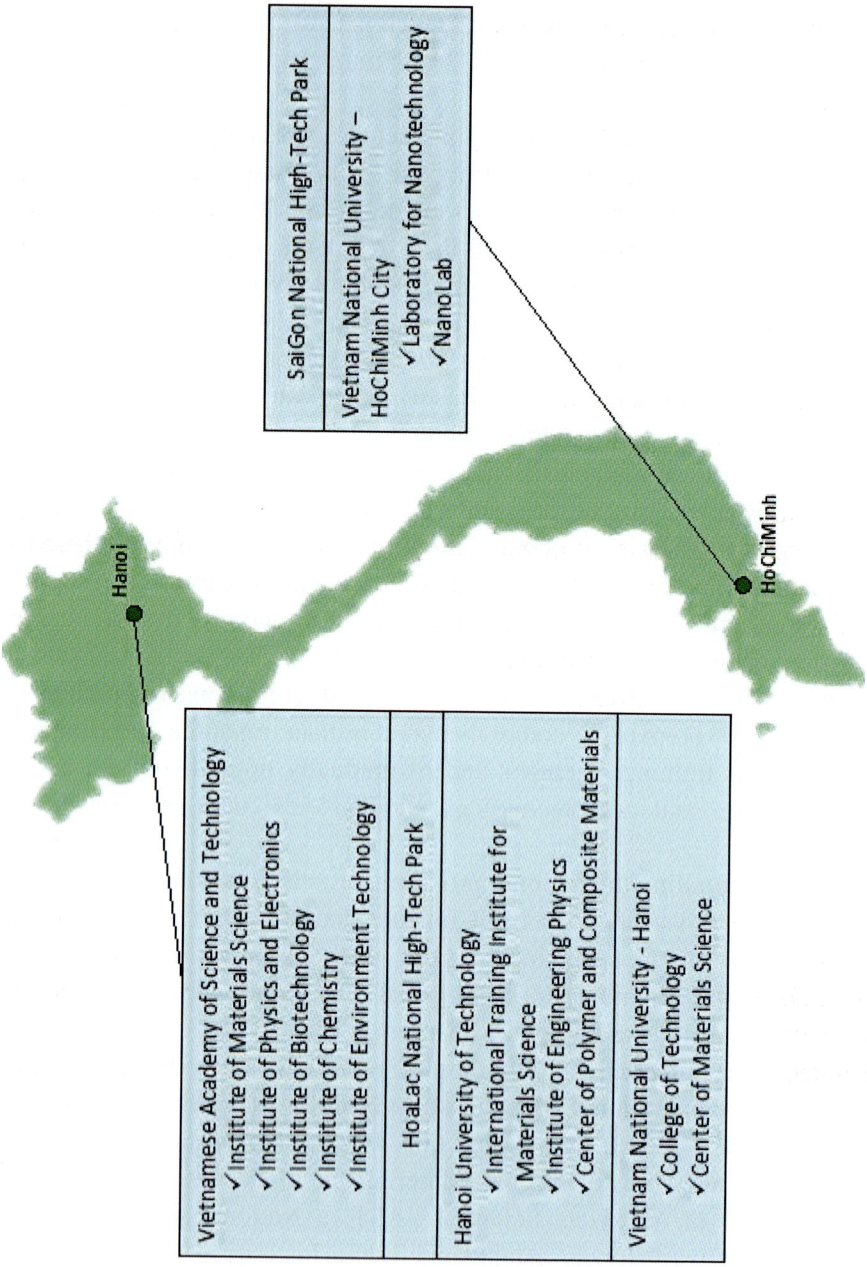

Fig. 1. R&D infrastructure and facilities in selected areas of science and technology in Vietnam.

equipped with the state of the art equipments for fundamental and application oriented research. The National Key Laboratories are operating as open labs[7]. Figure 1 shows the location of the main nanotechnology research centres in Vietnam.

Some other sources such as ODA and Ministerial level of 2-3 million $US each has been also allocated for improving the R&D infrastructure and facilities at various research institutions and laboratories of VAST and National Universities. For instance, recently a Laboratory for Nanotechnology (LNT) was officially established in September 2004 in Vietnam National University- HoChiMinh City. It is financed by the High Education Project of World Bank with the budget of 4.5 million $US. The LNT has the following missions: (i) Developing the scientific research in micro-nanotechnology and transferring research results from universities to industry, (ii) Supporting industrial sectors for conducting research on materials and microelectronics, (iii) Teaching the Micro-Nanotechnology in graduate and postgraduate programs, and (iv) Collaborating with foreign companies as joint-ventures for conducting research, training and technology transfer in microelectronics and other sectors[6]. Very recently, the Vietnam National University-Hanoi (VNU-Hanoi) provided funding to set up of the Department of MEMS and Microsystems. The main research topics of the department consist of micro-electro-mechanical systems (MEMS) for medical application, inertial navigation and automatic control systems. At the same time, VNU-Hanoi decided to establish a key laboratory for nanomaterials and nanodevices at the College of Technology-VNU. It is financed by the High Education Project of World Bank with the budget of 3.0 million $US. Nanomagnetic materials, multiferroics and magnetic nanoparticles, nanosilver and biochips are the main research topics of the laboratory[4]. A nanotechnology laboratory with the investment of about 10 million $US has come into operation staring 2008 in SHTP.

3.2. *Nanotechnology Research and Development Projects*

Main projects on nanoscience and nanotechnology are the following:

(1) Theoretical research on electronic and optical properties of quantum dots, quantum wells and super-lattice (Projects of National Program on Basic Science 1999-2003, 2004-2005, 2006-2007).
Principle Investigators: Group of Prof. Nguyen Van Hieu - Institute of Materials Science (IMS-VAST), Groups of Prof. Nguyen Ai Viet, Prof. Nguyen Toan Thang, Prof. Nguyen Ba An, Prof. Nguyen Van Lien and Prof. Doan Nhat Quang - Institute of Physics and Electronics (IPE, VAST).

(2) Experimental research on nanostructured semiconductors: preparation and investigation of electronic and optical properties of nanostructured semiconductors: silicon nanocrystals, silicon nanostructured thin films, porous silicon and Ge/Si quantum dots (Projects of National Program on Basic Research and Vietnam NCST - France NCSR S&T cooperation 1999-2003; Projects of National Program on Nanotechnology 2004-2005; projects of National Program on Basic Research 2006-2007).
Principle Investigators: Group of Prof. Phan Hong Khoi - IMS, VAST; Group of Ass. Prof. Nguyen Ngoc Long - Center of Materials Science, CMS-VNU-Hanoi; Group of Prof. Nguyen Duc Chien, International Training Institute for Materials Science (ITIMS, HUT).

(3) Experimental research on nanostructured magnetic materials: nanostructured magnetic materials, nanocrystalline perovskite oxides ABO_3, magnetic nanoparticles, giant magnetoresistance effects (Projects of National Program on Basic Research in Natural Science and National Program on New Materials 1999-2003; Projects of National Program on Nanotechnology 2004-2005; projects of National Program on Basic Research 2006-2007, Project of VAST 2006-2007).
Principle Investigators: Group of Prof. Nguyen Xuan Phuc - IMS, VAST; Group of Prof. Nguyen Chau - CMS, VNU; Groups of Prof. Than Duc Hien, Nguyen Phu Thuy - ITIMS, HUT, Group of Prof. Nguyen Hoang Nghi, Institute of Engineering Physics, HUT, and

group of Prof. Nguyen Huu Duc - College of Technology, VNU-Hanoi.

(4) Experimental research on nanophotonic materials and their applications: CdSe CdSe/ZnS (core/shell) quantum dots (QDs) and opal photonic crystals prepared by the colloidal chemical method for biological sensors and luminescence labeling applications; hydrophobic and hydrophilic thin films based on TiO_2-nano particles; rare earth-doped nanophosphor luminescence materials and OLED materials (Projects of National Program on Basic Research 1999-2003; Projects of National Program on Nanotechnology 2004-2005; projects of National Program on Basic Research 2006-2007 and Vietnam-Korean S&T cooperation Project 2006-2007).
Principle Investigators: Groups of Ass. Prof. Pham Van Hoi, Ass. Prof. Pham Thu Nga, Ass. Prof. Tran Kim Anh, Ass. Prof. Nguyen Quang Liem - IMS, VAST, and Prof. Nguyen Nang Dinh - ColTech, VNU.

(5) Experimental research on Carbon-based nanomaterials: High yield/low cost synthesis of multiwall Carbon nanotubes, vertical aligned carbon nanotubes and their applications for fabrication of field emission devices, nano coating and nanocomposite materials basing on CNTs; fabrication of Diamond-like thin films and Carbon-based nanomaterials by HF-CVD and MW-CVD methods; Field emission and thermal-field emission of Carbon nanotubes; Synthesis, properties and applications of single wall Carbon nanotubes (Project of VAST 2003-2004; Project of National Program on Nanotechnology 2004-2005, National Program on Basic Research 2006-2007; Vietnam-Korean S&T cooperation Project 2006-2007, Vietnam-French S&T cooperation Project 2008-2009).
Principle Investigators: Group of Prof. Phan Hong Khoi and Ass. Prof. Phan Ngoc Minh - IMS, VAST.

(6) Experimental research on nanocomposite materials: Fabrication and Applications of Nano Polymer Composite (Projects of National Program on New Materials 2003-2004, 2005-2006, Project of National Program on Nanotechnology 2004-2005).

Principle Investigators: Group of Ass. Prof. Nguyen Duc Nghia, Institute of Chemistry (IC, VAST); Group of Prof. Tran Vinh Dieu, Center of Polymer and Composite Materials (CPSM, HUT).

(7) Research on fabrication of Pt nanowires for glucose detection (Project of National Program on Biotechnology, 2008-2010).
Principle Investigator: Group of Ass. Prof. Dang Mau Chien, Nanolab, VNU- HoChiMinh.

(8) Thin magnetic films for biosensing applications (Project of NUH 2007-2008).
Principle Investigator: Group of Prof. Nguyen Huu Duc, ColTech, VNU.

(9) Research on preparation and application of Ag nanoparticles for environment treatment (Project of VAST 2005-2006, 2007-2008).
Principle Investigator: Group of Dr. Nguyen Hoai Chau, Institute of Environment Technology (IET, VAST).

A summary of main research topics of nanoscience and nanotechnology and corresponding research institutions are shown in Table 1.

3.3. *Nanotechnology Education*

Recognized the need of qualified manpower for research and application of nanoscience and nanotechnology in Vietnam, the MSc. and PhD. education programs in nanotechnology has been executed since 2003-2004 school years with a curriculum for the two years MSc and four years PhD education program on nanotechnology. Emphasis is placed on the ways how to ensure the multidisciplinary characteristics both from theoretical and practical points of view of these programs. The responsible institution for these programs is the Faculty of Physical Engineering and Nanotechnology of the College of Technology, Vietnam

Table 1. Current research topic of Nanoscience and Nanotechnology and responsible key researchers (selected)

Areas of research	Materials and topics of research	Research Institution	Responsible person and Contact address
Nano-physics	Theoretical research on electronic and optical properties of quantum dots, quantum wells and superlattices, quantum computers	IMS, VAST	Prof. Ac. Nguyen Van Hieu, nvhieu@iop.vast.ac.vn
		IPE, VAST	Prof. Nguyen Ai Viet, naviet@iop.vast.ac.vn
	Modeling of nanoelectronic and single electron devices	IPE, VAST	Ass. Prof. Nguyen Hong Quang, nhquang@iop.vast.ac.vn
Semi-conducting nano-materials	Porous silicon, Si nanoparticle, CdS, CdSe quantum dots	IMS, VAST	Prof. Phan Hong Khoi, phkhoi@ims.vast.ac.vn
	Compound and Oxide Semiconductors	CMS, VNU	Prof. Nguyen Ngoc Long, longngoc@hn.vnn.vn
	Si nanoparticles, micro-nano-sensors	ITIMS, HUT	Prof. Nguyen Duc Chien, chien@itims.edu.vn
	Nanoelectronic Devices, LED, biosensors	NanoLab-VNU-HoChi Minh	Ass. Prof. Dang Mau Chien, dmchien@vnuhcm.edu.vn
Magnetic nano-materials	Giant magnetoresistance materials, nanocrystalline perovskite oxides ABO_3, magnetic nanoparticles, magnetic thin films, multiferroics, spitronics; magnetic sensors.	CMS, VNU	Prof. Nguyen Chau chau@cms.edu.vn
		IMS, VAST	Prof. Nguyen Xuan Phuc phucnx@ims.vast.ac.vn Ass. Prof. Le Van Hong honglv@ims.vast.ac.vn
		ITIMS, HUT	Prof. Than Duc Hien hien@itims.edu.vn
		ColTech, VNU	Prof. Nguyen Huu Duc ducnh@vnu.edu.vn

Table 1 (continued)

Areas of research	Materials and topics of research	Research Institution	Responsible person and Contact address
Nano-photonic materials	CdSe QDs and CdSe/ZnS (core/shell) QDs, nanophotsphore luminescent materials, Nanophotsphore luminescent materials	IMS, VAST	Ass. Prof. Pham Van Hoi, hoipv@ims.vast.ac.vn Ass. Prof. Pham Thu Nga, phtnga@ims.vast.ac.vn Ass. Prof. Nguyen Quang Liem liemnq@ims.vast.ac.vn
	ITIMS, HUT	ITIMS, HUT	Prof. Nguyen Duc Chien; chien@itims.edu.vn
Carbon Nanotubes, Diamond thin films and carbon-based materials	MWCNTs, SWCNTs, Vertical aligned CNTs, Individual CNT, Diamond thin films: synthesis and applications in advanced composite, thermal dissipation materials, electron field emitter, etc.	IMS, VAST	Prof. Phan Hong Khoi phkhoi@ims.vast.ac.vn Ass. Prof. Phan Ngoc Minh minhpn@ims.vast.ac.cn
	Gas sensors using CNTs	ITIMS, HUT	Prof. Nguyen Duc Chien chien@itims.edu.vn
Nano particles	TiO_2	IMS, VAST	Ass. Prof. Pham Thu Nga, phtnga@ims.vast.ac.vn
	Magnetic nanoparticles for biomedicine application	IMS, VAST	Ass. Prof. Le Van Hong, honglv@ims.vast.ac.vn
	Silver nanoparticles	IET, VAST	Dr. Nguyen Hoai Chau
MEMS/ NEMS	Nanotips for scanning Microscopy	IMS, VAST	Ass. Prof. Phan Ngoc Minh, minhpn@ims.vast.ac.vn
	Pressure Sensors and Accelerometers	ITIMS, HUT	Ass. Prof. Vu Ngoc Hung hung@itims.edu.vn
	Micro-electro-mechanical systems for medical application, inertial navigation and automatic control systems	ColTech, VNU	Prof. Nguyen Phu Thuy thuynp@vnu.edu.vn
Nano-composite materials	Conducting polymer Nano composite materials	IC, VAST	Ass. Prof. Nguyen Duc Nghia ducnghia246@fpt.vn

National University, Hanoi. Practically these programs are a joint activity of several leading institutions in materials science, biotechnology, electronic and telecommunication technologies in Vietnam. They are Institute of Materials Science, Institute of Physics, Institute of Chemistry and Institute of Biotechnology of VAST as well as International Training Institute for Materials Science and Institute of Engineering Physics - Hanoi University of Technology, Laboratory for Nanotechnology - Vietnam National University in HoChiMinh City. In addition to enhance and improve the education in nanoscience and nanotechnology, joint-education programs for MSc and PhD study with the Institute of Materials Science and the College of Engineering Science, Osaka University (Japan) and with ColTech and University of Paris 11 (France) were established and have been executed.

Since 2002, to provide knowledge and explore opportunities to expand domestic expertise in nano- and micro-technologies and forge international ties to support the advancement of basic research, many National and International Workshops, Conferences, Schools and Seminars on Nanotechnology have been regularly organized in Vietnam. For instance, in July 2002, the first International Workshop on Nanophysics and Nanotechnology (IWONN-02) was successfully organized in Hanoi, Vietnam. This workshop was sponsored by the IMS-VAST (Vietnam), AOARD (USA), Tohoku University (Japan), and APCTP (Asian Pacific Center for Theoretical Physics). The workshop focused on examining regional and international efforts in nano- and micro-technology, reviewing opportunities of basic research support through international agencies and over-viewing basic research efforts and opportunities in Vietnam. Since then the IWONN Workshop has been organized regularly every two years in 2004, 2006 and 2008. From 20 to 24 January, 2003 the first Vietnam-France Training School on Micro-and Nanotechnologies, co-organized by the Laboratory of Electronics and Information Technology LETI-CEA (France), the Institute of Materials Science, the Institute of Physics–VAST and the Faculty of Technology–VNU was held in Hanoi. More than 150 participants from various institutions and universities in the whole country attended this school. The First Vietnam–Korea symposium on Chemistry and nano-structured materials organized by Vietnam National

Center for Natural Sciences and Technology was held in Halong Bay, Vietnam October 18-19, 2003. The second Vietnam-France Training School on Micro- and Nanotechnologies dealing with nanomagnetic materials and spintronics was successfully organized in Hanoi in April 2004. Many others International Workshops related to Nanoscience and Nanotechnology have been orgarnized such as "Nanophysics from Fundamentals to Applications" were organized in October 2004 and August 2005, "*The first International Workshop on Functional Materials-IWOFM-2006*" was held in Halong, Vietnam on December 2006 with more then 300 participants from 15 countries, "*The first International Workshop on Nanotechnology and Applications-IWNA2007*" was held in VungTau, Vietnam on November 2007. Recently, September 15-18, 2008, the APCTP–ASEAN Workshop on Advanced Materials and Nanotechnology (AMSN2008) was successfully organized in NhaTrang with about 250 participants, in which 150 from local, 60 from regional and 40 from international institutions.

4. Some Examples of Research on Nanotechnology

It is impossible to describe within this chapter the detail of all nanotechnology related research activities in Vietnam. In this session we describe some of the research highlights carried out at the Institute of Materials Science (IMS) (one of the largest research institution belongs to Vietnam Academy of Science and Technology (VAST)).

A typical nanomaterial developed at the IMS is quantum dots based on CdSe/ZnS. The quantum dots were produced by sol-gel technique with high controllability in size and shape. The materials can be mass produced and are applied in bio and other material labeling technologies. Using also the sol-gel technique, SiO_2 microspheres were prepared and three-dimensional structured for photonic crystals (Fig. 2)[8]. Research and development of TiO_2 nano coating on glass, ceramics and wood (using dip-coat technique) was carried out also at the IMS. This nano coating technology can be used in construction industry[9] to provide self-cleaning glass windows.

Many other excellent research work on magnetic materials for GMR and other application; magnetic nanoparticles for biomedicine;

nanocatalyst for environment treatment; toxic gas sensor based on ABO_3 Perovskite type materials; humidity and oxygen sensors based on porous alumina, SiC and so on are actively carried out at the IMS. The toxic gas sensors and related systems have been purchased by industry recently.

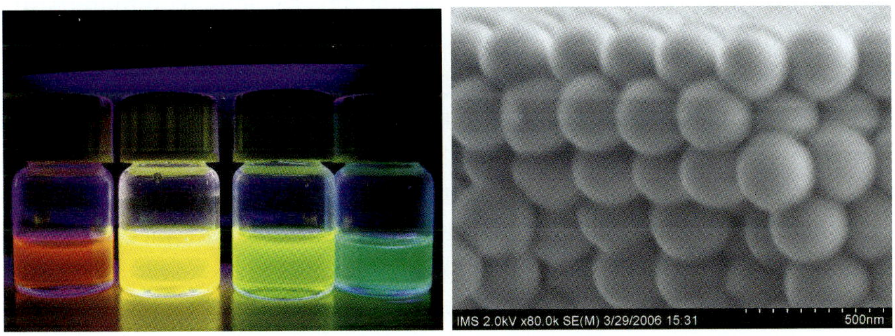

(a) (b)

Fig. 2. (a) CdSe/ZnS (core/shell) Quantum Dots prepared by the colloidal chemical method and (b) SEM image of stacked opal photonic crystal structure made from 287 nm silica spheres of the cleaved edge of opal crystal about 14 layers[8].

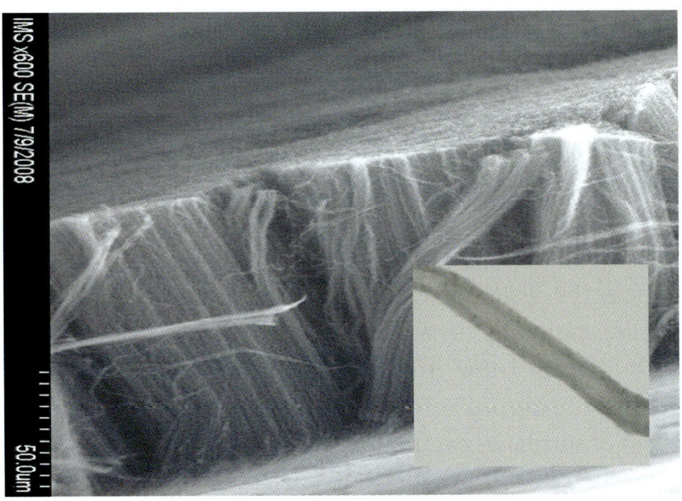

Fig. 3. Vertical-aligned CNTs growth on Si wafer using the Fe_3O_4 nanoparticle catalyst by thermal CVD. The insert is TEM image of 60 nm-diameter MWCNTs.

Institute of Materials Science in VAST is the pioneer in Vietnam in synthesizing and finding applications of the Carbon Nanotubes (CNTs) materials. Thermal chemical vapor deposition (CVD) techniques for the synthesis of carbon nano materials with several configurations such as carbon nanotubes (CNTs), nanorods, nanoballs have been well developed. Multiwalled CNTs of 10-50 nm diameter, several tens to hundreds μm length and 95% purity has been successfully synthesized (see Fig. 3). The system can be expanded to a large volume production of high quality and low cost CNTs by multiplying the number of the thermal furnace[10]. Vertically aligned MWCNTs and single walled CNTs are being extensively developed[11]. Several techniques for purification of the CNTs materials have been developed and the purity of the materials have been increased from 85.8% to more than 95%[12].

Several approaches in application of the grown CNTs materials are extensively carried out:

4.1. *Application of CNTs in Advanced Rubber*

It is well known that the CNTs materials are very attractive for nanocomposite application because of their nano size and excellent mechanical and electrical properties. We have utilized the grown MWCNTs as additive component to reinforce into natural rubber vulcanizates. Table 2 shows several mechanical properties of the examined natural rubber, rubber doped with Graphite powder, rubber doped with 7% and 10% MWCNTs. It is clear that the rubber doped with MWCNTs greatly enhanced the mechanical properties of the rubber. The advanced rubber (10%MWCNTs/NR) was cast into essential components in the water pumping systems and this has been practically used in water pumping systems in some places of Vietnam. It was confirmed by the user that life time of such components was two times improved. Such advanced rubber is also applicable for other field such as rubber wheel of airplane, car, and so on[13].

Table 2. Mechanical properties of the Natural rubber, rubber doped with graphite, MWCNTs of different contents

Parameters	Natural Rubber	Rubber+doped with Graphite Powders	Rubber doped with 7% MWCNTs	Rubber doped with 10% MWCNT
Hardness (Shore A)	42	54	53	53
Specific Mass (g/cm^3)	0.995	1.019	1.016	1.015
Broken durability (N/cm^2)	21.03	47.6	64.15	79.62
Broken length (%)	293.6	226.4	298.4	341.8
Arkon Abrasion (g/1.61 km)	4.5	2.4	1.7	1.3
Heat resistant (°C)	200	350	350	350

4.2. *Application of CNTs in Ni, Cr Coatings*

Besides applying the grown MWCNTs onto the natural rubber, we also utilized the grown MWCNTs in metallic based nanocomposite. Ni and Cr coating by electroplating was chosen for the research[14,15]. Ni and Cr coatings have many advantages and widely used in many industrial areas. To increase the hardness and other mechanical properties, it's essential to reinforce the other small, harder particles such as TiCN, TiC, SiC, B_4C, TiO_2 on the solution during electroplating process. Size of the particles was in the range of several ten nm to several ten micrometers. The difficulty, however, is the strong hydrogen release during the plating process that prevents the co-deposition of these particles into the coating. The larger size of the second phase particles have, the more difficult the co-deposition occurs. To overcome this problem, the MWCNTs were utilized as additives in the Cr and Ni electroplating solutions. It is found that the Ni/MWCNTs and Cr/MWCNTs plated films were approx. 1.5-2 times harder than the conventional Ni and Cr electroplated films and the corrosion resistance of the Ni/MWCNTs films were also greatly enhanced.

4.3. *Application of CNTs in Electromagnetic Absorption Materials and Conductive Paints*

The shielding of radio frequency radiation is important for many military and commercial applications. CNTs composite material was announced as a good candidate for making a light weight and effective EMI shielding media due to their geometric and electrical properties. The conductivity and shielding effectiveness of these composites are strongly correlated to the structure of the CNTs. We have tried to make the CNTs/epoxy composite and measured the radio frequency and Rada absorption efficiency. The measured results showed that the composites Epoxy/CNTs of 20% wt. can absorb more than 99% the electromagnetic wave in the Rada rage 8–12 GHz (X-band). The grown CNTs were also tested in making CNTs/Polyaniline conducting polymer. The electrical conductivity of the polyaniline/CNTs composite reached to the value of 30 S/cm that is comparable to the conductivity range of metallic materials. With this high electrical conductivity, polyaniline/CNTs material opened a promising way for application in making RF shielding materials and as well as protective paint, especially protective paint for sea-ship by eliminating the electro-chemical etching effect.

4.4. *Application of MWCNTs for Thermal Dissipation Media*

Carbon nanotubes were announced as the materials with highest thermal conductivity (2000 W/m.K compared to thermal conductivity of Ag 419 W/m.K). The CNTs therefore is an ideal candidate for making thermal dissipation media to improve the performance of computer processor and other high power electronic devices.

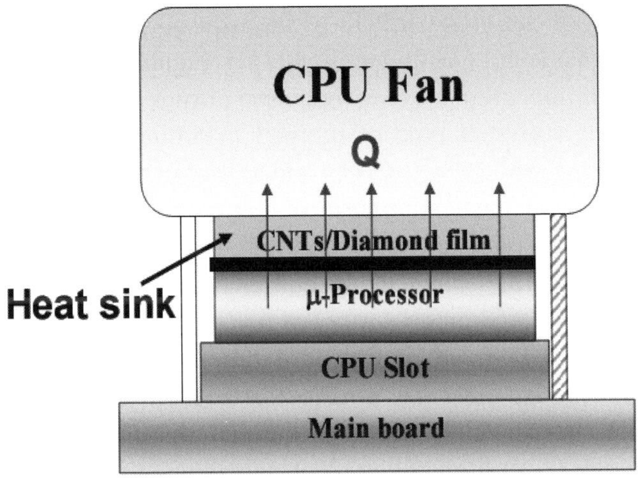

(a)

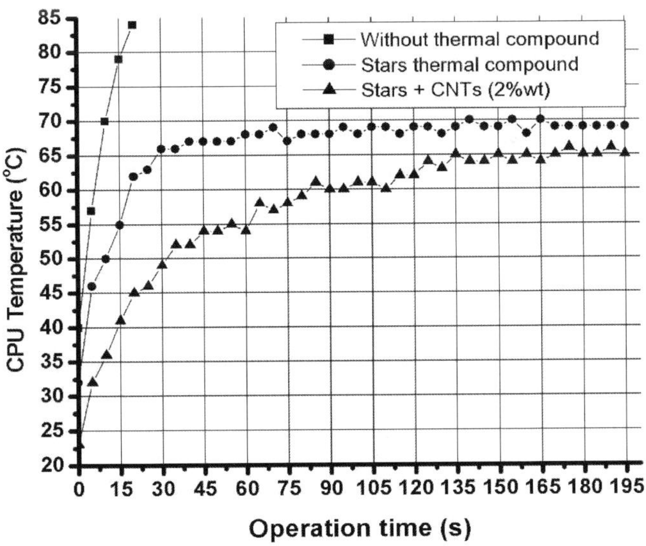

(b)

Fig. 4. Schematic of the thermal dispersive system using CNTs (a) and the measured temperature of the chip of a computer as a function of working time using different thermal matching media (b).

We employed the grown CNTs as the heat dissipation media in a μ-processor of a Personal Computer (Fig. 4(a)). Figure 4(b) is the measured temperature of the chip of the computer during the operation of the computer using different type of thermal matching media: commercial thermal matching and 2% wt. CNTs added thermal glues. The measured results showed that the temperature of the μ-processor decreased 5°C, and the time for increasing the temperature of the μ-processor was three times longer than that when using commercial thermal compound. In overclocking mode, the processor speed reached 3.8 GHz with 165 MHz of system bus clock speed. It was 1.24 times higher than that in non-overclocking mode. This is a direct evidence that confirmed the advantage of the MWCNTs for the thermal dissipation media for μ-processor of the PC or high power electronic devices in general.

4.5. *Application of CNTs in Electron Field Emission and Scanning Probe*

Individual CNT on a Tungsten (W) tip for an advanced miniature electron source and scanning tunneling probe was developed. It was announced that individual CNT can emit electrons as a pointed source at a low voltage and the energy spread of the emitted electrons by the CNTs is very narrow (0.1-0.2 eV). This means that the emitted electron beam can be easily focused into a small spot that is very essential for application such as in electron beam based instrumentation, parallel electron beam nanolithography, field emission information storage technology. The individual CNT has been successfully grown on the Fe coated Si tip formed by a standard microfabrication process. Detail of the synthesis technique was presented in Ref. 16. The electron field emission of the individual CNT on the Si tip was measured in comparison with the Si tip itself. It was shown that the individual CNT emit electrons at a threshold electric field of 4 V/μm that is five times lower than that of the Si tip itself[16]. The measured results suggested a possibility of utilizing individual CNT for ultra-small electron source. The current instability and lifetime of the emission process, however, is needed to be improved. One of the effective ways is to treat the structure in Hydrogen during the emission process[17]. In another approach, the CNTs were grown on W tip for an advanced electron emitter and scanning tunneling microscopy probes application[18].

5. Conclusion

Nanotechnology is recognized as a leading technology in the century. Vietnam is not an exception. The chapter is not meant to cover all the nanotechnology related activities in Vietnam but provide general overview on the current status of infrastructure, research and development of nanotechnology.

Many joint-research projects between industry, research institute and university on nanotechnology have started. With strong support of government, passion of nanotechnologists and the global cooperation, we expect to develop international competitive nanotechnology research and industry especially nanomaterial in Vietnam.

Acknowledgements

We wish to thank the National Program on Basic Research; Vietnam-Korea (2006-2007), Vietnam-France (2008-2009) Bilateral Research and AFOSR/AOARD 074030 projects for the financial supports related to CNTs works. We would like thanks the colleagues for their contribution to the works that we have data to report here.

References

1. http://www.most.gov.vn
2. http://www.vast.ac.vn
3. http://www.moet.gov.vn
4. http://www.vnu.vn
5. http://hut.edu.vn
6. http://www.hcmnlt.vnu.vn
7. Phan Ngoc Minh, Nguyen Xuan Phuc, "*Introduction and Collaboration Intentions of the National Key Laboratory for Electronic Materials and Devices*", The 1st International Workshop on Innovation and Commercialization of Micro/ Nanotechnology (ICMAN 2007), October 18-20, 2007, p. 46.

8. Pham Thu Nga, et al., "Experimental study of 3D self-assembled photonic crystals and colloidal core-shell semiconductor quantum dots", Asean J. on Science and Technology for Development 24 (2007) 161-170.
9. Pham Thu Nga, et al., Report of the Laboratory (unpublished).
10. Nguyen Van Chuc, et al., Journal of the Korean Physical Society, Vol. 52, No. 5, May 2008, pp. 1368-1371.
11. Ngo Thi Thanh Tam, et al., "Production of vertically aligned Carbon nanotubes by thermal chemical vapor deposition using Fe_3O_4 catalytic nanoparticles", Proc. IWNA-2007 Conference, Vungtau-Vietnam November 15-17, 2007, pp. 485-488.
12. Ngo Thi Thanh Tam, et al., "*Analyzing the purity of Carbon nanotubes by different methods*", Journal of the Korean Physical Society, Vol. 52, No. 5, May 2008, pp. 1382-1385.
13. Pham Huu Ly, Phan Hong Khoi, "*Preparation and Characterization of Carbon nanotubes/polymers composites*", Advances in Natural Science, Vol. 6, No. 4, 2005, pp. 339-361.
14. Bui Hung Thang, et al., "*Electrodeposition and characterization of Nickel – multi walled carbon nanotubes composite coatings*" Proceedings of the BME Conference, Hanoi-Vietnam July 25-27, 2007, pp. 308-311.
15. Nguyen Ngoc Khoai, et al., "*Initial results of making carbon nanotube reinforced chrome film by electroplating technique*" Proceedings of the BME Conference, Hanoi-Vietnam July 25-27, 2007, pp. 325-328.
16. Phan Ngoc Minh, et al., "*Carbon nanotube on a Si tip for electron field emitter*", Japanese Journal of Applied Physics, Vol. 41, Part 2, No. 2A, 2002, L1409-L1411.
17. Phan Ngoc Minh, et al., "*Schottky emitters with carbon nanotubes as electron source*", Technical Digest of the 13th International Conference on Solid-state Sensors, Actuators and Microsystems (Transducers' 05) Seoul-Korea, June 2005, 267-270.
18. Phan Ngoc Hong, et al., "*Fabrication and Characterization of Carbon nanotubes on Tungsten tips*", Journal of the Korean Physical Society, Vol. 52, No. 5, May 2008, pp. 1386-1389.

Author Biographies

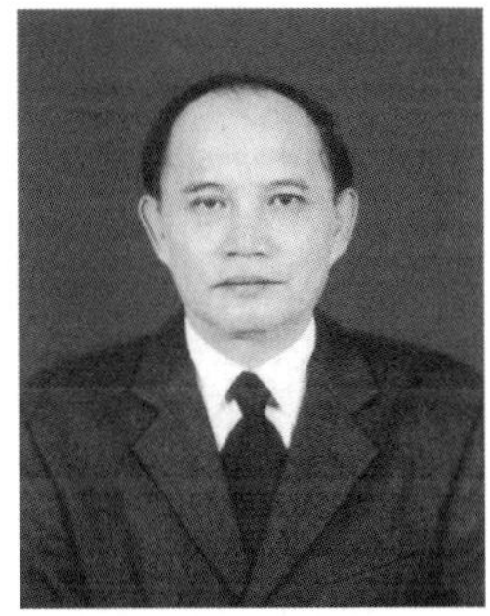

Professor Phan Hong Khoi is the President of Vietnam Physical Society, Chairman of the Scientific Council of the Institute of Materials Science (2003-2007), Vice-Chairman of the Scientific Council on Materials Science - Vietnam Academy of Science and Technology (VAST), Head of the Department of Nano-structured Semiconductors - College of Technology at Vietnam National University-Hanoi. He graduated at the Faculty of Physics, National University of Leningrad (now Sain-Peterbourg - Russia) in 1968 and got PhD degree in 1982. From 1997 to 2002 he was the Director of the Institute of Materials Science - National Center for Natural Science and Technology.

Prof. Khoi received research awards and honors worldwide, which include: Fellow of Institute of Physics (IOP) - United Kingdom; Foreign Member of the Russian Academy of Natural Sciences - Russia; Fellow of Ecole Normale Superieure, Paris France; Visiting Professor of University AIX Marseile 1, CRMC2, France; Visiting Professor of the Military University, Munchen, Germany; Visiting Professor of Osaka University; Co-Director of the International College on Real-Time Microprocessor (3rd World Academy of Science, Triest, Italy); Member of the Editorial Board of Journal Communication in Physics; Member of the Management Board of National Program on New Materials and National Program in Fundamental Research on Physics. In 2005 he received an outstanding state award on Science and Technology.

His scientific research interests focus on (1) Physics and Technology of silicon materials and Devices; (2) Nano-structured materials for optoelectronic and photonic applications (Porous Silicon, Silicon Nano-crystals, Carbon Nanotubes and Nano-structured Diamond-like Films).

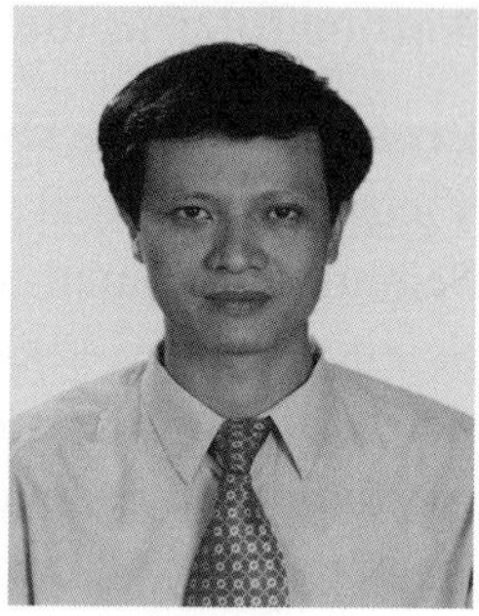

Associate Professor Phan Ngoc Minh was born in QuangBinh, Vietnam in 1969. He received BS degree in Physics from Hanoi University, PhD degree in Physics from Institute of Physics, Hanoi, Vietnam in 1996 and PhD degree in Engineering from Tohoku University, Japan in 2001. From 2001-2004 he worked as a Post-doctoral Researcher and then as an Assistant Professor at the Graduate School of Engineering, Tohoku University in the field of MEMS/NEMS, CNT and diamond materials.

He returned to Vietnam from April 2004 and he currently is a Vice Director of the Institute of Materials, Vietnam Academy of Science and Technology. He is Director of the National Key Laboratory on Electronic Materials and Devices. He is also a teaching Professor at the Department of Nano-structured Semiconductors - College of Technology at Vietnam National University-Hanoi.

He is a member of organizing/program committee of international conferences: IEEE Sensor-2004, IEEE Sensor-2006, IWOFM and IWONN-2006, IWONA-2007, NDNC-2008, AMSN and IWONN-2008, IEEE Sensor-2008. He is a referee of international journals: Nanotechnology, Journal of Electrochemical Society, Sensors and Actuators A, Applied Optics, Diamond and Related Materials. He has been awarded hornor including prize from Minister of Education and Training for young resercher-1991; DFG scholarship - Germany (1994), Monbusho scholarship - Japan (1997-2001), C. N. Yang prize 2004.

His current interests are Physics, Technology and Application of Nano-structured materials for electronic, photonic and optoelectronic applications; Carbon based nanomaterials; Micro/Nano Electro Mechanical Systems.